Future City

AF806262

Volume 20

As of 2008, for the first time in human history, half of the world's population now live in cities. And with concerns about issues such as climate change, energy supply and environmental health receiving increasing political attention, interest in the sustainable development of our future cities has grown dramatically.

Yet despite a wealth of literature on green architecture, evidence-based design and sustainable planning, only a fraction of the current literature successfully integrates the necessary theory and practice from across the full range of relevant disciplines.

Springer's *Future City* series combines expertise from designers, and from natural and social scientists, to discuss the wide range of issues facing the architects, planners, developers and inhabitants of the world's future cities. Its aim is to encourage the integration of ecological theory into the aesthetic, social and practical realities of contemporary urban development.

Giuseppe E. Scarascia-Mugnozza •
Vicente Guallart • Fabio Salbitano •
Giovanna Ottaviani Aalmo • Stefano Boeri
Editors

Transforming Biocities

Designing Urban Spaces Inspired by Nature

Springer

Editors
Giuseppe E. Scarascia-Mugnozza
University of Tuscia
Viterbo, Italy

Vicente Guallart
Institute for Advanced Architecture of Catalonia
Barcelona, Spain

Fabio Salbitano
University of Sassari
Sassari, Italy

Giovanna Ottaviani Aalmo
Norwegian Institute of Bioeconomy Research
(NIBIO)
Ås, Norway

Stefano Boeri
Politecnico di Milano
Milano, Italy

ISSN 1876-0899 ISSN 1876-0880 (electronic)
Future City
ISBN 978-3-031-29468-6 ISBN 978-3-031-29466-2 (eBook)
https://doi.org/10.1007/978-3-031-29466-2

This Springer imprint is published by the registered company Springer Nature Switzerland AG
The registered company address is: Gewerbestrasse 11, 6330 Cham, Switzerland

Foreword

Biocities: From a Concept to a New Urban Reality

Cities represent the good, the bad, and the ugly of our world. They showcase some of our greatest challenges—but also offer some of our greatest opportunities for leading the transformation towards a climate-neutral and nature-positive economy.

We built our first cities many thousands of years ago, and since then cities have shaped human civilisation. However, urbanisation as a megatrend is rather new. It was only during this century that, for the first time, half of the world's population lived in urban areas. Two hundred years ago, only 7% of the world's population lived in cities and towns. Since then, urbanisation has accelerated: every day our cities add around 200,000 more people and by 2050 more than two-thirds of the global population will live in urban areas.

Given cities are our economic and innovation hubs, and also the major consumer of energy and resources, it is crucial that we reflect on why and how cities grow, and the consequences of such rapid urbanisation for sustainable development.

Cities emerged because they are the most efficient system for self-organising ourselves in social networks that optimise our social interactions and the exchange of ideas and information and support wealth creation based on the division of labour, specialisation, and innovation. They enable all this while minimising the transaction and infrastructure costs. Cities, therefore, are the most efficient system for creating social and economic capital. But what are the implications for our natural capital and for our environment, and for the relationship between humans and nature which together form the basis for sustainable development?

The visionary physicist Geoffrey West considered these implications in his book *Scale*, where he looked at the fundamental difference between how cities grow compared to biological systems. In biological systems, the amount of energy available for growth continuously decreases with increasing size until a point where growth stops. Biological systems/organisms grow sublinearly. With cities, the bigger the city, the more resources can be allocated for its socio-economic growth, and the faster it grows. The bigger the city, the more the average individual systematically owns, produces, and consumes in terms of goods, resources, and ideas. Cities grow superlinearly with increasing returns to scale, provided the energy and resources are available.

This explains why urbanisation did not accelerate until the Industrial Revolution started, once there was access to massive and affordable fossil energy and materials. It also explains why at a global level we only reached 50% of urban population this century, after experiencing the greatest global economic acceleration ever taking place in the last 30 years. In this period of time, the urban global population has doubled, but the global GDP and the global middle class have tripled. Clearly, economic growth and urbanisation mutually accelerate each other. The problem is also that the environmental problems related to the existing fossil-based economy accelerate too.

After 200 years of unprecedented urbanisation and economic growth based on a fossil-based economy, we have arrived at a tipping point. Our urbanised world has become too big for our planet. This is clearly exemplified by climate change, biodiversity loss, and the degradation of our natural resources.

We now need a new way of thinking, as a basis for a new economic paradigm for our urbanised world. A paradigm where cities, our economic and innovation hubs, take the lead in rethinking our economy and its relationship to nature in order to ensure it prospers within the renewable boundaries of our planet. This new paradigm should be based on a synergistic relationship between nature and society, economy and ecology, and rural and urban areas, to develop a circular bioeconomy centred around life and not consumption.

As this book argues, cities need to lead this change, not only in replacing fossil energy by renewable energy but also by taking the lead in replacing non-renewable materials like plastics, steel, or concrete with renewable biobased materials, and replacing grey infrastructures with green ones, making nature a basic urban infrastructure. Here, trees, forests, and wood have a crucial role to play.

The book also highlights why cities using wood in construction become carbon capture and storage infrastructures and how urban forests and the strategic placements of trees around buildings decrease the energy consumption in buildings for heating and cooling. But they also reduce the increasing problem of the urban heat island effect and play a major role in human health and wellbeing.

Transitioning to biocities is a challenge for truly transdisciplinary research and for transformative approaches that combine urban and landscape planning, medical science, architecture, forestry, ecology, biology, chemistry, sociology, agriculture, landscape architecture, industrial design, engineering, economics, governance, and social sciences. It also requires political leadership, and the active participation of urban and rural citizens.

To accelerate this transition, the European Forest Institute launched its Biocities Facility, aiming to create an informed dialogue on how trees, forests, and wood can rethink and form the backbone of climate smart cities: *Biocities*. Connecting the dots between different disciplines, sectors, and actors, the Facility generates and communicates scientific knowledge on the potential of the circular bioeconomy concept to rethink urban areas, particularly based on forest-based solutions.

Together with the recently published *Research Agenda—Biocities of the future*, this book will frame the activities of EFI's Biocities Facility in the coming years, to enable it to develop a new and holistic conceptual framework for the use of green infrastructures and biobased solutions in urban environments.

European Forest Institute, Joensuu, Finland

Marc Palahí

References

Hurmekoski E (2017) How can wood construction reduce environmental degradation? European Forest Institute.

Leskinen P et al (2018) Substitution effects of wood-based products in climate change mitigation. From Science to Policy 7. European Forest Institute. https://doi.org/10.36333/fs07

Palahí M et al (2020) Investing in Nature as the true engine of our economy: A 10-point Action Plan for a Circular Bioeconomy of Wellbeing. Knowledge to Action 02, European Forest Institute. https://doi.org/10.36333/k2a02

West G (2017) Scale. The universal laws of growth, innovation, sustainability, and the pace of life in organisms, cities, economies, and companies. New York, Penguin Press

Wilkes-Allemann J, van der Velde R, Kopp M, Bernasconi A, Karaca E, Coleman Brantschen E, Cepic S, Tomicevic-Dubljevic J, Bauer N, Petit-Boix A, Cueva J, Živojinović I, Leipold S, Saha S (2022) Research Agenda – Biocities of the future. European Forest Institute. https://doi.org/10.36333/rs4

Contents

Towards the Development of a Conceptual Framework of BioCities

Vicente Guallart, Michael Salka, Daniel Ibañez, Fabio Salbitano, Silvano Fares, Arne Sæbo, Stefano Boeri, Livia Shamir, Lucrezia De Marco, Sofia Paoli, Maria Chiara Pastore, Jerylee Wilkes-Allemann, Evelyn Coleman Brantschen, and Ivana Živojinović

1 Introduction

This introductory chapter will evaluate how we have reached the current point in the history of world urbanity, its relationship with nature, and why a fusion between the two is now necessary. In order to define BioCities as cities which follow the principles of natural ecosystems to promote life, we will refer to the extensive knowledge of the history of urban science, the need for cities to be reinvented based on ecological principles, and new methods of analysing and measuring reality through digital systems. This vision of the main functions and traits of BioCities will also serve as a thread and reference for the subsequent chapters which will highlight

V. Guallart (✉) · M. Salka · D. Ibañez
Institute for Advanced Architecture of Catalonia (IAAC), Barcelona, Spain

F. Salbitano
University of Sassari, Sassari, Italy

S. Fares
National Research Council of Italy, Institute for Agriculture and Forestry Systems in the Mediterranean, Naples, Italy

A. Sæbo
Norwegian Institute of Bioeconomy Research (NIBIO), Ås, Norway

S. Boeri · L. Shamir · L. De Marco · S. Paoli
Stefano Boeri Architetti (SBA), Milan, Italy

M. C. Pastore
Politecnico of Milan (PoliMi), Milan, Italy

J. Wilkes-Allemann · E. C. Brantschen
Bern University of Applied Sciences (BFH), Bern, Switzerland

I. Živojinović
Bern University of Applied Sciences (BFH), Bern, Switzerland

G. E. Scarascia-Mugnozza et al. (eds.), *Transforming Biocities*, Future City 20,
https://doi.org/10.1007/978-3-031-29466-2_1

and elaborate on the different properties of the BioCity vision. The final chapter will draw from this vision the constituting principles of the BioCity and will outline possible pathways of transition towards BioCities.

2 A Brief History of Urbanisation

Scientific evidence shows us that the actions of humanity on the planet are producing global warming, due in large part to the effects of greenhouse gas (GHG) emissions, and that these are produced mainly by urban areas; as Geoffrey West argues in his inspiring book, *Scale* (2018), 'we live in the age of the Urbanocene, and globally the fate of the cities is the fate of the planet'.

The history of cities reveals that they have always had complementary and conflicting relationships with nature. Since the creation of the first cities in the Levant more than 5000 years ago, cities were conceived as places different from nature—places from which to organise agricultural practices, manage human activity, and provide security against war. These cities were permanent settlements of previously nomadic, hunter-gatherer communities. This transition accelerated the already ongoing erosion of biodiversity during the Holocene epoch (the last 12,000 years of Earth's history), marking the beginning of the profound transformation of what we call the environment, which began to be assigned the fundamental function of systematically supplying human life through agriculture, water, biomass, and other elements that guaranteed the prosperity of the cities. Throughout the centuries, the history of humanity is a sequence of historical events, technological advances, social organisations, and economic models that have ultimately led us to become a predominantly urban species (UN DESA 2018).

Until the start of the industrial revolution that ushered in the modern era, dynamics between the urban, rural, and natural spheres were still, by and large, balanced. This meant that population growth was intrinsically linked to the capacity of the hinterland to support its inhabitants, a constraint epitomised by the theory of the Malthusian trap. The modern era, with the introduction of fossil fuels such as coal, oil and gas, and the development of communications, facilitated an explosion in the urbanisation process, drastically and irreversibly altering this prior balance, with cities rapidly expanding into the rural or traditional agricultural landscape and into the 'natural' environment. Attracted by perceptions of greater job opportunities, better living conditions, and socio-political rights, freedoms, and security rural populations left the countryside for urbanised areas, whilst the gradual innovations of science and medicine promoted longer life expectancies. Currently, this process—with much broader and more exponentiated proportions than ever before—is underway in the countries of South America, Asia, and Africa, amplified by the climatic crises that will soon make entire regions inhospitable (Xu et al. 2020).

The forms and functions of cities have evolved throughout history in response to the civilisations that built them, based on the challenges of each era, using the technologies and knowledge available at that time. For example, at the time in history when the main function of the city was to defend the population from invasions, cities were transformed into castles or fortresses; when the main function was religious, cities were organised around places of worship; and, when the main function was to house an industrial workforce, cities became great machines of manufacturing. However, in all of these stages, with variations dependent upon their respective environmental contexts, cities tended to develop on sites with nutrient-rich and productive soils, with good water supplies, with game-rich forests that provided building materials and energy, and with easy access to large rivers, lakes, or to the sea (Diamond 1997; Bosker 2021). Therefore, urban sprawl in many places causes great losses from the perspective of biodiversity or food production. Today, cities and their metropolitan regions, in many cases, form continuous urban areas stretching many kilometres, conserving 'nature' elsewhere as that which is 'not urban'.

At the expected rate, the world population is projected to reach approximately 9.7 billion people by 2050, 68% of which will be concentrated in cities (UN DESA 2018, 2019). This is an alarming prospect, considering that today only 55% of the world's population is urban, yet that portion is already responsible for 70% of global CO_2 emissions (UN Habitat 2020). Put otherwise, more than 2.4 billion people will become urbanised in the next 30 years, which implies building the equivalent to a city of more than 6 million people every month. This is an unbearable pace for the planet if cities continue to be built in the same ways as they are today, and an urgent call for a global transformative approach to designing urbanised areas and for improving the quality of life of citizens, as is proposed by the BioCity vision and concept.

In fact, the actual scale of urbanisation, population growth, and lifestyle changes generates increasing pressures on the natural environment in terms of environmental contamination, GHG emissions, climate change, traffic congestion, air quality, lack of affordable housing, deterioration of biodiversity, loss of ecosystem services (ES), and resource depletion, even though carbon footprint and land consumption may differ from densely populated city centres to sparsely built-up suburban areas (UN Habitat 2022). Cities may well count amongst our most powerful tools for accommodating growing populations with contemporary habits whilst mitigating impacts on land uptake and carbon footprint given the per capita efficiencies achieved through scaling laws by densified human habitations, the sharing of communal resources, and the networking of infrastructures, as compared to lower density developments of detached homes reliant upon personal vehicles (West 2018). However, the plea for a transition to BioCities acknowledges that these good aspects of urbanisation must be made better still. From a demographic and social perspective, the current mass urbanisation process physically translates into overcrowded informal settlements (e.g. slums and favelas) in many cities of the world, sprawling doorway or dormitory districts on the border of metropolises, and people living in cars and trucks in prestigious, high-value cities like San Francisco. Thus urbanisation, in many cases, generates inequalities and social tensions, especially when accompanied by growing migratory flows to urban areas. With regard to the natural environment, the profound impacts of cities extend far beyond urban borders; such is the case of the food and timber supply chains through which current consumer culture has far-reaching consequences in distant regions or continents (Pendrill et al. 2019). Resulting deforestation, extreme increases in land use for intensive agriculture and ranching, soil consumption and degradation, as well as air and water pollution, are prominent examples. As shown by the seminal book, *The Limits of Growth* (Meadows et al. 1972), commissioned 50 years ago by the Club of Rome, our planet is physically limited, and humanity cannot continue to use more physical resources nor generate more emissions than nature is capable, respectively, of supplying or removing in a sustainable manner. This concept of a human ecological footprint that is overshooting the carrying capacity of the planet has been further elaborated by Rockström et al. (2009a) through the proposal of a set of 'planetary boundaries' which, taken together, delineate a 'safe operating space' for humanity; and yet, recent data have shown that four out of these nine planetary boundaries—namely, biodiversity loss, damage to nutrient cycles, climate change, and land use—have already been crossed, sliding the biosphere into or beyond the 'uncertainty zone' (Steffen et al. 2015).

A deep rethinking of the use of natural resources is overdue. Fatefully, cities are also the crucibles of invention and progress (Glaeser 2011; Kern 2019), so it is certain that, if solutions are to be found, many will come from urban, rather than rural, areas.

A new way of inhabiting the planet should be imagined, a new way of existence for the human species on Earth, which must start from a rediscovered and rethought relationship with the natural sphere. The task is not just to increase the presence of urban forests, trees, and greenery within cities, although this is important for liveability; there are also deep changes necessary in the socio-economic, cultural, natural, institutional, and technological spheres. Advancement in one area alone is grossly insufficient. To adequately address multifaceted urban challenges and promote urban resilience, progress is needed in all areas simultaneously.

Since the beginning of the mid-nineteenth century, approximately every 50 years a new paradigm for city construction has been defined in Europe and around the world, [1] in order to respond to the technological, economic, and social changes of the corresponding era. Such shifts have always occurred after major conflicts, crises, or epidemics forced widespread re-evaluation of the ways we live and how the economy develops locally and globally. The year 2020 may mark another of these historical inflection points, given the COVID-19 pandemic, and the majority recognition that climate change is a reality which must be faced immediately, including development of a new urban model based on the protection and valuation of the environment. Indeed, a unified human, technological and natural environment, which recognises that

[1] After the modernisation of European cities in mid-nineteenth century, symbolised by the Haussmann's renovation of Paris, and after the Garden City Movement of 1898, the Athens Charter brought together the essential principles of functionalist urbanism of the Modern Movement, as drafted in 1933 by Le Corbusier, Gropius, Aalto and other famous architects of the time; in 1977 followed the Charter of Machu Picchu, insisting on a more organic growth of human settlements; in 1996 the Congress for the New Urbanism; and in 2022 the Charter of Rome was elaborated within the programme of the New European Bauhaus.

these distinctions have only ever been illusory, and is based on the scientific principles of ecology, will likely prove requisite to our collective survival.

The terms 'ecology' and 'urbanisation' were coined in practically the same year, in the middle of the nineteenth century, at a time of scientific development without equality in Europe. Ernst Haeckel defined ecology in 1866 in his book, *Generelle Morphologie der Organismen*, as a science that studies the relationship of living beings with their environment, whilst Ildefonso Cerdà minted the term urbanisation in his *General Theory of Urbanisation* in 1867, as a science which should allow the rational construction of human settlements (Haeckel 1866; Cerdà 2018).

It was also at that time the first modern urban revolution took place, related to industrialisation and the massive influx of populations from the countryside to the city, which initially grew on itself, within the defensive bounding walls inherited from past epochs. However, epidemics, the science of hygienism, new transport systems based on the steam engine (such as the train), and the realisation that new ballistic technologies made the city walls militarily obsolete, prompted a process of demolition of these walls and development of urban extension projects throughout the European continent. In this manner, the construction of large cities manifested new pressures for the exploitation of natural resources, as these cities had to create the first truly immense energy, water, and sanitation networks whose impacts surpassed their urban limits.

This early modern urbanisation process took on a new form at the beginning of the twentieth century, with the so-called 'garden cities' concept advocated by Ebenezer Howard, who in his book, published in 1898, *To-Morrow: A Peaceful Path to Real Reform*, proposed a new model of industrial cities in greater harmony with nature (Howard 1898).

The second great transition occurred exactly 100 years ago, just after the First World War, which redefined the political map of Europe with important consequences for global economic relations. Also, due to the so-called Spanish flu (1918–1920), more than 40 million people perished, begetting redefinition of how to build cities and houses based, again, on a new hygienism. In response, the Bauhaus School was founded to champion advancement of the modern city agenda, relying on the automobile, subway, and aviation, and with them the ubiquitous use of oil. In this urban model, defined by Le Corbusier in 1924 in *Vers une Architecture*, the buildings were conceived of as 'a machine to inhabit' (Le Corbusier 1924). In the new neighbourhoods, the housing blocks were oriented to the south for solar exposure and separated amongst vast green spaces. The mantra 'form follows function' was applied to the city such that the segregation of urban functions into specialised districts (e.g. housing, industry, commerce, and leisure) made the continuous movement of the inhabitants of the cities mandatory. Cities have long been guided by the optimism of technological progress. Relatedly, concrete reinforced with steel became the new material system of choice for construction, and the freeing of façades from structural or ornamental purposes was favoured, which allowed more light to enter the interior of buildings. This model, which has been employed massively throughout the twentieth century, is still being followed in many locales. Despite certain positive aspects of modernism's functional cities, such as social housing blocks, green spaces, and the development of expansive public transportation networks; other traits contributed, albeit unintentionally, to the current climate crisis, such as the exclusive segregation of functions and the ensuing need for individual cars and their infrastructures, fossil fuel consumption, and the excessive use of carbon-intensive building materials like concrete, steel, and aluminium (Architecture 2030 2018).

Fifty years later, several simultaneous crises of supreme relevance to cities occurred, notably, the social and cultural crises culminating in May 1968 in Paris, the oil crisis of 1973, and the crises in urban centres fostered by abandonment of city cores due to the pull towards life in the suburbs; or, at a global scale, the unprecedented growth of (mega) cities, and the pervasive pollution of air, soils, and water. This was also the period during which the use of plastics innovated by the petrochemical industry began to gain traction, and that today we recognise as having devastating effects on our oceans. It was, moreover, the moment when environmental movements started to emerge throughout the world to combat the loss of biodiversity and the expansion of nuclear energy. From an urban point of view, Aldo Rossi's 1966 book, *L'architettura della Città*, is a canonical text that catalysed the newfound interest in urban regeneration, and in the city centre, which has prevailed in recent decades (Rossi 1966) with outstanding examples visible in Barcelona of the 1980s or through the ripple effect of the Guggenheim Museum in Bilbao and the resulting boom of urban tourism from the 1990s onwards.

In 1992, at the Rio Summit, the concept of 'sustainable development' was affirmed, which has since enabled reconsideration of the goals of unlimited growth and the exploitation of planetary resources. Shortly afterwards, in 1996, the concept of the 'ecological footprint' was presented, which quantitatively analysed patterns of resource consumption and waste production (Wackernagel and Rees 1996). From the first data, the need to develop a novel model of resource production and consumption (Fig. 1) became blatantly evident, resulting in the evolving conception of the 'circular economy' or 'circular bioeconomy' (Boulding 1966; Stahel and Reday Mulvey 1981; D'Amato et al. 2017; EMF 2021).

Subsequently, the UN-FCCC Kyoto Protocol and the Paris Agreement represented key instances in the acceptance of global climate change as a universal challenge, based on scientific data collected in previous decades which foreshadows a catastrophe for humanity if the current trends do not change (IPCC 2022).

Along these lines, in 2020, the European Union (EU) asserted its aspiration to become climate neutral by 2050, hosting an economy with net-zero GHG emissions (EC 2020). This goal is at the heart of the European Green Deal, and is in keeping with the EU's commitment to global climate action under the Paris Agreement.

3 Why Is Now the Time for BioCities?

The preceding history makes clear why it is necessary to define a new urban model in order to face the grand challenges of our time, starting with the application of the tenets of the circular bioeconomy to the urban reforms typically required of

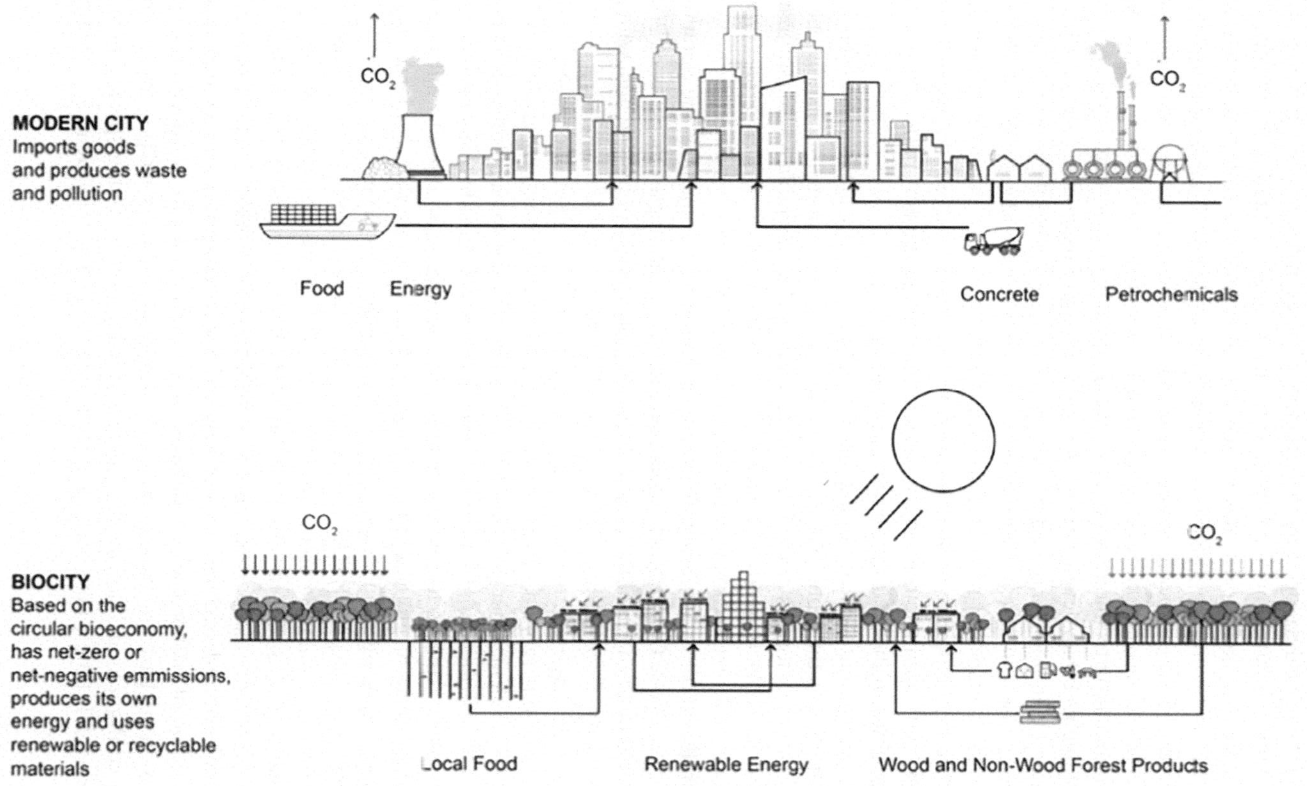

Fig. 1 Through intentional engagement with trees and forests both beyond and within the city, the BioCity achieves net CO_2 absorption. ©Vicente Guallart (adapted from Guallart Architects 2019)

European or North American cities, and to the creation of new or expanded settlements which will occur mainly in Latin America, Asia, and Africa.

BioCities can be developed by merging the sciences of ecology and urbanisation, using systems and solutions developed with information technologies. In fact, the methodologies applied for decades by the natural sciences aimed at analysing the behaviour and evolution of natural systems have already been fundamental in inspiring new processes being applied to cities via what has come to be known as the 'smart cities' movement, leveraging contributions from the field of information and communication technologies (ICT) towards the creation of more intelligent and more sustainable urban developments (United for Smart Sustainable Cities-U4SSC). Information technologies applied to cities have made possible the use of empirical data to recognise and assess phenomena such as urban heat island effect (UHI), air pollution, the collapse of mobility, the pollution of rivers, or the impacts of urban activities on the natural systems that surround and support the densest cities on the planet, and will continue to be indispensable in defining the new paradigms within which the cities of the near future must learn to operate.

Urban centres and their communal networks have been permitted to grow almost without limits, creating conurbations spanning hundreds of kilometres. For instance, the Northeast megalopolis of the USA, elaborated by the geographer Jean Gottmann in his 1961 book, *Megalopolis: The Urbanised Northeastern Seaboard of the United States,* as a vast metropolitan region approximately 970 km long stretching from Boston in the north to Washington DC in the south, also containing the populous cities of New York City, Philadelphia, and Baltimore, amongst others, as of 2010 concentrated nearly 17% of the country's peoples on just 2% of its land area with an average density of 390 people/km^2 (in contrast to the national average of 31 people/km^2) and is the largest in the world in terms of economic output (Gottmann 1961). The pressures this type of urbanisation has put on the environment are becoming increasingly evident. Therefore, in the search for genuine sustainability, it is high time to advance the definition of the BioCity beyond the idea of unifying two different spheres (that is to say, the natural and the urban), in order to mitigate or reverse the effects of global climate change and inequity, as well as to rise to the many related challenges embodied in the 17 Sustainable Development Goals (SDGs) of the United Nations (UN) (UN DESA 2015). It is imperative to begin to conceive of, and enact, a sustainable urban growth model with 'nature' broadly defined at its core, rather than as a state of 'otherness' to be discretely conserved as in a museum: an urban settlement deeply rooted in the natural ecosystem and capable of multiplying, through nature-based solutions (NBS) and ecosystem services (ES), the values and potentials of nature itself. Similarly, it is timely to begin to conceive of the BioCity as being based on the translation of natural mechanisms and principles into a spatially configured and holistically integrated habitat for all living species, including humans. It is timely now because there is no time to lose (UN General Assembly 2019).

4 BioCities Manifesto

BioCities are cities that follow the principles of natural ecosystems to promote life. Elaborated further, BioCities emulate the principles of social-ecological systems to better connect humans with nature, and to contribute to the solutions of environmental crises and of global climate change, within the framework of the Earth system's planetary boundaries. Ecological principles focus on the flow of energy and the cycling of matter through ecosystems as well as on the crucial role of information embedded in biodiversity, as basic concepts of the general theory of ecology (Scheiner and Willig 2008; O'Connor et al. 2019).

Cities, like ecosystems, are complex adaptive systems characterised by a dynamic network of interactions in which the behaviour of the ensemble may not be predictable according to the behaviour of the components (West 2018). Therefore, BioCities are cities that strive to approximate ecosystem's functioning (see also chapter "Towards BioCities: The Pathway to Transition" for forest ecosystem's analogue), particularly their network interactions such as the harnessing and flow of renewable energy, the storage of carbon, the cycling of biomaterials or other matter, and the conservation of evolutionary information as a fundamental feature of ecological as well as sociological systems. The processing of information, which includes biodiversity at all scales, allows components of living systems to interact with environmental conditions and to adapt to their anticipated future states (O'Connor et al. 2019). Although a number of definitions have been recently proposed for cities to be designed considering sustainability and environmental impacts (see chapter "Towards BioCities: The Pathway to Transition"), the key aspect of the BioCity concept resides in a paradigm change, focused on identifying solutions to climate and health crises by mimicking natural systems, rather than being the cause of these problems due to historically exceptional urban development. BioCities should be considered as social-ecological systems (Holling 2001) where the technosphere responsible for the production of goods and services is embedded and integrated within the biophysical constraints of the surrounding biosphere, the latter providing the needed flows from primary sources of energy, matter, and biodiversity, both on the supply and the sink side (Giampietro 2019).

Hence, the societies of BioCities are composed of inhabitants, civic leaders, and public and private actors committed to principles and practices that can accurately be described as nature based and supportive of social and environmental justice.

BioCities are also places where a particularly dense concentration of projects and configurations evidencing these principles can be found. Qualitatively, BioCities are 'cities hosted by nature', rather than 'nature hosted by cities'. In BioCities, green and blue components are understood to be assets, rather than costs. BioCities are not static objects, but amalgamations of dynamic processes. Accordingly, being recognised as a BioCity involves cumulative thresholds continuously evolving throughout the process of urban development—it is a journey as much as a destination. The transformation of existing urban areas into BioCities, in response to shifting environmental feedback, mirrors the progressive evolution and succession of natural species and ecosystems.

To realise this vision, we must look to nature and ecological functioning. The biomimicry approach promotes a new way of addressing economic and social developments of global territories, starting from the implementation of NBS, the optimisation of ES, and the advancement of the circular bioeconomy. This approach is, by definition, an interdisciplinary endeavour synthesising the fields of forestry, ecology, biology, chemistry, sociology, agriculture, urban planning, architecture, landscape architecture, industrial design, engineering, economics, governance, medical science, and social science amongst others.

Since the first surviving writings on the interactions between animals and their environments by Aristotle from the fourth century BC, many have ventured to define the key operational criteria of resilient social-ecological systems (Ramaley 1940; Berkes et al. 2002). For the purposes of the following manifesto, it will suffice for the reader to bear in mind the following criteria elaborated in the 2015 book by Reinette Biggs, *Principles for Building Resilience: Sustaining Ecosystem Services in Social-Ecological Systems*. Resilient social-ecological systems: (1) maintain diversity and redundancy; (2) manage connectivity; (3) manage slow variables and feedbacks; (4) foster complex adaptive systems thinking; (5) encourage learning; (6) broaden participation; and (7) promote polycentric governance systems (Biggs 2015).

Reinterpreting the above to specifically address the urban context, if a city behaved as a resilient, nature-based social-ecological system (practically rather than metaphorically), it would respect the following 10 key functional traits:

1. *The BioCity as a Carbon Sink*

   The BioCity has no net emissions of carbon dioxide (CO_2) and other greenhouse gasses (GHGs) but rather net absorption (Fig. 1) (see also chapter "Mitigation and Adaptation for Climate Change: The Role of BioCities and Nature Based Solutions"), as a forest ecosystem does (Harris et al. 2021). In this way, the challenge of creating net-zero emissions cities posed by the European Commission (EC) (EC 2021), is achieved. The BioCity interacts intentionally with trees and forests within and beyond the urban boundary to benefit from the goods and services they sustainably provide both during life and whilst incorporated within building materials. Equally, the actions of urban dwellers are made to benefit trees and forests. Vegetation is not merely decorative, but a critical infrastructure of the urban system thoroughly integrated in BioCity design and planning. Nature is not artificialised, instead naturalness is reintroduced to the city and expanded, maximising benefit whilst minimising maintenance. The BioCity goes beyond employing technocratic NBS, as 'designing…and managing ecosystems in very intrusive ways… to mitigate city warming and clean polluted air' (Type 3 of NBS) according to Eggermont et al. (2015), by adopting holistic nature-based thinking (NBT) (Randrup et al. 2020).

2. *The Self-Sufficient BioCity*
   Whereas natural ecosystems do not import resources from beyond the scope of their natural environmental fluxes, the BioCity produces within its own buildings, neighbourhoods, urban areas, and local BioRegion (see chapter "From BioCities to BioRegions and Back: Transforming Urban-Rural Relationships" for the definition of BioRegion) the derivative resources underpinning its operation (Fig. 2) (Guallart 2014). It produces energy through its own renewable systems, extracts water from its own natural basins or subsoils, and grows food and biomass for its own population (Fig. 3). All of these resources are used in such a way that they remain, cascading and recycling through interconnected, decentralised networks, within the localised system and hinterlands in near perpetuity, accessible to all inhabitants who need them, without surpassing ecosystem thresholds nor planetary boundaries (see also chapter "The Enabling Environment for BioCities") (Rockström et al. 2009b). Sufficiency supersedes the goal of efficiency.
3. *The Multi-Layered BioCity*
   Like forests (see also chapter "Towards BioCities: The Pathway to Transition"), which capitalise on vertical stratification (Fig. 4) to enhance biodiversity, light and rainfall interception, and primary productivity, the BioCity must be organised such that each of its layers, from the subsoil to the ground, the central body, and the roofs, can develop different, mutually reinforcing functions and provide resources using elements of green, blue, brown, and grey infrastructures (see Glossary) to service the BioCity as a whole (Fig. 3) (Silva et al. 2020).
4. *The Healthy Living BioCity*
   Just as in a natural community, where most individuals and populations thrive when living in stable and sustainable ecosystems, so too the health of individuals and populations living in a BioCity (see also chapter "BioCities as Promotors of Health and Wellbeing") benefits from the highest possible quality of life, a sustainable environment (WHO 1946), and a strongly connected, mutually supportive community (Fig. 4) (Adkins 2015). The BioCity exceeds characterisation as a collection of human settlements, instead people are understood to be part of an ecosystem. Since BioCities are necessarily urban areas that promote a wide spectrum of life (*bios*); human well-being and biodiversity are fostered by the same multi-scalar strategies as in natural ecosystems (Fig. 5). This is achieved by using the entire palette of biophysical structures and functions to aid the provision of ES demonstrated by how forests reach equilibrium (Brockerhoff et al. 2017). The removal of atmospheric and substrate pollutants, the mitigation of periods of extreme heat or cold, the reimagining of urban mobility to encourage active lifestyles, the improved access to green or blue spaces for exercise or recreation, the implementation of natural systems, and the enhancement of biodiversity improve the quality of people's lives, along with health and well-being, whilst simultaneously bolstering non-human communities.
5. *The Circular Bioeconomy BioCity*
   Paralleling the natural phenomena of trophic cascades and biodegradation, the circular bioeconomy makes the BioCity a vibrant, regenerative system (see

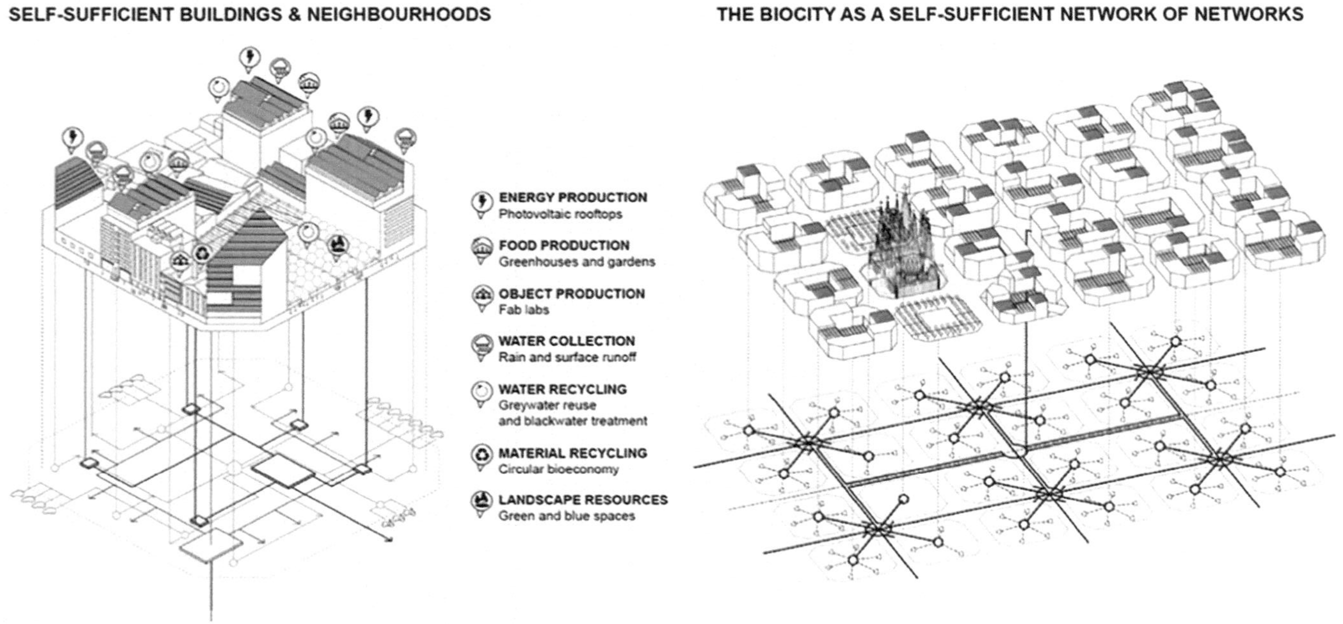

Fig. 2 By networking self-sufficient buildings and neighbourhoods, the BioCity becomes a self-sufficient network of networks. ©Vicente Guallart (adapted from Guallart 2014)

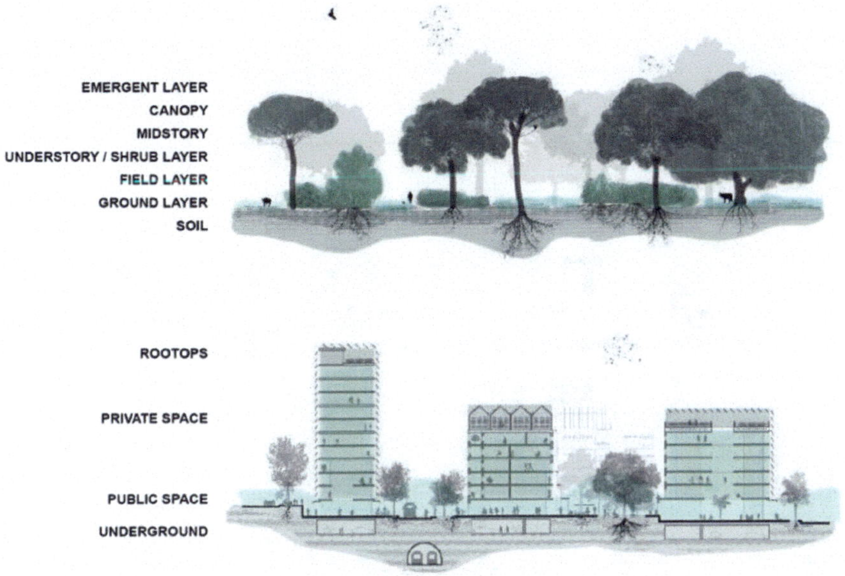

Fig. 3 Like a forest, the BioCity is composed of vertical strata which perform diverse, mutually reinforcing functions. ©IAAC (adapted from Guallart et al. 2021)

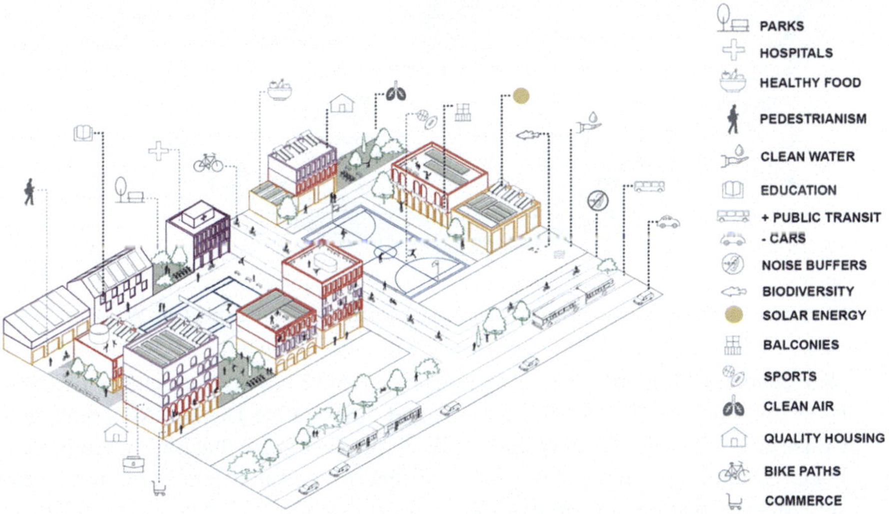

Fig. 4 Connected communities in sustainable environments support the healthy living of citizens in the BioCity. ©IAAC (adapted from Guallart et al. 2021)

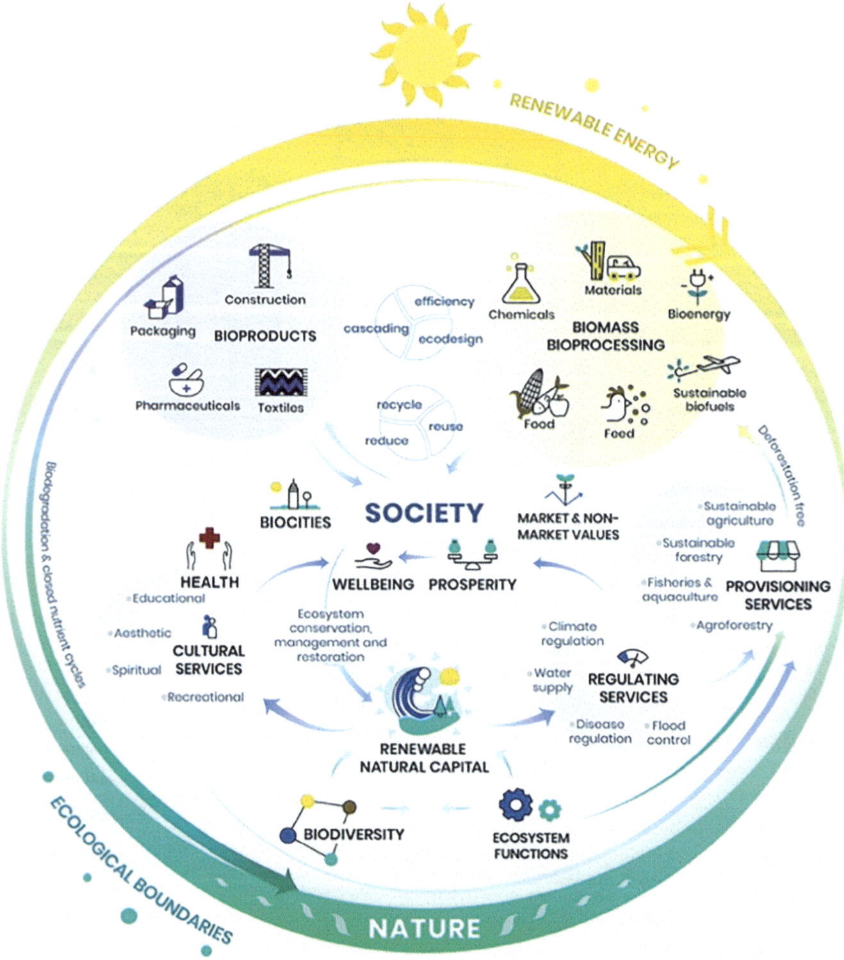

Fig. 5 The circular bioeconomy of well-being. ©EFI (adapted from Palahí et al. 2020)

also chapters "Forests, Forest Products and Services to Activate a Circular Bioeconomy for City Transformation" and "Innovative Design, Materials, and Construction Models for BioCities"). BioCities recognise that highly productive linear economic models, based on fossil energy and non-renewable materials, should be converted into circular bioeconomy systems (Fig. 5) which replace the 'end-of-life' concept with recovering and reusing materials and biomaterials in production processes operating at different economic levels, from the local to the macro-scale (EC 2018).

However, as argued by Giampietro (2019), the circular bioeconomy model should also be analysed according to a wider thermodynamic narrative considering that every loop in the reusing and recycling pathways 'creates dissipation and entropy, attributed to losses in quantity (physical material losses, by-products) and quality (mixing, downgrading). New materials and energy must be injected into any circular material loop, to overcome these dissipative losses' (Cullen 2017). Therefore, a sustainable circular bioeconomy should necessarily comply with the biophysical constraints governing the ecological processes which provide inputs of natural resources and absorb wastes (Giampietro 2019).

At the same time, the continuous processing of circulating energy and matter within complex social-ecological systems, as urban ecosystems are, creates an integrated system of flows, combining primary flows of natural resources, secondary flows of industrial transformations, and services and tertiary flows of recycled products, connecting the techno- and biosphere which ground the self-organising, open systems. For these complex metabolic systems to survive, they must learn and adapt to changes in their environmental boundary conditions as well as anticipate potential future changes (Giampietro 2019). Likewise, these systems generate dynamic hierarchies of multiple activities, services, and production value chains, which undergo constant reinvention, and spawn ample aligned job opportunities through the use and development of local bio-based materials and recycled materials to manufacture, maintain, and improve the products demanded for the proper functioning of the BioCity (Silliman and Angelini 2012). BioCity products are designed to be repurposed as resources at the end of their period of initial use, thereby ensuring their constituent matter remains within the manufacturing ecosystem as long as possible. Otherwise, they are designed to be compostable, ending up as necessary nutrients for building sound and productive soils. Accordingly, waste is non-existent in the BioCity. Compelling incentives to redesign not just what we make, but also how, will spark a new wave of fruitful creativity.

6. *The Low-Mobility Connected BioCity*

In a natural ecosystem, many animals' mobility is bound to the radii containing their vital sources of food and water, along with potential mates (Fig. 6). Reflecting this principle, the low-mobility BioCity promotes changes in the movement habits of its population. Through functional reorganisation of an urban area, all basic services necessary to live are made readily available within the radius of a 15-minute walk or cycle (Moreno et al. 2021). This localised concentration of activity reduces overreliance on motorised transport through network redundancy, which more than makes up for any increase in resources invested in the repetition of services by greatly reducing the amount of resources demanded for costly vehicular infrastructures, in addition to achieving numerous health and well-being, economic, and environmental improvements associated with favouring multimodal human-powered transport. Such network redundancy, modelled after the mycorrhizal networks (Fig. 7) (Simard et al. 2012) which serve as the distributed 'brain' of a forest ecosystem, is also manifested by the BioCity's physical and information infrastructures. The connected BioCity enables individuals to exchange goods and information as a superlinear function of community size (Schläpfer et al. 2014); in such a way it

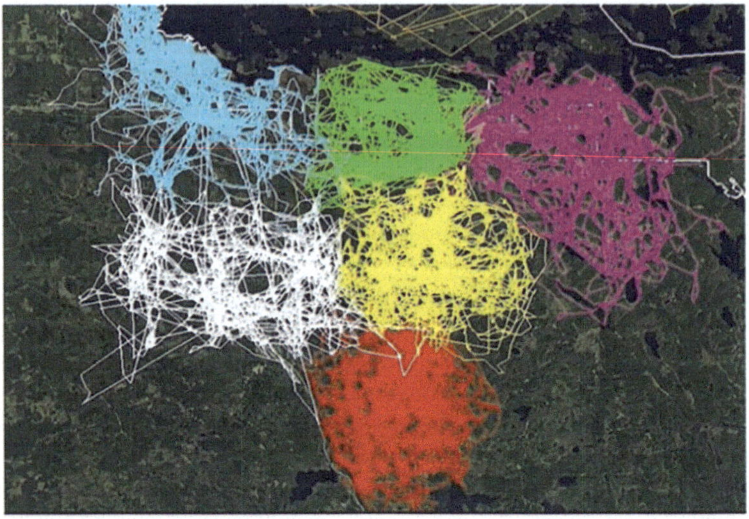

Fig. 6 The tracked movements of wild wolves demonstrate their territoriality, and how an individual's mobility patterns are limited to the radii containing their vital necessities. ©Voyageurs Wolf Project (Gable 2018)

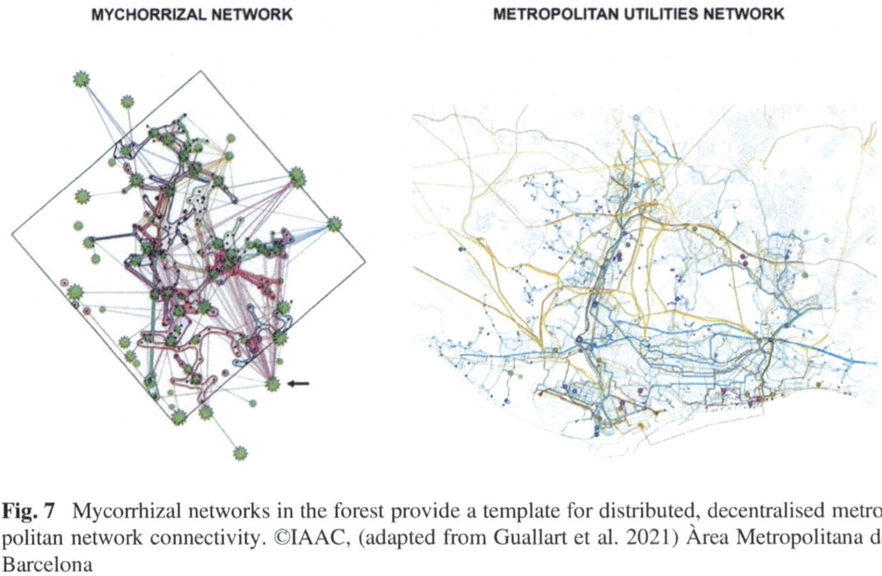

Fig. 7 Mycorrhizal networks in the forest provide a template for distributed, decentralised metropolitan network connectivity. ©IAAC, (adapted from Guallart et al. 2021) Àrea Metropolitana de Barcelona

will allow society to function, flow, and progress together in the most sustainable, efficient, and ecological manner. Every BioCity subcomponent has a specific role (or, whenever possible, combines multiple roles), adding up to a whole greater than the sum of its parts. All actors are connected by

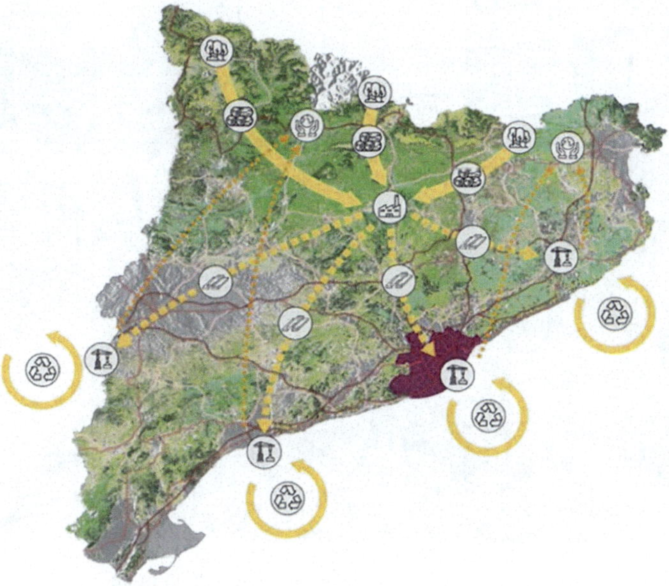

Fig. 8 Scheme of nature-based value chains for engineered timber in the region of Catalonia serving the Metropolitan Area of Barcelona. ©IAAC (adapted from Guallart et al. 2021)

communication and collaboration, resulting in a widespread awareness of, and mutual responsibility for, all processes. Moreover, the BioCity itself is connected to other BioCities and BioRegions as a unique node within a compounded network.

7. *The Urban-Rural Balanced BioCity*

   Just as soft, blurred, gradated, fluid, and reciprocal boundaries between discrete natural ecosystems (ecotones) optimise health and function, unbiased symbioses and dialogues between the BioCity and its encompassing BioRegion (see chapter "From BioCities to BioRegions and Back: Transforming Urban-Rural Relationships") enable urban systems to work in harmony with the natural systems of their territorial environments. This balance thus fuels both the urban and rural economies, through the growth of robust, regionalised, nature-based value chains (NBVCs) (Yahner 1988; Ibañez et al. 2022). Urban populations are bestowed with regional supplies, and the regional populations they depend upon are economically enlivened (Fig. 8). Environmentally arbitrary managerial borders are re-evaluated with deference to the functional extents of natural systems.

8. *The Participatory Local Culture BioCity*

   Natural ecosystems are typified by characteristic ecological communities, or groups of actual or potentially interacting species bound together in a network of influence and a shared environment (Whittaker 1970). Likewise, the BioCity is not only adaptive to its environment and to a changing climate, at local and global scales, but also promotes a material, cultural and social identity based on

Fig. 9 Communal urban gardens enable citizens to assert bottom-up self-determination by co-creating the ecology of the BioCity. ©Justin Pickard (Creative Commons: https://creativecommons.org/licenses/by-sa/2.0/legalcode)

its unique local history and traditions via continuous exchange with the broader world through physical and information networks. Moreover, the interweaving of nature and culture can have such a positive effect on biodiversity that the concept of *biocultural diversity* has been authored to express the crucial role of knowledge, innovations, and practices of local communities in conservation and sustainability (MAB 2017). Similarly, it is evident that place-based and intangible knowledge, cultural heritage, and vernacular practices are pivotal for understanding and shaping the enormously diverse landscapes of the world. Through an integrated governance ecosystem incorporating top-down and bottom-up decision-making with communal rights, local residents and communities are proactively engaged through participatory approaches in self-determining the realities and networks of influence of their BioCity, coming to see its spaces as shared property, whilst tacit knowledge leads to insightfully attuned nature-based interventions (Fig. 9).

9. *The Resilient BioCity*

   Newly established or disrupted ecosystems proceed through sequential seral stages along the process of ecological successions, before attaining a sustainable, 'mature' climax community, scaffolded by the accomplishments of prior phases, culminating with concurrently high biodiversity, productivity, and stability (Bai et al. 2004). According to the more recent ecological and socio-economic theory of panarchy, natural as well as social-ecological systems are interlinked in never-ending adaptive cycles of growth, accumulation, restructuring, and renewal whilst the equilibrium state is only temporary or,

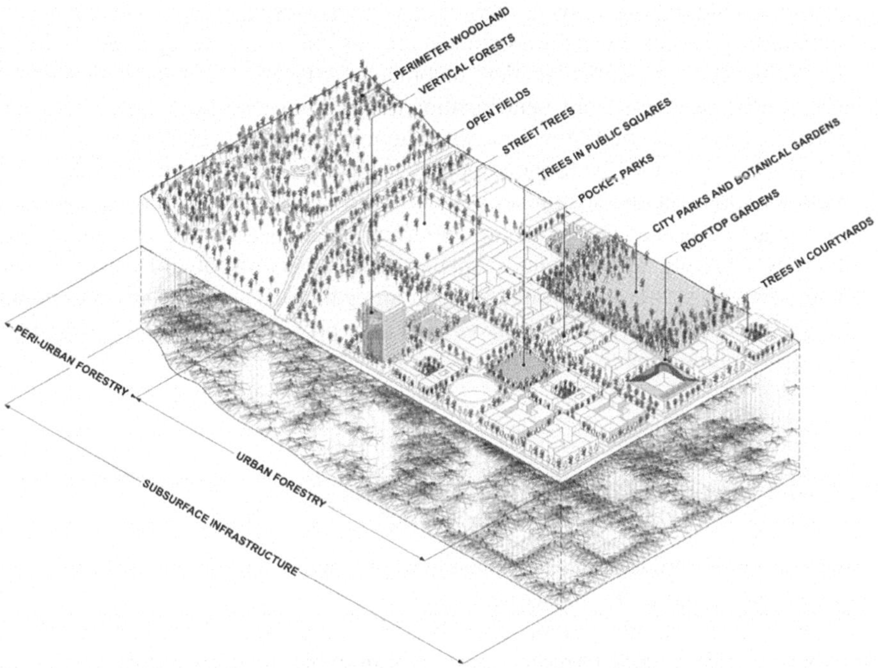

Fig. 10 A diversity of trees, forests, and other green and blue spaces and infrastructures empower the BioCity with resilience. ©IAAC (adapted from Guallart et al. 2021)

more accurately, composed of a complexity of dynamic states of equilibria (Walker et al. 2004; Stanger and Beauchamp 2015). In a mature BioCity, publicly accessible urban blue and green nature in the forms of forests, meadows, individual trees, rivers, creeks, lakes, ponds, and waterfronts, provide a diverse population of citizens with opportunities to lead productive, stable, healthy, meaningful lives (Fig. 10). Such public, accessible places provide democratic realms in accordance with the justice perceptions of all affected stakeholders and globally accepted standards for human rights. In doing so, they perpetuate the value of past human and natural heritage, as well as form the infrastructures which will be required to absorb the shocks and meet the challenges of tomorrow. The preserved historical dimension, combined with a safeguarded basis for future resilience, enables innovation whilst maintaining the resources posterity will rely on. In this way, the BioCity balances continuity with prosperity.

10. *BioCities for All*

In nature, competitive exclusion amongst similar species is alleviated by variation in tailored environmental niches, defined as the range of conditions necessary for persistence of the species, and its ecological role in the ecosystem (MacArthur 1958; Junker et al. 2019). Within the BioCity, biodiversity is prioritised not only in terms of sheltering a variety of species, but also in terms of maximising accessibility for all citizens, regardless of ability, age, race, ethnicity, religion, occupation, gender, income, or education, whilst undermining forced displacement from gentrification, with commensurate variation in tailored niches of the built environment. The involvement of citizens is natural at all levels, from locally founded activities and management to planning and policy-making at the overall city scale. Ultimately, the BioCity for all will eliminate systemic and structural environmental inequalities and injustices, thereby ensuring all residents reap the Gestalt benefits of diverse exchange and common stewardship founded upon an egalitarian sense of urban community. Hence, BioCities will be universal in their provision of attainable resources for every demographic population.

5 Outcomes and Concluding Remarks

From these principles, we can extrapolate multiple guidelines and initiatives that will be further elaborated in the subsequent chapters.

Governance and administration in the BioCity entail a shift in the mindset of city administrators from short-term economic favourability in decision-making facilitated primarily by technological solutions, towards a commitment to nature as a means for solving many of society's contemporary challenges. Instigating this new paradigm will require radical approaches to city planning, management, and co-governance prioritising long-sighted approaches, iterative processes, and citizen participation. A local and place-based focus is needed to establish contextualised, engaging, and inclusive environments for all. These aspects will be developed particularly in chapters "Towards the Development of a Conceptual Framework of BioCities", "BioCities as Promotors of Health and Wellbeing" and "The Social Environment of BioCities".

Governance mechanisms should protect the equilibrium between urban areas and their associated rural regions and resources, emphasising their reciprocal interests. The nexus between BioCities and their BioRegions is especially elaborated in chapter "From BioCities to BioRegions and Back: Transforming Urban-Rural Relationships".

The BioCity places special emphasis on blue and green infrastructures (Fig. 10). This will help to ensure that the potential to serve urban residents and visitors is secured in the longer term. A sample of specific goals include: (1) reforming urban surfaces and limiting the impermeable materials of streets and facades to ameliorate rapid runoff and mitigate the urban heat island (UHI) phenomenon; (2) renaturing rivers, canals, and wetlands in the interiors of cities, and constructing water retention systems for evaporative cooling, irrigation, and biodiversity; (3) creating connective, multifunctional green infrastructures, natural corridors, and ecological networks for people and wildlife to freely cross habitats and move between the city and hinterland. Chapters "Biodiversity and Ecosystem Functions as Pillars of BioCities", "Green Infrastructure and Urban Forests for BioCities: Strategic and Adaptive Management", and "Mitigation and Adaptation for Climate Change: The Role of BioCities and Nature Based Solutions" specifically deal with nature-based solutions (NBS), with trees and forests within urban areas, and with their interactions with the urban climate and environment.

Developing urban forestry at multiple scales, from peri-urban forests to green belts, urban parks, pocket parks, green roofs, green facades, gardens, and street trees is essential. These provide refuge for both people and wildlife, and play a key role in generating and providing nature's contributions to people. Urban forestry developments must also be paired with the cessation of deforestation to ensure sustainability at a global scale.

Cities are also the ideal testing grounds in which to drive the global transition towards a new circular bioeconomy: their characteristic concentrations of actors, data, and capital make policy changes easier to implement than at the territorial or national scale, because they can adapt more quickly and demonstrate the results of implementations faster and with immediate and visible benefits. A circular economic model is based on three principles: (1) design out waste and pollution; (2) keep products and materials in use; and (3) regenerate natural systems (EMF 2021). A circular bioeconomy extends these principles to use renewable natural capital to comprehensively transform and manage land, food, health, and industrial systems with the goal of achieving sustainable well-being (EFI 2020). Chapter "Forests, Forest Products and Services to Activate a Circular Bioeconomy for City Transformation" and "Innovative Design, Materials, and Construction Models for BioCities" develop the issue of circular bioeconomy in BioCities as well as the role of bio-based materials, primarily wood, in the transition towards BioCities.

The overarching challenge of the new urban system of BioCities will be to depart from a global degenerative and exploitative fossil-fuelled culture, dependent upon depleting non-renewable sources of materials and energy, and to adopt instead a restorative and regenerative culture based on renewable biomaterials, adapted to the available sources, and contributing to the improvement of planetary health. Meeting this challenge will necessitate programmes and platforms encouraging and enabling investment in discrete projects adding up to a circular bioeconomy. In light of urgent social and climatic pressures, it will prove critical to confirm definitive plans and timescales for investment and establish financial schemes guaranteeing capital commitments as soon as possible, as described in chapter "The Enabling Environment for BioCities".

To adequately address the multifaceted and variable existential environmental, social, and economic challenges facing contemporary civilisation, there is no singular set of categorical rules BioCities can follow without adapting to their unique local contexts. Responsive methods and strategies for implementing the principles of BioCities (as shown in the final chapter "Towards BioCities: The Pathway to Transition" in differing world regions and circumstances must therefore be well developed. Whilst nature-based thinking (NBT) represents a universally valid approach, its delivery and manifestation should rightly differ from city to city, reflecting the particular needs of the respective citizens, biodiversity, cultural history, and geography.

References

Adkins S (2015) ESS relevant—pedagogical implications of class background. Association for Environmental Studies and Sciences 2015 annual conference: confronting frontiers, borders and boundaries, San Diego. aessonline.org/wp-content/uploads/2015/12/AESS_2015_PROGRAM_CONFERENCE_FINAL.pdf

Architecture 2030 (2018) New buildings: embodied carbon. architecture2030.org/new-buildings-embodied/

Bai Y, Han X, Wu J, Chen Z, Li L (2004) Ecosystem stability and compensatory effects in the Inner Mongolia grassland. Nature 431(7005):181–184. https://doi.org/10.1038/nature02850

Berkes F, Colding J, Folke C (eds) (2002) Navigating social-ecological systems: building resilience for complexity and change. Cambridge University Press. https://doi.org/10.1017/CBO9780511541957

Biggs R (2015) Principles for building resilience: sustaining ecosystem services in social-ecological systems. Cambridge University Press

Bosker M (2021) City origins. Reg Sci Urban Econ 103677. https://doi.org/10.1016/j.regsciurbeco.2021.103677

Boulding K (1966) The economics of the coming spaceship earth. In Jarrett H (ed) Environmental quality in a growing economy. Resources for the Future/Johns Hopkins University Press, pp 3–14. www.google.com/url?sa=t&rct=j&q=&esrc=s&source=web&cd=&ved=2ahUKEwjY4tCS7f_5AhWBYMAKHWZ4BIsQFnoECAkQAQ&url=http%3A%2F%2Fwww.zo.utexas.edu%2Fcourses%2Fthoc%2FBoulding_SpaceshipEarth.pdf&usg=AOvVaw16dXp8Y2Ej7TtdTu_-E9ht

Brockerhoff EG, Barbaro L, Castagneyrol B, Forrester DI, Gardiner B, González-Olabarria JR, Lyver PO, Meurisse N, Oxbrough A, Taki H, Thompson ID, van der Plas F, Jactel H (2017) Forest biodiversity, ecosystem functioning and the provision of ecosystem services. Biodivers Conserv 26(13):3005–3035. https://doi.org/10.1007/s10531-017-1453-2

Cerdà I (2018) General theory of urbanisation 1867. Guallart V (ed) Actar Publishers Institute for Advanced Architecture of Catalonia. actar.com/product/general-theory-of-urbanization-1867/

Cullen JM (2017) Circular economy: theoretical benchmark or perpetual motion machine? J Ind Ecol 21(3):483–486. https://doi.org/10.1111/jiec.12599

D'Amato D, Droste N, Allen B, Kettunen M, Lähtinen K, Korhonen J, Leskinen P, Matthies BD, Toppinen A (2017) Green, circular, bio economy: a comparative analysis of sustainability avenues. J Clean Prod 168:716–734. https://doi.org/10.1016/j.jclepro.2017.09.053

Diamond JM (1997) Guns, germs and steel: the fates of human societies. Jonathan Cape

Eggermont H, Balian E, Azevedo JMN, Beumer V, Brodin T, Claudet J, Fady B, Grube M, Keune H, Lamarque P, Reuter K, Smith M, van Ham C, Weisser WW, Le Roux X (2015) Nature-based solutions: new influence for environmental management and research in Europe. GAIA Ecol Perspect Sci Soc 24(4):243–248. https://doi.org/10.14512/gaia.24.4.9

European Commission (2018) Guidance on cascading use of biomass with selected good practice examples on woody biomass. Publications office of the European Union. https://data.europa.eu/doi/10.2873/68553

EFI (2020) Circular bioeconomy offers game-changing solutions. European Forest Institute. efi.int/articles/circular-bioeconomy-offers-game-changing-solutions

EMF (2021) What is a circular economy? Ellen MacArthur Foundation. www.ellenmacarthurfoundation.org/circular-economy/concept

European Commission (2020) 2050 long-term strategy. ec.europa.eu/clima/eu-action/climate-strategies-targets/2050-long-term-strategy_en
European Commission (2021) EU missions: concrete solutions for our greatest challenges. Publications Office of the European Union. https://doi.org/10.2777/500470
Gable T (2018) Wolf territoriality [map]. Voyageur Wolf Project. www.voyageurswolfproject.org/animations
Giampietro M (2019) On the circular bioeconomy and decoupling: implications for sustainable growth. Ecol Econ 162:143–156. https://doi.org/10.1016/j.ecolecon.2019.05.001
Glaeser EL (2011) Triumph of the City. Macmillan. cam.ldls.org.uk/vdc_100039975747.0x000001
Gottmann J (1961) Megalopolis: the urbanized northeastern seaboard of the United States. https://doi.org/10.7551/mitpress/4537.001.0001
Guallart Architects & Built by Associative Data (2019) The forest city project [Masterplan]. www.worldbuildingsdirectory.com/entries/the-forest-city-project/
Guallart V (2014) The self-sufficient city—internet has changed our lives but it hasn't changed our cities, yet. Actar Publishers. actar.com/product/the-self-sufficient-city/
Guallart V, Ibañez D, Salka M, Capra Ribeiro F (2021) Biociudad: Modelo para la aplicación de los principios de la bioeconomía circular en la Área Metropolitana de Barcelona. Institute for Advanced Architecture of Catalonia
Haeckel E (1866) Generelle Morphologie der Organismen. Georg Reimer. books.google.co.uk/books?id=dthOAAAAMAAJ&printsec=frontcover&redir_esc=y#v=onepage&q&f=false
Harris NL, Gibbs DA, Baccini A, Birdsey RA, de Bruin S, Farina M, Fatoyinbo L, Hansen MC, Herold M, Houghton RA, Potapov PV, Suarez DR, Roman-Cuesta RM, Saatchi SS, Slay CM, Turubanova SA, Tyukavina A (2021) Global maps of twenty-first century forest carbon fluxes. Nat Clim Chang 11(3):234–240. https://doi.org/10.1038/s41558-020-00976-6
Holling CS (2001) Understanding the complexity of economic, ecological, and social systems. Ecosystems 4(5):390–405. https://doi.org/10.1007/s10021-001-0101-5
Howard SE (1898) To-morrow: a peaceful path to real reform. Swan Sonnenschein books.google.co.uk/books/about/Tomorrow.html?id=gtf0nAEACAAJ&redir_esc=y
Ibañez D, Guallart V, Salka M (2022) On pedagogical prototyping of advanced ecological buildings and biocities at Valldaura labs. AGATHÓN: International Journal of Architecture, Art and Design, 11
IPCC (2022) Climate change 2022: impacts, adaptation and vulnerability: summary for policymakers. Cambridge University Press. www.ipcc.ch/report/ar6/wg2/downloads/report/IPCC_AR6_WGII_SummaryForPolicymakers.pdf
Junker RR, Lechleitner MH, Kuppler J, Ohler LM (2019) Interconnectedness of the Grinnellian and Eltonian niche in regional and local plant-pollinator communities. Front Plant Sci 10. https://doi.org/10.3389/fpls.2019.01371
Kern K (2019) Cities as leaders in EU multilevel climate governance: embedded upscaling of local experiments in Europe. Environ Polit 28(1):125–145. https://doi.org/10.1080/09644016.2019.1521979
Le Corbusier (1924) Vers une architecture (Nouv. éd. revue et augmentée). G. Crès et Cie. www.worldcat.org/title/vers-une-architecture/oclc/3756638
MAB (2017) A new roadmap for the man and the biosphere (MAB) Programme and its world network of biosphere reserves. UNESCO unesdoc.unesco.org/ark:/48223/pf0000247418
MacArthur RH (1958) Population ecology of some warblers of northeastern coniferous forests. Ecology 39(4):599–619. https://doi.org/10.2307/1931600
Meadows D, Meadows D, Randers J, Berens, WI (1972) The limits to growth. Potomac Associates. www.clubofrome.org/publication/the-limits-to-growth/
Moreno C, Allam Z, Chabaud D, Gall C, Pratlong F (2021) Introducing the "15-minute city": sustainability, resilience and place identity in future post-pandemic cities. Smart Cities 4(1): 93–111. https://doi.org/10.3390/smartcities4010006

O'Connor MI, Pennell MW, Altermatt F, Matthews B, Melián CJ, Gonzalez A (2019) Principles of ecology revisited: integrating information and ecological theories for a more unified science. Front Ecol Evol 7. https://doi.org/10.3389/fevo.2019.00219

Palahí M, Pantsar M, Costanza R, Kubiszewski I, Potočnik J, Stuchtey M, Nasi R, Lovins H, Giovannini E, Fioramonti L, Dixson-Declève S, McGlade J, Pickett K, Wilkinson R, Holmgren J, Trebeck K, Wallis S, Ramage M, Berndes G (2020) Investing in nature as the true engine of our economy: a 10-point action plan for a circular bioeconomy of wellbeing. European Forest Institute. https://doi.org/10.36333/k2a02

Pendrill F, Persson UM, Godar J, Kastner T, Moran D, Schmidt S, Wood R (2019) Agricultural and forestry trade drives large share of tropical deforestation emissions. Glob Environ Chang 56:1–10. https://doi.org/10.1016/j.gloenvcha.2019.03.002

Ramaley F (1940) The growth of a science. University of Colorado Studies 26:3–14

Randrup TB, Buijs A, Konijnendijk CC, Wild T (2020) Moving beyond the nature-based solutions discourse: introducing nature-based thinking. Urban Ecosyst 23(4):919–926. https://doi.org/10.1007/s11252-020-00964-w

Rockström J, Steffen W, Noone K, Persson Å, Chapin FSI, Lambin E, Lenton TM, Scheffer M, Folke C, Schellnhuber HJ, Nykvist B, de Wit CA, Hughes T, van der Leeuw S, Rodhe H, Sörlin S, Snyder PK, Costanza R, Svedin U, Foley J (2009a) Planetary boundaries: exploring the safe operating space for humanity. Ecol Soc 14(2). https://doi.org/10.5751/ES-03180-140232

Rockström J, Steffen W, Noone K, Persson Å, Chapin FSI, Lambin EF, Lenton TM, Scheffer M, Folke C, Schellnhuber HJ, Nykvist B, de Wit CA, Hughes T, van der Leeuw S, Rodhe H, Sörlin S, Snyder PK, Costanza R, Svedin U, Foley JA (2009b) A safe operating space for humanity. Nature 461(7263):472–475. https://doi.org/10.1038/461472a

Rossi A (1966) L'architeturra della Città. Marsilio

Scheiner SM, Willig MR (2008) A general theory of ecology. Theor Ecol 1(1):21–28. https://doi.org/10.1007/s12080-007-0002-0

Schläpfer M, Bettencourt LMA, Grauwin S, Raschke M, Claxton R, Smoreda Z, West GB, Ratti C (2014) The scaling of human interactions with city size. J R Soc Interface 11(98):20130789. https://doi.org/10.1098/rsif.2013.0789

Silliman BR, Angelini C (2012) Trophic cascades across diverse plant ecosystems. Nat Educ Knowl 3(10):44

Silva I, Rocha R, López-Baucells A, Farneda FZ, Meyer CFJ (2020) Effects of forest fragmentation on the vertical stratification of neotropical bats. Diversity 12(2):67. https://doi.org/10.3390/d12020067

Simard SW, Beiler KJ, Bingham MA, Deslippe JR, Philip LJ, Teste FP (2012) Mycorrhizal networks: mechanisms, ecology and modelling. Fungal Biol Rev 26(1):39–60. https://doi.org/10.1016/j.fbr.2012.01.001

Stahel WR, Reday-Mulvey G (1981) Jobs for tomorrow: the potential for substituting manpower for energy. Vantage Press

Stanger N, Beauchamp J (2015) Panarchy, transformation, and place: exploring social change and resiliency through an ecological lens. Trumpeter 31(1):14–38

Steffen W, Richardson K, Rockström J, Cornell SE, Fetzer I, Bennett EM, Biggs R, Carpenter SR, de Vries W, de Wit CA, Folke C, Gerten D, Heinke J, Mace GM, Persson LM, Ramanathan V, Reyers B, Sörlin S (2015) Planetary boundaries: guiding human development on a changing planet. Science 347(6223):1259855. https://doi.org/10.1126/science.1259855

UN DESA (2015) THE 17 GOALS | Sustainable Development. sdgs.un.org/goals

UN DESA (2018) 2018 Revision of World Urbanization Prospects | Multimedia Library—United Nations Department of Economic and Social Affairs. www.un.org/development/desa/publications/2018-revision-of-world-urbanization-prospects.html

UN DESA (2019) World Population Prospects Highlights, 2019 Revision Highlights

UN Habitat (2020) World Cities Report 2020—The Value of Sustainable Urbanization. unhabitat.org/World%20Cities%20Report%202020

UN Habitat (2022) World Cities Report 2022: Envisaging the Future of Cities. unhabitat.org/wcr/

UN General Assembly (2019) Only 11 years left to prevent irreversible damage from climate change, speakers warn during general assembly high-level meeting. Meetings coverage and press releases. press.un.org/en/2019/ga12131.doc.htm

Wackernagel M, Rees W (1996) Our ecological footprint: reducing human impact on the Earth. New Society Publishers

Walker B, Holling CS, Carpenter S, Kinzig A (2004) Resilience, adaptability and transformability in social–ecological systems. Ecol Soc 9(2). https://doi.org/10.5751/ES-00650-090205

West G (2018) Scale: the universal laws of life and death in organisms, cities and companies. Weidenfeld & Nicolson

Whittaker RH (1970) Communities and ecosystems. Communities and Ecosystems. www.cabdirect.org/cabdirect/abstract/19740615709

WHO (1946) Constitution of the world health organization. https://www.google.com/url?sa=t&rct=j&q=&esrc=s&source=web&cd=&cad=rja&uact=8&ved=2ahUKEwiolJiuv4D6AhUOS8AKHZgcCHQQFnoECBgQAQ&url=https%3A%2F%2Fwww.who.int%2Fabout%2Fgovernance%2Fconstitution&usg=AOvVaw2vCe7z8J3d1tCeewCn3nB0

Xu C, Kohler TA, Lenton TM, Svenning JC, Scheffer M (2020) Future of the human climate niche. Proc Natl Acad Sci 117(21):11350–11355. https://doi.org/10.1073/pnas.1910114117

Yahner RH (1988) Changes in wildlife communities near edges. Conserv Biol 2(4):333–339. https://doi.org/10.1111/j.1523-1739.1988.tb00197.x

Urban Sustainable Futures: Concepts and Policies Leading to BioCities

Giovanni Sanesi, Fabio Salbitano, Giovanna Ottaviani Aalmo, Wendy Chen, Silvija Krajter Ostoic, Jerylee Wilkes-Allemann, and Clive Davies

1 Introduction

BioCities are strongly linked to evolving contemporary concepts that have become an integral part of the urban discourse around multi-scale policy making. Urban communities are searching for new ideas for developing a sustainable future (Calhoun 2012). This includes concepts such as urban forest, ecosystem services, green infrastructure, ecosystem-based adaptation, nature-based solutions (FAO 2016; Pauleit et al. 2017; Escobedo et al. 2019), nature's contributions to people (Managi et al. 2022), and nature-based thinking (Randrup et al. 2020). Many of these concepts have been adopted in the European Union (EU) urban policy framework

G. Sanesi (✉)
Italian Society of Silviculture and Forest Ecology (SISEF), Viterbo, Italy
e-mail: giovanni.sanesi@uniba.it

F. Salbitano
University of Sassari, Sassari, Italy

G. O. Aalmo
Norwegian Institute of Bioeconomy Research (NIBIO), Ås, Norway

W. Chen
Hong Kong University, Hong Kong, China

S. K. Ostoic
Croatian Forest Research Institute, Jaska, Croatia

J. Wilkes-Allemann
Bern University of Applied Sciences (BFH), Bern, Switzerland

C. Davies
University of Newcastle, Callaghan, NSW, Australia

G. E. Scarascia-Mugnozza et al. (eds.), *Transforming Biocities*, Future City 20,
https://doi.org/10.1007/978-3-031-29466-2_2

based on the declared recognition of *bringing nature back to cities and reward community action.*[1]

Cities are seen as both the source of, and solution to, today's economic, environmental, and social challenges. The pathways to achieve these solutions may have different approaches, with alternative modalities and needs, but they share a common thread to address universal global challenges. Some cities already provide models of this change process, such as Vancouver and Melbourne, and they focus attention, at least initially, on transformative governance according to ecological, social, and economic (ESE) sustainability criteria. On the other hand, it is precisely in large urban areas that the symptoms of environmental discomfort and the worsening of the quality of life appear critical. Rapid experimentation of new socio-environmental solutions in cities, adapted to local conditions, can promote urban environments as champions and drivers of a sustainable green transition with the perspective of cities as socio-ecological systems (Frank et al. 2017).

Europe's urban areas are home to over two-thirds of the EU's population, and they account for about 80% of energy use and generate up to 85% of Europe's GDP (Eurostat 2016). These urban areas are the engines of the European economy and act as catalysts for creativity and innovation throughout the continent. They are also places where persistent problems, such as unemployment, segregation, social inequality, and poverty, are at their most severe. Increasing environmental and health problems are illuminated by the planetary crisis of climate change and habitat depletion. Urban policies have wide cross-border international significance, which is why urban development is central to European policy and those of many other regions globally. In addition, the urban footprint has substantially changed the characteristics and functioning of ecosystems, not only in the immediate proximity to cities but also in remote environments far from urban centres. The evolution of the European landscape is linked to the urbanisation process, particularly over the last two centuries, and the dynamic relationship between the city and the rural environment has often been based on hegemonic and transformative policies of the urban elite. It is, therefore, necessary to explore the development of policies that, particularly in Europe, are conscious of environmental problems generated by the consumption-focused "way-of-life" in cities. To address this issue includes not only economic and technological transformations to limit the impacts of human activities on natural systems, but will also cause profound social and cultural changes. Redefining concepts of the human role in the world environment and how that impacts our urban and rural lifestyles, based on our values and vision of what society can and should be, are the beginning of the transitional process towards creating more sustainable cities and, definitively, to Biocities. This involves, in a central way, the acquisition of the fact that future transformation must have processes and functions built on the foundation of nature.

[1] Biodiversity Strategy for 2030: Bringing nature back into our lives (2020). Communication from the European Commission to the European Parliament, The Council, The European Economic and Social Committee, and the Committee of the Regions.

2 The Journey Towards BioCities

For more than 99% of human history, we have co-evolved with nature. We developed biologically through an adaptive response to our interactions with nature, more so than the "artificial" ecosystem that we have created in cities (including all urban areas) and that has developed in recent generations (Kellert and Calabrese 2015; el-Baghdadi and Desha 2017). Stemming from this evolutionary history, the term *biophilia* was first used by Erich Fromm (1973) and subsequently popularised by Edward O. Wilson (1984), referring to *humans' innate tendency to affiliate with natural and life-like systems and processes*. Biophilia highlights our fundamental, genetically based attraction to and interest in nature.

Even though the precise kinds of nature available in a particular city varies depending on its unique environmental setting, climate, and development context, many forms of urban nature are usually found across a range of scales and degrees of human management. Three categories of urban nature have been defined, including *remnant nature* such as rivers or large urban forests which defy urban development, *accidental nature* such as diverse plants which spontaneously colonise (Clément 2015) vacant or abandoned urban lands (e.g. brownfields), and *human-constructed nature* such as green roofs or vertical gardens, alongside urban parks and green spaces which are actively maintained (Beatley 2020; Hoyle 2020). Newman and Dale (2013) stressed that urban nature is different from our collective concept of pristine nature and wildness, yet worthy of celebrating, since urban nature is intrinsically valuable.

At the species level, nature provides us with food, energy, shelter, soil, water, and air, what the United Nations Millennium Ecosystem Assessment (MEA 2005) defines as "ecosystem services" (ESS) (i.e. the benefits people obtain from ecosystems). ESS are vital to our survival (Thomas and Xing 2021). Even if we were able to replace nature with artificial substitutes, humanity would still be spiritually impoverished simply because those artificial substitutes provide neither biophysical wonder nor comfort, and they would not be able to satisfy our spiritual, psychological, cognitive, imaginative, or emotional needs (Clowney 2013). Though there is nothing comparable to being outdoors in a green space, viewing photographs of nature has positive psychological impacts on people. This concept was illustrated during the COVID-19 pandemic when hospital recovery was delayed for those with limited physical access to nature (Spano et al. 2021).

Regardless of its form and size, urban nature provides abundant benefits to those inhabiting metropolitan areas as well as cities and smaller urban settlements (McDonald and Beatley 2021), as described in detail in chapters "Mitigation and Adaptation for Climate Change: The Role of BioCities and Nature Based Solutions" and "BioCities as Promotors of Health and Wellbeing". Urban trees, for instance, remove air pollutants such as nitrous oxides and particulate matter (Nowak et al. 2014), mitigate urban heat island (Livesley et al. 2016), reduce urban flooding, and improve surface water quality (Armson et al. 2013). Moreover, having nature nearby delivers immense emotional and mental health benefits (Kaplan 1983; Beatley

2016). Contact with nature is a powerful way to release tension and anxiety, relax tired minds, support recovery from mental fatigue, and prevent depression, eventually boosting happiness and creativity (Van den Berg et al. 2017). Indeed, a wealth of scientific evidence has been recently generated showing compellingly benefits and the value of urban nature.

3 Nature-Based Solutions

3.1 Background

The concept of nature-based solutions (NBS) was introduced in the first decade of this new millennium by the World Bank (MacKinnon et al. 2008) and the International Union for Conservation of Nature (IUCN 2009) to highlight the importance of biodiversity for mitigating the impacts of climate change. NBS can be defined as "actions to protect, conserve, restore, sustainably use and manage natural or modified terrestrial, freshwater, coastal and marine ecosystems, which address social, economic and environmental challenges effectively and adaptively, while simultaneously providing human well-being, ecosystem services and resilience and biodiversity benefits" (UNEA 2022). It has recently emerged as a key issue for offsetting climate issues in the urban setting (Kabisch et al. 2016, 2017). The European Commission, in delineating general environmental policies that are specifically addressed to cities, has characterised NBS as actions that are inspired and supported by natural processes, or that reproduce their functions to increase and simulate natural processes (Davis et al. 2017).

The NBS concept highlights the potential of adopting alternative strategies to the so-called conventional grey solutions (Anderson et al. 2022) that have characterised the transformations of cities and landscapes at the end of the last century. In particular, NBS highlight the importance of biodiversity conservation for the mitigation and adaptation to climate change, but also include a series of actions based on the replication of natural systems in the management of critical environmental issues with regard to urban landscapes.

NBS were proposed by IUCN in the context of the Paris climate summit negotiations of 2015 to mitigate and adapt to climate change, secure water, food, and emphasise the resilience of landscapes to disturbances and disasters induced by natural and/or anthropogenic factors and actions. The principles articulating the concept of NBS refer primarily to ecological disciplines, but integrate substantial aspects of socio-economic processes. NBS, therefore, whilst becoming extremely popular in various research fields and in the activation of specific policies (as is the case for the innovative environmental policies promoted by the European Commission), remain strongly action oriented. In this sense, the concept of NBS includes cost efficiency, access, and use of an integrated system of public and private financial support through clear and robust partnerships, the development of well-structured and feasible communication programmes, and the wide and concrete participation of the public (van Ham and Klimmek 2017).

3.2 Global Policies

To address long-term targets and pathways, countries are mainstreaming NBS by developing enabling policies supported by science as they update and fulfil their commitments towards the Paris Agreement and the United Nations' (UN) Sustainable Development Goals (SDGs). In 2015, the UN adopted the 2030 Agenda for Sustainable Development as a "universal call to action to end poverty, protect the planet, and ensure that by 2030 all people enjoy peace and prosperity". The UN Agenda 2030[2] specified 17 SDGs with 169 associated targets.

NBS, as well as urban forests and green and blue infrastructures, are essential in delivering, directly or indirectly, a wide variety of ecosystem services that can decisively aid in achieving the targets of the UN SDGs (FAO 2016). In principle, whilst the NBS contribute to all SDGs, they are also key in interconnecting the urban and rural parts of the landscape, both physically and ecologically, as well as raising awareness and educating urban communities on the importance of nature and natural resources. Figure 1 illustrates the relative importance of NBS and urban forests in achieving each of the SDGs. These qualitative judgements were expressed according to pre-defined classes (irrelevant, moderately important, very important, extremely important, primary).

SDG 11 specifically addresses the sustainability of cities and communities to encourage inclusivity, safety, and resilience. The targets of SDG 11 articulate socio-ecological linkages, economic and livelihood support, and health and quality of life improvement in urban societies. Two of the targets illustrate the role that cities play in sustainability and environmental resilience:

- *11.4 Strengthen efforts to protect and safeguard the world's cultural and natural heritage.*
- *11.7—By 2030, provide universal access to safe, inclusive, and accessible, green and public spaces, in particular for women and children, older persons, and persons with disabilities.*

The New Urban Agenda (NUA) (UN 2017), introduced in 2016 at the UN Conference on Housing and Sustainable Urban Development in Quito, Ecuador, explores an "upside-down" paradigm where it is necessary to configure, think, govern, and experience cities of the future from a bottom-up rather than from a top-down perspective. Even if the concept eventually proves to be overly optimistic, there is a belief that well-planned and well-managed urbanisation can be a powerful tool for sustainable development for both developing and developed countries.

[2] https://www.un.org/sustainabledevelopment/

Fig. 1 The relative importance of urban forests and NBS in supporting SDGs; as developed by the authors of this chapter

The NUA depicts the cities of tomorrow as socio-ecological systems, strongly based on sustainability and nature. The vision of future urban development looks to the protection of ecosystems and biodiversity, and it is based on the adoption of "healthy lifestyles in harmony with nature, by promoting sustainable consumption and production patterns, by building urban resilience, by reducing disaster risks and by mitigating and adapting to climate change". In the framework of commitments within the NUA, ecosystem services, NBS, accessible green public spaces, and forest sustainable management and protection are envisioned as key tools for achieving the sustainability goals in urban areas, as well as in minimising the potential impacts that cities can cause to terrestrial and marine ecosystems.

The NUA, by institutional mandate, cannot introduce effective and binding policies to directly support cities or countries, but refers to a series of multilateral funds supporting the concrete activation of regional, national, and local policies that can contribute to the implementation of NBS in urban environments. The Quito declaration explicitly refers to "the Green Climate Fund, the Global Environment Facility, the Adaptation Fund, and the Climate Investment Funds, among others, to secure resources for climate change adaptation and mitigation plans, policies, programmes and actions for subnational and local governments, within the framework of agreed procedures".

Given the unique specifics of urban policies, which are frequently not included in dedicated overall national strategies, a global role in implementing actions related to NBS, green infrastructure, ecosystem services, urban forestry and agriculture, and urban public greenspaces, have been assumed by partnerships of cities. Examples of such networks include:

- Local Governments for Sustainability (ICLEI), consisting of 2500 cities, towns, and regions hosting approximately 25% of the global urban population.
- C40 Cities, a network of urban mayors of nearly 100 world-leading cities collaborating to deliver the action needed to counteract the climate crisis.

Figure 2 illustrates ICLEI Development Pathways (ICLEI 2021), many of which are linked to the adoption of NBS and related issues.

4 European Policy Framework for NBS

NBS became a key issue in European policies on research, development, and innovation during the last decade. The European Commission multiplied the efforts of mainstreaming the concept of NBS across policy sectors accompanying it with other tools such as sustainable urban planning, green infrastructure implementation, urban forests and urban greenspaces improvement, and ecosystem services (de Luca et al. 2021). Local urban governments, however, typically do not have direct binding commitments for the implementation of such strategies, even though they are asked to develop dedicated tools and plans (e.g. plans for green infrastructure, urban

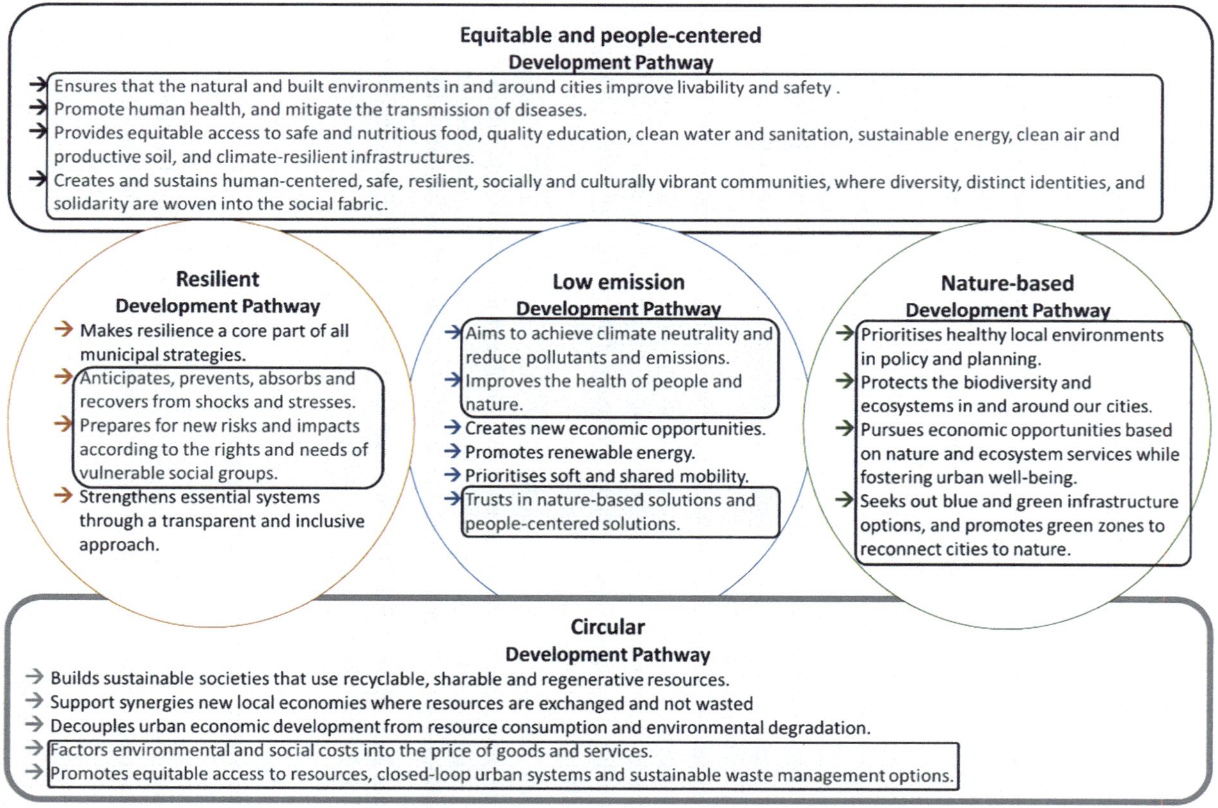

Fig. 2 ICLEI development pathways. Adapted by the authors from ICLEI Malmö Commitment and Strategic Vision 2021–2027 (ICLEI 2021)

greening and urban forestry, and climate adaptation), highlighting the need for increased coordination and a strong response from urban areas.

The Convention on Biological Diversity explicitly supports ecosystem-based approaches (Secretariat of the Convention on Biological Diversity 2004), setting associated targets and recently adopting voluntary guidelines for their design and effective implementation. The UN also adopted the Sendai Framework for Disaster Risk Reduction 2015–2030 (UNISDR 2015), which encourages "ecosystem-based approaches to build resilience and reduce disaster risk".

As part of the European Green Deal,[3] the European Commission adopted the EU Biodiversity Strategy 2030,[4] which recognises the role of ecosystem restoration and NBS as key contributions to both climate change mitigation and adaptation. The Green Deal promotes the integration of NBS and ecosystem restoration into urban planning. In June 2022, the European Commission proposed a new nature restoration law with binding targets on wetlands, peatlands, rivers, forests, marine ecosystems, and urban areas.

The new EU Soil Strategy for 2030[5] is a key deliverable of the EU Biodiversity Strategy for 2030. It will contribute to the objectives of the European Green Deal. Healthy soils are essential for achieving climate neutrality, a clean and circular economy, and reducing the risks of desertification and land degradation. They are also essential to reverse biodiversity loss whilst providing healthy and nutritious food and safeguard human health. For these reasons, NBS are closely linked to the implementation of the soil strategy. The EU Strategy on Adaptation to Climate Change puts a strong emphasis on NBS as a cross-cutting priority. The 2019 review of the Green Infrastructure Strategy[6] highlights the economic, social, and cultural co-benefits arising from green infrastructure and ecosystem-based solutions.

4.1 *Green Infrastructure*

The EU strategy on green infrastructure promotes climate adaptation and mitigation through the deployment of a network across Europe to help attain the decarbonisation of cities (Rosenzweig and Solecki 2018). This transition requires linking research priorities to local needs, since capacity and requirements differ between cities. Transnational networks on climate actions driven by cities (i.e. C40 Cities and 100 Resilient Cities) have the potential to accelerate the development of new strategies to integrate climate change mitigation and adaptation activities.

A growing number of urban centres in Europe have signed up with the "Tree Cities of the World" programme of the UN Food and Agriculture Organisation of the

[3] https://commission.europa.eu/strategy-and-policy/priorities-2019-2024/european-green-deal_en

[4] https://environment.ec.europa.eu/strategy/biodiversity-strategy-2030_en

[5] https://environment.ec.europa.eu/publications/eu-soil-strategy-2030_en

[6] https://eur-lex.europa.eu/legal-content/EN/TXT/HTML/?uri=CELEX:52019DC0236&from=EN

United Nations (FAO) and the Arbor Day Foundation, promoting cooperation and the exchange of best practices amongst European towns and cities on supporting urban forestry. Networks of scientists and practitioners, principally the European Forum on Urban Forestry (EFUF), play a key role in sharing knowledge and experience across the continent and beyond. Research and innovation projects, such as the Horizon 2020 Clearing House Project (2019–2023), are producing results on how urban forests can act as a nature-based solution.

The Urban Agenda for the EU provides a new framework for involving cities in the development and implementation of EU policy. As an example of cross-border cooperation, "Eurocities" is a network of major European cities that, for the most part, are already committed to climate neutrality by 2050. The European Green Deal aims to drastically reduce greenhouse gas emissions and design resilient cities. Taken as a whole, Europe is well positioned in the policy arena, along with its research and innovation agenda and existing networks, to provide global leadership in the role that urban forests can play in the transition towards climate objectives. This effort requires coordination, however, spanning policy, planning practice, research, innovation, management, and visionary leadership. This coordination has yet to be fully defined, and whilst some elements are already present, there continue to be opportunities.

4.2 *Ecosystem-Based Adaptation*

Ecosystem-based adaptation (EbA) can be defined as a strategy to adapt to climate change and its consequences (CBD 2009), such as extreme weather events. Through harnessing nature-based solutions and ecosystem services, urban areas can increase their resilience towards heat waves, droughts, and floodings, as well as increasing sea levels and erosion. A vulnerability assessment (VA) of current conditions is an essential tool for determining effective climate change adaptation strategies, but power relations across all constituencies are central to determining these vulnerabilities and therefore EbA are highly influenced by the local context.

EbA can be of great interest for urban systems (Geneletti and Zardo 2016) contributing to the mitigation of urban heat islands (UHI), the reduction of soil sealing, and the enhancement of water storage capacity in the urban watershed (Grimsditch 2011). EbA in the urban context is considered part of the wider concept of nature-based solutions since EbA primarily refers to climate change adaptation, and is hence more limited in scope than NBS (Pauleit et al. 2017). Moreover, EbA are strictly related to planning green and blue infrastructures and designing their components (e.g. urban green spaces, green roofs and walls, urban forests, ponds, and swales). Because of contrasting soil imperviousness, EbA is expected to provide both general and specific adaptation to the effects of climate change (Roberts et al. 2012).

The Global EbA Fund[7] provides a funding mechanism that supports innovative approaches to EbA to climate change. The Global EbA Fund, based on partnerships between the IUCN and UNEP fund managers and local project implementers, supports catalytic initiatives to help overcome identified barriers to upscaling EbA. It prioritises filling in gaps in planning, knowledge, and resourcing, with a broad thematic focus on innovation and urgency, and encourages creative solutions and partnerships. By supporting catalytic interventions, the Global EbA Fund addresses research gaps, pilots innovative EbA approaches, engages in strategic EbA policy mainstreaming, and incentivises innovative finance mechanisms and private sector EbA investment.

4.3 Urban Forestry

First mentioned as early as the nineteenth century, the term "urban forestry" has been experiencing a renewed interest since the 1960s with the gradual recognition of its potential and substantial role in making cities more liveable and sustainable in the long term. Whilst having its roots in the world's oldest civilisations, the concept and practices of planting and managing trees in human settlements have followed different trends through the ages, depending on cultures and regions.

The Guidelines on Urban and Peri-urban Forestry (FAO 2016) provides a synthesis of the evolution of discourses, governance, policies, and actions that link urban forests to the implementation of BioCities at global level. Extending the concept of urban forest to peri-urban areas, has major implications for integrating environmental policies beyond the city boundaries, whilst strengthening the role that forest-based solutions can have in reducing the urban footprint. The urban and peri-urban forest (UPF) is included in not only dense cities, but also the less-dense regions surrounding cities. Dijkstra and Poelman (2012) highlight the major issues about the required enabling environment, the policies and strategies, and the concrete actions to take in term of planning, design, and management, to strengthen the role that urban forests, green infrastructure, and NBS should have in the cities of tomorrow. Professionals, practitioners, and researchers are joining forces in creating forums at a continental and global level. In 1998, for instance, the European Forum on Urban Forestry (EFUF) held its first annual meeting in Wuppertal, Germany, and has met annually ever since. Similarly, the first Asia-Pacific Urban Forestry Meeting was held in 2016 in Zhuhai, China, the first World Forum on Urban Forestry was held in 2018 in Mantova, Italy, and the first African Forum on Urban Forests was held online in 2021. These forums have facilitated the flow of knowledge to a degree unimaginable to previous generations.

Today, a growing number of communities are applying a technical and scientific approach to urban tree placement and maintenance. However, UPF can still be

[7] https://globalebafund.org/

considered a fledgling discipline. Efforts must be made to continue raising awareness amongst policymakers and the wider public about its crucial contributions, with emphasis on developing countries. Traditionally, urban forestry has often focused on how to educate politicians and local communities about the importance of urban trees and urban woodland. We have come to realise, however, that urban forestry needs to move beyond this, and find ways of including a wide set of stakeholders into decision-making and management (Sheppard et al. 2017).

5 Emerging Concepts on Future Sustainable Cities

Over the last few decades, we have been witnessing a rapidly expanding movement in support of naturalising cities. The key assumption has been that *nature hosts the city and not the opposite*. The implications of this assumption are reflected not only in some components of the city (e.g. urban green spaces, parks and urban forests, and green infrastructure elements), but in the overall urban policies and specific ways of thinking, governing, designing, planning, and managing the cities of tomorrow.

Numerous experiences of "green practices" at urban level have established themselves in different geographical, social, and cultural contexts, but all of them have contributed to delineating a trajectory that is leading towards the concept of BioCities. In some cases, the green experiences are limited to a district, whilst in others they are extended to the whole city. Various concepts, policies, and practices have been devised:

5.1 The Green City

The Green City is a very broad concept that focuses on achieving an environmental balance in supporting human activities by considering the carrying capacities of the natural environment. It follows a worldwide conceptualisation of "green" as a reference to what makes a place (any place) more sustainable, resilient, liveable, and, very generally, "natural". The green city concept has been widely used by governments across the globe and European countries have been pioneering in this area and contributed substantially to grow the idea of Green Cities (Keane and Davies 2020; Beatley 2012). More recently, authors developed and elaborated on tools for measuring green performance over time, setting targets and tracking achievements (Brilhante and Klaas 2018). The green city concept is a key reference for some award schemes, notably the Green Capital of Europe and the Green Leaf.

In 2020, FAO launched the Green Cities Initiative, which focuses on "improving the urban environment, strengthening urban-rural linkages and the resilience of urban systems, services and populations to external shocks".[8] The Green Cities Initiative aims at ensuring access to a healthy environment and healthy diets from sustainable food systems, adopting solutions in the domains of urban agriculture and agro-forestry, urban and peri-urban forestry. A declared goal is also contributing to climate change mitigation and adaptation and sustainable resource management. The goal of the Green Cities Initiative is to "improve the livelihoods and well-being of urban and peri-urban populations in at least 100 cities around the world in the next three years, looking to have 1000 cities join by 2030". The Green Cities Network was created for cities of all sizes to share experiences, including successful cases and practices, with a goal of building city-to-city cooperative opportunities.

5.2 *The Resilient City*

The multiple risks and impacts of climate change on cities (e.g. food insecurity, heat waves, and windstorms) have acted as a catalyst for planning resilient cities, based on the design and implementation of NBS and green-blue infrastructure to reduce vulnerabilities and improve urban resilience. The UNISDR (2012) defined a resilient city by its capacity to withstand or absorb the impact of a hazard through resistance or adaptation, and which enables it to maintain certain basic functions and structures during a crisis and bounce back or recover from an event (Newman et al. 2017). This concept was also explored by the Organisation for Economic Co-operation and Development (OECD 2015), which has investigated how to measure and increase resilience. Many city networks have been born and developed around the concept of urban resilience.

5.3 *The EcoCity*

The term "EcoCity" or "ecological city" was coined in 1975 by a group of architects and ecologists in California, USA (Roseland 1997). An EcoCity is a place where people can live in harmony with nature whilst reducing their ecological footprint. Conceptually, the EcoCity is an ecologically sound, compact, and vigorous settlement that co-exists with nature and enables the society and ecological environment to develop in harmony. Many cities across the globe have developed different types of EcoCity projects, including Tianjin (China), Berkeley (USA), and Copenhagen (Denmark) (Caprotti 2014; Li et al. 2019). In some respects, the EcoCity can be regarded as a precursor to the concept of the BioCity.

[8] https://www.fao.org/green-cities-initiative/en

5.4 The Sponge City

A sponge city is based on a green and sustainable strategy that envisions a city's water management system functioning as a "sponge", absorbing, storing, infiltrating, and purifying rainwater, and releasing it for use/reuse when needed. It emphasises "design with nature", where sealed surfaces are replaced by permeable ones; and green roofs, wetlands, urban forests, and meadows are encouraged; and sunken green fields are used to facilitate rainwater infiltration, storage, and purification. Since 2014, about 130 cities across Mainland China have formulated plans to transform themselves into sponge cities (Nguyen et al. 2020).

5.5 The Urban Food City

Cities around the world are engaged in agriculture to produce food, including urban gardens as a strategy to mitigate climate change, create sustainable urban foodscape, and provide alternative food networks for consumers (Maye 2019). A notable example is the City of Havana, Cuba, where urban agriculture is a marquee green "solution". This was initially born out of a need for swift and urgent food security reform following the end of the USSR, upon which Cuba was highly dependent. It has since become recognised as a highly creative sustainable solution, including by the World Wildlife Fund (WWF) for multiple accrued benefits including the higher resilience of food chains, reduced energy use, employment creation, and biodiversity conservation, amongst others.[9]

5.6 The Smart City

A smart city is an urban area that uses different types of technologies and sensors to collect data to manage assets, resources, and services efficiently. Citizens also contribute to the collection of data, which is processed and analysed to monitor and manage transport and traffic systems, power plants, public services, water supply networks, waste, crime detection, information systems, schools, libraries, hospitals, and other community resources and services, such as urban green areas (Meijer and Bolívar 2016).

In one respect, the Smart City is the antithesis of the BioCity and has come under much criticism, prompting lead urban designer Dan Hill to say that "...the smart city was the wrong idea pitched in the wrong way to the wrong people".[10] Indeed, what

[9] https://wwf.panda.org/wwf_news/?204427/Havana-urban-farming

[10] https://www.theguardian.com/cities/2014/dec/17/truth-smart-city-destroy-democracy-urban-thinkers-buzzphrase

has perturbed others is the idea that running the city like a computer network reduces human agency, and further reduces the stature of "non-digital" species that share the city with human to little more than an algorithm (Keeton 2021). The idea that the "smart" and "bio" components of a city could be combined, however, is an attractive notion in managing future cities, but it remains to be effectively explored.

5.7 *The BiodiverCity*

Initially, the notion of BiodiverCity was explored by the European Parliament with the aim of enhancing the use of urban green infrastructure to augment urban socio-ecological systems, providing benefits for both people and nature. BiodiverCities were meant to encourage civil society participation in local and urban policy and governance processes. The ultimate aim was to raise public awareness and to build a joint vision of future, more sustainable cities, inclusive of people and nature (see Box 1). More recently, the concept of BiodiverCity was also developed by the Institute von Humboldt in Bogotà, Colombia, to launch an urban strategy and provide a framework of analysis of urban transformation to 2030, including a network of cities adopting "biodiversity" as the major driver of urban development in the next decade (Mejía and Amaya-Espinel 2022).

Box 1 European Policy for BiodiverCities (Extracted and adapted from Maes et al. 2021)

Between 2019 and 2022, the European Commission promoted a number of initiatives to protect the environment and minimise risks to climate, human health, and biodiversity. In December 2019, the European Commission presented the European Green Deal, which includes a set of proposals to make the EU's climate, energy, transport, and taxation policies consistent with reducing net greenhouse gas emissions by at least 55% by 2030, compared to 1990 levels. In May 2020, the European Commission adopted the new Biodiversity Strategy for 2030, a comprehensive, ambitious, and long-term plan to protect nature and reverse the degradation of ecosystems. In June 2021, the European Commission adopted the European Climate law, which translates into a law the goals set in the European Green Deal, setting a legally binding target of net zero greenhouse gas emissions by 2050. In May 2022, the European Commission adopted the 8th Environmental Action Plan to guide the European environmental policy until 2030. In June 2022, the Commission adopted a Proposal for a Nature Restoration Law, which aims at restoring damaged ecosystems, bringing nature back across Europe, from agricultural land and seas to forests and urban environments. In July 2022, the European Commission proposed the introduction of ecosystem accounts amending the

(continued)

Box 1 European Policy for BiodiverCities (Extracted and adapted from Maes et al. 2021) (continued)
regulation (EU) No 691/2011. These initiatives have implications for the plan and management of EU urbanised areas, which should also contribute to fulfil their objectives.

6 The Route of European Policies and Actions for Future Sustainable Cities

In Europe, local authorities are at the forefront of change and their commitment to approaching sustainability has increased over time. A wide range of EU policies and strategies regulate, drive, and/or support this commitment in many cities of various sizes. The issue of urban sustainability is framed by EU policies regarding the environment, forest, landscape, green infrastructure, and cultural capital, as well as the EU Research and Innovation (R&I) agenda and the Green Deal. Nevertheless, urban policy is not an EU-level responsibility under the treaties of the European Union. In the Lisbon Treaty of 2009, the notion of territorial cohesion appeared for the first time, including a focus on urban areas that reflect a steady progression over the past quarter century to reinforce a continental urban and territorial agenda. A visual summary of the major EU policies concerning and/or influencing urban government and strategies is reported in Fig. 3.

The first steps towards addressing urban sustainability policy occurred in 1989, when the Urban Pilot Project targeted the financing of primarily area-based actions. The objectives of the Urban Pilot Projects were to contribute to economic and social cohesion in urban areas through supporting urban regeneration and planning activities, recognising that European cities must face the challenge of integrating economic, environmental, and employment considerations within a logic of sustainable urban development.

In the following decade, during the 1990s, the EU realised that the improvement of urban quality strengthen social cohesion, economic accountability, and environmental sustainability. Several programmes and projects were created to improve urban quality in problematic and declining areas at many spatial scales—from districts and neighbourhoods to cities and regions. In 1994, 80 municipalities signed the Aalborg Charter at the European Conference on Sustainable Towns (held in Denmark), and started the Campaign for European Sustainable Cities, aiming to reach a consensus amongst local communities on Local Agenda 21, a voluntary process for community consultation. In 1998, the EU review of the 5th Environmental Action Programme resulted in commitments to develop "a comprehensive approach to urban issues with special emphasis on the assistance required to support actions by local authorities to implement the Programme and Local Agenda 21". Later in 1998, the European Commission released a communication memo entitled *Sustainable Urban Development in the EU: A Framework for Action.*

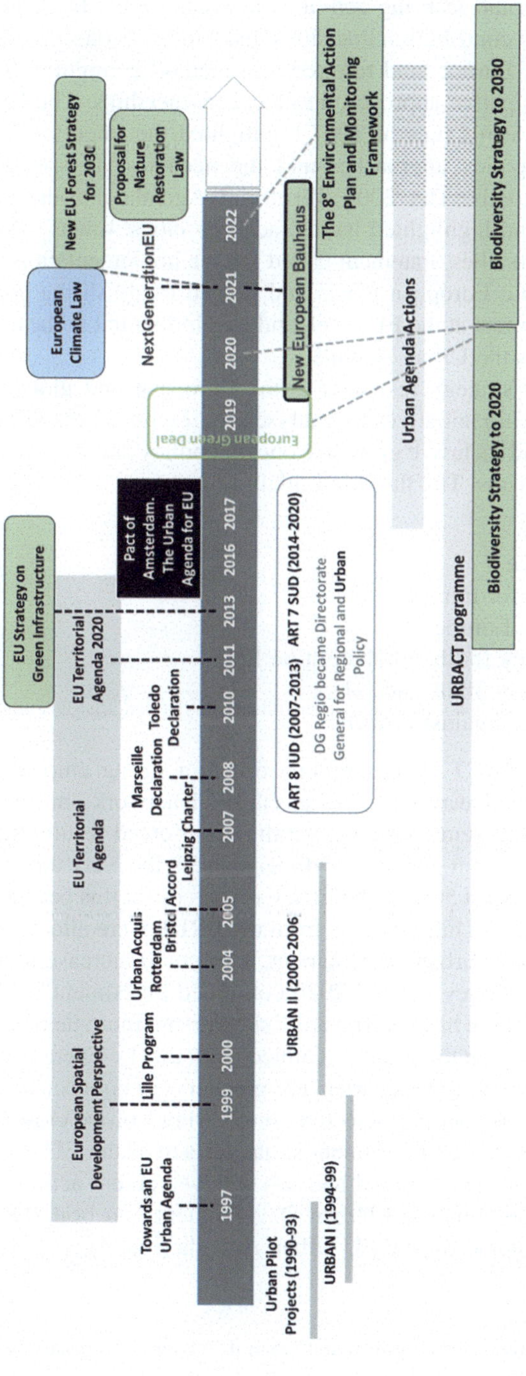

Fig. 3 Timelines of major EU policies, strategies, and programmes concerning cities or environmental issues where the role of cities is mentioned or substantial. Adapted and expanded from Fioretti et al. (2020)

Into the new millennium, the URBACT Programme was launched to support the knowledge exchange and learning activities between cities. In 2004, the *URBAN Acquis* recognised the contribution that cities make to the economic, environmental, and social success of Europe, and referred to a method combining the area-based, integrated, and participative approach into local partnerships. The Leipzig Charter on Sustainable European Cities of 2007 highlighted the importance of integrated urban development policy approaches and the need to pay special attention to deprived neighbourhoods. The 2007 Territorial Agenda introduced the idea of territorial cohesion and highlighted issues faced by cities, towns, and urban areas. And in 2008, the Marseilles Statement called for the implementation of the Leipzig Charter on Sustainable European Cities and helped establish the concept of integrated urban development at the EU level, and was influential in the development of EU initiatives such as the Urban Agenda.

The Europe 2020 strategy responds to the European and global challenge by proposing seven flagship initiatives to catalyse progress under the priority themes of smart, sustainable, and inclusive growth. Cohesion policy and its structural funds are key delivery mechanisms. The flagship initiatives are:

- Innovation Union
- Youth on the Move
- A Digital Agenda for Europe
- Resource Efficient Europe
- An Industrial Policy for the Globalisation Era
- An Agenda for New Skills and Jobs
- European Platform Against Poverty

The 2014–20 URBACT III programme brings a strong emphasis on capacity building, knowledge and learning exchanges through networks, and a renewed focus on capitalisation and dissemination set within a reinforced results framework. The 2014–2020 period has put the urban dimension at the very heart of European Cohesion Policies. At least 50% of the ERDF resources for this period were invested in urban areas. Around 10 billion euros from the ERDF were allocated to integrated strategies for sustainable urban development. This could increase in the future.

The EU Cohesion Policy beyond 2020 continued investment in all regions and the European Commission has put forward a simpler and more flexible framework to better reflect the reality on the ground (Cunico et al. 2021). There is a focus on five policy objectives around a (1) smarter, (2) greener, (3) connected, and (4) social Europe, and a new cross-cutting objective to (5) bring Europe closer to citizens by supporting locally developed investment strategies across the EU.

In the framework of EU regional and urban development actions, the Commission set up a topic on Cities and Urban Development[11] to bridge knowledge and actions towards the themes of the EU Urban Agenda (Fig. 4).

[11] https://commission.europa.eu/eu-regional-and-urban-development/topics/cities-and-urban-development_en

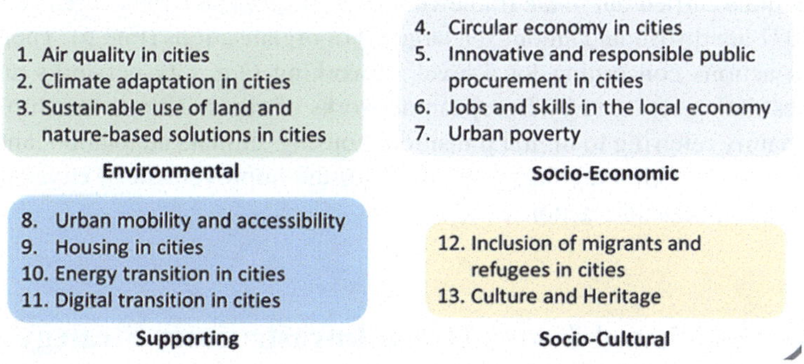

Fig. 4 European Commission Priority themes for EU cities in the frame of cities and urban development policies. Adapted and regrouped from https://ec.europa.eu/info/eu-regional-and-urban-development/topics/cities-and-urban-development_en

7 The New European Urban Agenda

The Urban Agenda for the EU was launched in May 2016 alongside the Pact of Amsterdam. It represents a new multi-level working method promoting cooperation between member states, cities, the European Commission, and other stakeholders, in order to support growth, liveability, and innovation in the cities of Europe and to identify and successfully tackle social challenges. The Pact of Amsterdam (2016) affirms that:

1. The Urban Agenda for the EU aims to realise the full potential and contribution of urban areas towards achieving the objectives of the EU and related national priorities in full respect of subsidiarity and proportionality principles and competences.
2. The Urban Agenda for the EU strives to establish a more effective integrated and coordinated approach to EU policies and legislation with a potential impact on urban areas and also to contribute to territorial cohesion by reducing the socio-economic gaps observed in urban areas and regions.
3. The Urban Agenda for the EU strives to involve urban authorities in the design of policies, to mobilise urban authorities for the implementation of EU policies, and to strengthen the urban dimension in these policies. By identifying and striving to overcome unnecessary obstacles in EU policy, the Urban Agenda for the EU aims to enable urban authorities to work in a more systematic and coherent way towards achieving overarching goals. Moreover, it will help make EU policy more urban friendly, effective, and efficient.
4. The Urban Agenda for the EU will not create new EU funding sources, unnecessary administrative burden, nor affect the current distribution of legal competences and existing working and decision-making structures and will not transfer competences to the EU level (in accordance with Articles 4 and 5 of the Treaty on European Union).

The actions carried out in the framework of the European Urban Agenda up to the end of 2021 are distributed amongst a large set of organisations (Fig. 5). There is still a lack of actions concerning local-level networking (e.g. city networks and local authorities) being replaced by European networks of cities. The partnership categories are mainly referring to digital transition, housing, climate adaptation, and socio-economic issues, which highlights the need for the improvement of environmental, cultural, and governance actions.

8 Cities as Viewed from a Green Infrastructure Strategy or Biodiversity Strategy

8.1 Urban Greening Platform

Green urban spaces, from parks and gardens to green roofs and urban farms, provide a wide range of benefits for people and the planet. They provide vital space for physical and mental well-being and a very important habitat for nature, including for birds and pollinators. Green space helps reduce air, water, and noise pollution, provides protection from flooding, droughts, heat waves, and much more.

Whilst protection of some urban green spaces has increased, green spaces often lose out to development in the competition for land, as the share of the population living in urban areas continues to rise. The EU Biodiversity Strategy for 2030 aims to reverse these trends, and to protect and restore our precious urban ecosystems. As part of the biodiversity strategy (i.e. bringing nature back to cities and rewarding community action), the Commission called on European towns and cities of at least 20,000 inhabitants to "…develop ambitious urban greening plans", including "measures to create biodiverse and accessible urban forests, parks and gardens, urban farms, green roofs and walls, treelined streets, urban meadows, and urban hedges".

The urban greening platform[12] aims at assisting and supporting local authorities in achieving this objective. It has been developed in collaboration with Eurocities and ICLEI, and is based on discussions with many local authorities that have already gone through the process of developing and implementing successful urban greening plans. It stresses the importance of the collaborative process of developing an urban greening plan, including the need for working with citizens and other stakeholders, and for cross-departmental working, and the integration of the greening plan with other aspects of urban development, from mobility and health, air and water, to energy and climate adaptation.

[12] https://platformurbangreening.eu/

A

Type of organisation \ Action category	D & I	G & H	Gov	PP	PR	Stra	TK	TOT
City	12	16	5	3	10	7	6	60
City network			1		1			2
European network/association URBACT	4	5	2		8	2	2	23
Local authority/Union of Communities		1			1		1	1
Region		1	1			1	1	5
National authority	4	8	3	1	2		1	20
Council of Europe Development Bank (CEB)					1			1
EU Agency					1			1
European Commission	1	3	1		2	2	1	10
European Investment Bank		1					2	3
TOT	21	35	13	4	27	13	13	129

Legenda Action category

D & I	Data and indicators	PR	Policy recommendations
G & H	Guidance documents and handbooks	Stra	Strategy/Plan/Roadmap
Gov	New governance body/structure	TK	Toolkit
PP	Policy preparation/Impact assessment		

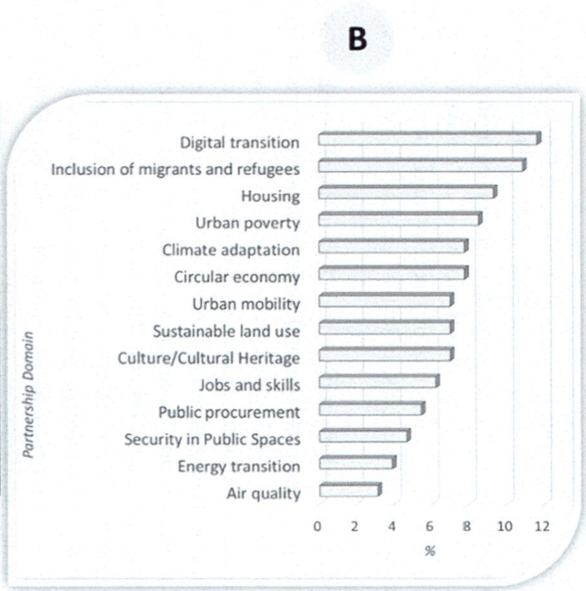

Fig. 5 Actions in the framework of the European Urban Agenda from 2017 to 2021: (**a**) is the number of projects by the type of organisation and action category; (**b**) is the share of actions per partnership category. Source: Monitoring MTA September.xls available at https://futurium.ec.europa.eu/en/urban-agenda/monitoring-actions/monitoring-table/table-actions-update-september-2021, downloaded on 30-11-2022

8.2 Green Deal and Related Urban Challenges

The European Commission has put forward a "European Green Deal" as the first priority of EU Political Guidelines for the new Commission. The agenda includes the goal of making Europe the first climate-neutral continent, ensuring a just transition, and moving towards zero pollution by putting forward a "cross-cutting strategy to protect citizens' health from environmental degradation and pollution, addressing air and water quality, hazardous chemicals, industrial emissions, pesticides, and endocrine disrupters".

In the European Union, actions within the "Green Deal" strategy also aim to achieve carbon neutrality and preserve and restore ecosystems and biodiversity. Cities of more than 20,000 inhabitants are asked to develop ambitious Urban Greening Plans, with the research programme "Horizon Europe" financing research on urban nature to realise future green and sustainable cities. The IUCN plays an important role. Their focus spans issues of governance and policy, gender, NBS, water, heritage forests, and many others. The IUCN, as well as other European projects related to urban development, consider nature in cities as a major factor to increase the sustainability, biodiversity, and livability of cities. The EU's biodiversity strategy for 2030 "is a comprehensive, ambitious and long-term plan to protect nature and reverse the degradation of ecosystems".

Engineered green infrastructures are a strategic response to the UN Paris Agreement, making significant contributions to a "cleaner and more efficient energy system" and favouring the process of energy transition (Asarpota and Nadin 2020). An example of a green infrastructure project to mitigate UHI was implemented in Padua, Italy, by the Interreg Central Europe Programme and co-financed by the European Regional Development Fund (Musco et al. 2016). This project aims to ameliorate consequences of UHI through the development of transnational heat mitigation and adaptation strategies and their utilisation as urban planning tools.

NBS are also an important topic on the EU Research and Innovation policy agenda. The Horizon 2020 and Horizon Europe programmes financially supported the URBAN GreenUP Project, which aims to mitigate the effects of climate change by alleviating the UHI effect and improving air quality and water management, as well as increasing the sustainability of cities through innovative NBS. The Horizon 2020 Clearing House Project, which is a Sino-Europe collaboration, is looking specifically at how urban forests act as an NBS.[13]

The biophysical green infrastructure provided by trees, shrubs, grasslands, and water is a potential solution to be adopted within the urban environment as a means to re-nature the city (Breuste 2021). Biophysical green infrastructure is almost always a NBS, but can be spatially challenging in dense consolidated cities more than in dispersed multi-centered agglomerations, of which the Ruhrgebeit, in Germany, is an outstanding example. Today, the management option of urban

[13] www.clearinghouseproject.eu

reforestation has been consolidated in the actions of policymakers and through the participation of urban communities in the region's Covenant of Mayors,[14] which promoted a series of significant actions in the adaptation plans developed by associated municipalities.[15]

Air quality also benefits from green infrastructure as plants absorb carbon and emit oxygen. Emission standards set by the EU encourage member states to reduce emissions by 2030 (NECD 2016; De Marco et al. 2019). Article 9 of standards support the monitoring of negative impacts of air pollution on ecosystems including freshwater, forests, and natural and semi-natural habitats. Some European countries (i.e. Italy with its "climate decree") responded by adopting measures aimed at improving air quality, reducing waste, and improving soil quality. Actions aimed at increasing urban air quality include direct subsidies for cities to increase urban forest coverage.

Together with other international organisations, the World Meteorological Organisation (WMO) created an initiative, called Integrated Urban Hydrometeorological, Climate and Environmental Services (IUS), to develop science-based services to support safe, healthy, resilient, and climate-friendly cities (WMO 2021). It supports the role of NBS as essential urban service in cities. "It is important to foster green design over a city to activate secure pathways for fragile populations, to furnish warnings (including climate watch advisories) and to design a proper texture of the city itself (for example, where to place hospitals, schools or commercial centres". The fundamental role of green infrastructure must therefore be considered an integral part of the meteo-climatic services with which a city must be equipped in the project, urban regeneration, and transition phases.

8.3 New Bauhaus and Next Generation EU

The New European Bauhaus[16] is a creative and interdisciplinary initiative that connects the European Green Deal to our living spaces and experiences. The New European Bauhaus initiative calls on all of us to imagine and build together a sustainable and inclusive future that is beautiful for our eyes, minds, and souls. Beautiful are the places, practices, and experiences that are:

- Enriching, inspired by art and culture, responding to needs beyond functionality.
- Sustainable, in harmony with nature, the environment, and our planet.
- Inclusive, encouraging a dialogue across cultures, disciplines, genders, and ages.

[14] https://www.covenantofmayors.eu/

[15] https://unfccc.int/topics/adaptation-and-resilience/workstreams/national-adaptation-plans; https://climate-adapt.eea.europa.eu/knowledge/tools/urban-ast/step-5-2

[16] https://new-european-bauhaus.europa.eu/index_en

The role and impact of forests to achieve SDG 11 targets have been explored by Devisscher et al. (2019). Forest-based solutions, both in purely urban contexts and in relationships of sustainable provision of ecosystem services, are highlighted as fundamental components of the future of cities. The political implications of these assumptions are reflected in programmes recently developed by European policies relating to the New Bauhaus and the Next Generation EU. The policy support for the use of sustainably produced wood (see chapter "Innovative Design, Materials, and Construction Models for BioCities") and the programmes of planting trees in urban and peri-urban contexts (see chapters "Mitigation and Adaptation for Climate Change: The Role of BioCities and Nature Based Solutions", "BioCities as Promotors of Health and Wellbeing", "Forests, Forest Products and Services to Activate a Circular Bioeconomy for City Transformation", "The Social Environment of BioCities") are currently happening at local to global scales, with effective policies, raised community awareness, and provide data for living lab research (e.g. H2020 CONEXUS[17] promoting NBS and urban forest-based solutions). "In the quest for location-based responses, radical change is understood as being strongly dependant on the evolution of city administrations' routines away from the traditional silo-based approach and towards a cross-cutting and citizen-driven way of operating" (Marchigiani and Bonfantini 2022).

The Next Generation EU[18] (NGEU) is the EU's unprecedented response to the COVID-19 crisis. The Commission is empowered to borrow up to 806.9 billion euros between 2021 and 2026 to drive Europe's recovery from the pandemic via a combination of loans and grants to member states and centrally managed EU programmes (European Commission 2022). "Make it Green" is one of the primary programmes in support of the EU goal for Europe to become the first climate-neutral continent by 2050. NGEU expects national governments to invest in environmentally friendly technologies, roll out greener vehicles and public transport, and make our buildings and public spaces more energy efficient. Actions supporting the improvement of knowledge on urban sustainability (e.g. National Centres of Research on Biodiversity) and the implementation of ecosystem restoration opportunities (e.g. Urban and Peri-urban Forests implementation campaigns) are widely financed in the framework of NGEU, and complement multiannual financial support schemes. Natural resources, the environment, and resilience are at the heart of the programme, and cities are tasked with being at the forefront of building healthier and sustainable futures for Europe (Nieuwenhuijsen 2021).

[17] https://www.conexusnbs.com/

[18] https://next-generation-eu.europa.eu/index_en

9 Case Studies: Green City Policies in Action

"Going green" appears as a common path taken up in multi-scalar urban policies and governance in terms of addressing the present and future challenges of cities. Indeed, the "green city" is a concept widely and frequently referred to and used in the institutional sphere as well as by the media and in the communication and dissemination field. It is both a current and contemporary idea in politics, planning, science, and public opinion.

Each discipline, however, attaches a different meaning and relevance to the concept. The Green City itself carries a positive message, but needs to be specified in a local and firm context. Citizens and their representatives, as well as the media and politicians, advocate for the goals of a green city on a national, regional, and local scale, but generally in an imprecise way. The Green City must therefore establish firm green credentials at a local scale for the community and stakeholders to "feel" it is real at various levels. Hence, the concept should not only be seen as a vision, but also expressed through a realistic delivery programme. So what should be in this programme? According to Breuste et al. (2020), the Green City is a city where all forms of nature—living organisms, biocoenoses, and their habitats—are highly significant components of green infrastructure. In a Green City, these forms of nature are preserved, maintained, and extended for the benefit of the City's residents.

1. Vancouver, Canada

When the City of Vancouver administration launched its first *Greenest City Action Plan* in 2011, it was soon evident worldwide that this was the start of an innovative transition towards green issues in the overall governance of forward-looking contemporary cities. In the following 10 years, many cities developed strategies and action plans where green concepts are the engine of urban changes (i.e. the City of Melbourne developed its *Green Our City Strategic Action Plan* in 2017).

The contents of the strategies concerning green cities go far beyond the planning, design, and management of green spaces, urban forests, and green infrastructure components. The goal areas of Vancouver's *Greenest City Action Plan*[19] include:

- Climate and renewables
- Green buildings
- Green transportation
- Zero waste
- Access to nature
- Clean water
- Local food
- Clean air
- Green economy
- Lighter footprint

[19] https://vancouver.ca/green-vancouver/greenest-city-action-plan.aspx

The action plan for Vancouver is constantly monitored and works through implementation projects. The Vancouver *Greenest City Action Plan* is integrated in the overall urban planning process and works back-to-back with the City's *Climate Emergency Action Plan*.

2. Barcelona, Spain

In the last decades, Barcelona declared the need for more greenery and adopted a vision about green issues, combined with a new sense of joint responsibility of citizens. The strategy to achieve this is based on the idea of a connected network of green spaces, conceived as a green infrastructure forming part and parcel of the city, serving environmental and social functions. To work towards a more sustainable and resilient city, an urban transformation was planned with the focus on increasing green areas, in particular in the less equipped neighbourhoods, to ensure a fair distribution of the ecosystem services and benefits that greenery and biodiversity can provide.

3. Turin, Italy

Turin's historical urban development, mainly from the mid-1500s to the 1800s, continues to influence the asset value of the entire central city, strongly limiting the adaption of the urban core to modern challenges and the development of new infrastructures. At the same time, the incontrovertible growth of industry in the twentieth century was the driver of rapid expansion of the city and a massive influx of migration in only a few decades. It has led to an unprecedented scale of soil sealing and green spaces loss, whilst creating major social tensions as new residents struggled to integrate into the local sociocultural fabric. Based on these characteristics, the City Council acknowledged the ecological importance of urban greenery and defined strategies for the enhancement of greenery, increase of biodiversity and ecological connectivity, and for the quantification and strengthening of ecosystem services. The *Strategic Green Infrastructure Plan*, however, does not broadly address existing green infrastructure, but it has elaborated a municipal *Corporate Forest Plan 2020* as a tool for the sustainable management of the city's urban forest.

10 Outcomes and Concluding Remarks

Europe, and especially the Europe of cities, is facing epochal challenges. The health, climate, economic, and energy crises are defying the future of the places in which we live. European cities have the potential and the character to carry out a fundamental transformation of the urban environment, to reduce the urban footprint on the landscape, and become leading examples in introducing more sustainable, equitable, and ecofriendly processes in the Urban Millennium. In fact, the quality and direction of transformation go beyond some environmental, sociological, and technological solutions: it requires a complex and systemic ecological approach translated into sound and complete policies, where the performance of urban habitats, the improvement of healthy living conditions for all, the sustainable energy and mobility challenges, and a just and equal growth interact. The policy framework towards BioCities recognises that nature is the best ally of cities. The commitment and articulation of global, European, and local policies are developing, in an ever more convincing way, strong options to support this transformation and thus activate a real revolution of paradigms, which sees Biocities as the main players in our common future.

10.1 Key Messages

- In Europe, the European Commission's policy framework that promotes, supports, and can implement Biocities is already well developed with accompanying and supportive measures.
- In Europe, there are many inputs and overlaps in the policy measures coming from different sectors at the national, regional, and local levels. A decisive effort to follow the BioCity concept can lead to the integration of devolved policies within the framework of the European Urban Agenda, hence strengthening capacity building for decision takers and awareness raising by the community.
- There are differing speeds in adopting the various relevant components that contribute to a BioCity (e.g. green infrastructure and urban forestry), according to national measures and the still persisting absence of a specific urban policy in many countries.
- Major cities and multi-centered urban regions are generally favoured in this process whilst there is still a need to include more small-medium size cities. Some issues are already declared but there is a concrete need of supporting smaller cities in achieving the sustainability and resilience goals towards BioCities, potentially through sharing agreements and mentoring.
- A future priority is to attain a natural balance by protecting biodiversity and bringing nature closer to people whilst not being blind to potential disservices. Prioritising trees in most urban contexts is paramount in this process, as pointed out from different policy perspectives, since they improve the quality of the living environment by reducing pollution and noise, and lowering the ambient temperature.

References

Anderson CC, Renaud FG, Hanscomb S, Gonzalez-Ollauri A (2022) Green, hybrid, or grey disaster risk reduction measures: what shapes public preferences for nature-based solutions? J Environ Manag 310:114727. https://doi.org/10.1016/j.jenvman.2022.1147277

Armson D, Stringer P, Ennos AR (2013) The effect of street trees and amenity grass on urban surface water runoff in Manchester. UK Urban Forestry and Urban Greening 12(3):282–286. https://doi.org/10.1016/j.ufug.2013.04.001

Asarpota K, Nadin V (2020) Energy strategies, the urban dimension, and spatial planning. Energies 13:3642. https://doi.org/10.3390/en13143642
Beatley T (2012) Green cities of Europe. Island Press, Washington, DC
Beatley T (2016) Handbook of Biophilic City planning and design. Island Press, Washington, DC
Beatley T (2020) Biophilic cities. In: Loftness V, Haase D (eds) Sustainable built environments. Encyclopedia of Sustainability Science and Technology Series. Springer, New York, NY, pp 275–292
Breuste JH (2021) The green city: from a vision to a concept from national to European perspectives. In: Arcidiacono A, Ronchi S (eds) Ecosystem services and green infrastructure. Springer, Cham, pp 29–43
Breuste J, Artmann M, Ioja C, Qureshi S (eds) (2020) Making green cities. Cities and nature series. Springer Nature. https://doi.org/10.1007/978-3-030-37716-8_1
Brilhante O, Klaas J (2018) Green city concept and a method to measure green city performance over time applied to fifty cities globally: influence of GDP, population size, and energy efficiency. Sustainability 10:2031
Calhoun C (2012) The roots of radicalism: tradition, the public sphere, and early nineteenth-century social movements. University of Chicago Press, Chicago
Caprotti F (2014) Critical research on eco-cities? A walk through the Sino-Singapore Tianjin Eco-City. Cities 36:10–36
CBD (2009) Connecting biodiversity and climate change mitigation and adaptation: report of the second ad hoc technical expert group on biodiversity and climate change. CBD technical series 41. Convention of biological diversity, Montreal
Clément G (2015) The planetary garden and other writings. University of Pennsylvania Press, Philadelphia
Clowney D (2013) Biophilia as an environmental virtue. J Agric Environ Ethics 26:999–1014
Cunico G, Aivazidou E, Mollona E (2021) Beyond financial proxies in cohesion policy inputs' monitoring: a system dynamics approach. Eval Program Plann 89:101964. https://doi.org/10.1016/j.evalprogplan.2021.101964
Davis M, Abhold K, Mederake L, Knoblauch D (2017) Nature-based solutions in European and national policy frameworks. Deliverable 1.5, NATURVATION. Horizon 2020 Grant Agreement No 730243, European Commission, 42 pp
De Luca C, Naumann S, Davis M, Tondelli S (2021) Nature-based solutions and sustainable urban planning in the European environmental policy framework: analysis of the state of the art and recommendations for future development. Sustainability 13:5021. https://doi.org/10.3390/su13095021
De Marco A, Proietti C, Anav A, Ciancarella L, D'Elia I, Fares S, Fornasier MF, Fusaro L, Gualtieri M, Manes F, Marchetto A, Mircea M, Paoletti E, Piersanti A, Rogora M, Salvati L, Salvatori E, Screpanti A, Vialetto G, Vitale M, Leonardi C (2019) Impacts of air pollution on human and ecosystem health, and implications for the National Emission Ceilings Directive: insights from Italy. Environ Int 125:320–333. https://doi.org/10.1016/j.envint.2019.01.064
Devisscher T, Konijnendijk C, Nesbitt L, Lenhart J, Salbitano F, Cheng Z, Lwasa S, Van den Bosch M (2019) SDG 11: sustainable cities and communities – impacts on forests and forest-based livelihoods. In: Katila P, Pierce Colfer C, De Jong W, Galloway G, Pacheco P, Winkel G (eds) Sustainable development goals: their impacts on forests and people. Cambridge University Press, Cambridge, pp 349–385. https://doi.org/10.1017/9781108765015.013
Dijkstra L, Poelman H (2012) Cities in Europe: the new OECD-EC definition. Regional Focus 1
El-Baghdadi O, Desha C (2017) Conceptualising a biophilic services model for urban areas. Urban Forestry Urban Greening 27:399–408
Escobedo FJ, Giannico V, Jim CY, Sanesi G, Lafortezza R (2019) Urban forests, ecosystem services, green infrastructure and nature-based solutions: nexus or evolving metaphors? Urban Forestry Urban Greening 37:3–12. https://doi.org/10.1016/j.ufug.2018.02.011

European Commission (2022) EU budget policy brief: the EU as an issuer: the next generation EU transformation. Directorate-General for Budget, Publications Office of the European Union, European Commission. https://doi.org/10.2761/111076

Eurostat (2016) Urban Europe statistics on cities, towns, and suburbs, 2016 Edition. Luxembourg

Fioretti C, Pertoldi M, Busti M, Van Heerden S (eds) (2020) Handbook of sustainable urban development strategies. Publications Office of the European Union, Luxembourg. https://doi.org/10.2760/020656

FAO (2016) Guidelines on urban and peri-urban forestry, by F. Salbitano, S. Borelli, M. Conigliaro and Y. Chen. FAO Forestry Paper No. 178. Food and Agriculture Organization of the United Nations, Rome

Frank B, Delano D, Caniglia S (2017) Urban systems: a socio–ecological system perspective. Sociol Int J 1(1):1–8. https://doi.org/10.15406/sij.2017.01.00001

Fromm E (1973) The anatomy of human destructiveness. Holt, Rinehart, and Winston, New York

Geneletti D, Zardo L (2016) Ecosystem-based adaptation in cities: an analysis of European urban climate adaptation plans. Land Use Policy 50:38–47

Grimsditch G (2011) Ecosystem-based adaptation in the urban environment. In: Otto-Zimmermann K (ed) Resilient cities: cities and adaptation to climate change - proceedings of the global forum 2010. Springer, Dordrecht

Hoyle H (2020) What is urban nature and how do we perceive it? In: Dempsey N, Dobson J (eds) Naturally challenged: contested perceptions and practices in urban green spaces. Springer, Cham, pp 9–35

ICLEI (2021) ICLEI in the urban era: our vision for a sustainable urban world. Local Governments for Sustainability, Bonn

IUCN (2009) Position paper for UNFCCC COP15, Copenhagen. International Union for the Conservation of Nature, Gland

Kabisch N, Frantzeskaki N, Pauleit S, Naumann S, Davis M, Artmann M, Haase D, Knapp S, Korn H, Stadler J (2016) Nature-based solutions to climate change mitigation and adaptation in urban areas: perspectives on indicators, knowledge gaps, barriers, and opportunities for action. Ecol Soc 21(2). https://www.jstor.org/stable/26270403

Kabisch N, Korn H, Stadler J, Bonn A (2017) Nature-based solutions to climate change adaptation in urban areas – linkages between science, policy and practice. Springer International Publishing, Gewerbestrasse. https://doi.org/10.1007/978-3-319-56091-5

Kaplan R (1983) The role of nature in the urban context. In: Altman I, Wohlwill JF (eds) Behavior and the natural environment. Human behavior and environment (Advances in theory and research), vol 6. Springer, Boston, MA, pp 127–161

Keane A, Davies P (2020) Green cities. In: Rogers D, Keane A, Alizadeh T, Nelson J (eds) Understanding urbanism. Palgrave Macmillan, Singapore, pp 179–194

Keeton R (2021) When smart cities are stupid. http://www.newtowninstitute.org/spip.php?article1078. Accessed August 18

Kellert S, Calabrese E (2015) The practice of biophilic design. www.biophilic-design.com

Li Y, Commenges H, Bordignon F, Bonhomme C, Deroubaix JF (2019) The Tianjin eco-city model in the academic literature on urban sustainability. J Clean Prod 213:59–74

Livesley SJ, McPherson EG, Calfapietra C (2016) The urban forest and ecosystem services: impacts on urban water, heat, and pollution cycles at the tree, street, and city scale. J Environ Qual 45: 119–124. https://doi.org/10.2134/jeq2015.11.0567

MacKinnon K, Sobrevila C, Hickey V (2008) Biodiversity, climate change, and adaptation: nature-based solutions from the World Bank portfolio. World Bank Group, Washington, DC. http://documents.worldbank.org/curated/en/149141468320661795/Biodiversity-climate-change-and-adaptation-nature-based-solutions-from-the-World-Bank-portfolio

Maes J, Quaglia A, Martinho Guimaraes Pires Pereira A, Tokarski M, Zulian G, Marando F, Schade S (2021) BiodiverCities: a roadmap to enhance the biodiversity and green infrastructure of European cities by 2030. Publications Office of the European Union, Luxembourg. ISBN: 978-92-76-38642-1. https://doi.org/10.2760/288633 (online), JRC125047

Managi S, Islam M, Saito O, Stenseke M, Dziba L, Lavorel S, Pascual U, Hashimoto S (2022) Valuation of nature and nature's contributions to people. Sustain Sci 17:701–705. https://doi.org/10.1007/s11625-022-01140-z
Marchigiani E, Bonfantini B (2022) Urban transition and the return of neighbourhood planning: questioning the proximity syndrome and the 15-minute city. Sustainability 14:5468. https://doi.org/10.3390/su14095468
Maye D (2019) 'Smart food city': conceptual relations between smart city planning, urban food systems and innovation theory. City Cult Soc 16:18–24
McDonald R, Beatley T (2021) The urban century. In: McDonald R, Beatley T (eds) Biophilic cities for an urban century. Palgrave Pivot, Cham. https://doi.org/10.1007/978-3-030-51665-9_1
MEA (2005) Ecosystems and human well-being: a framework for assessment. Millennium Ecosystem Assessment, Island Press, Washington, DC
Meijer A, Bolívar MPR (2016) Governing the smart city: a review of the literature on smart urban governance. Int Rev Adm Sci 82(2):392–408. https://doi.org/10.1177/0020852314564308
Mejía MA, Amaya-Espinel JD (eds) (2022) BiodiverCities by 2030: transforming cities with biodiversity. Instituto de Investigación de Recursos Biológicos Alexander von Humboldt, Bogotá, Columbia, p 288
Musco F, Fregolent L, Ferro D, Magni F, Maragno D, Martinucci D, Fornaciari G (2016) Mitigation of and adaptation to UHI phenomena: the Padua case study. In: Musco F (ed) Counteracting urban Heat Island effects in a global climate change scenario. Springer, Cham, pp 221–256
NECD (2016) Directive (EU) 2016/2284 of the European Parliament and of the Council of 14 December 2016 on the reduction of national emissions of certain atmospheric pollutants, amending Directive 2003/35/EC and repealing Directive 2001/81/EC. NEC Directive reporting status 2015. European Union, Strasbourg
Newman L, Dale A (2013) Celebrating the mundane: nature and the built environment. Environ Values 22(3):401–413
Newman P, Beatley T, Boyer H (2017) Resilient cities: overcoming fossil fuel dependence. Island Press, Washington, DC
Nguyen TT, Ngo HH, Guo W, Wang XC (2020) A new model framework for sponge city implementation: emerging challenges and future developments. J Environ Manag 253:109689
Nieuwenhuijsen MJ (2021) New urban models for more sustainable, liveable and healthier cities post Covid-19; reducing air pollution, noise and heat Island effects and increasing green space and physical activity. Environ Int 157:106850. https://doi.org/10.1016/j.envint.2021.106850
Nowak DJ, Hirabayashi S, Bodine A, Greenfield E (2014) Tree and forest effects on air quality and human health in the United States. Environ Pollut 193:119–129. https://doi.org/10.1016/j.envpol.2014.05.028
OECD (2015) The metropolitan century: understanding urbanisation and its consequences. OECD Publishing, Paris. https://doi.org/10.1787/9789264228733-en
Pauleit S, Zölch T, Hansen R, Randrup TB, van den Bosch KC (2017) Nature-based solutions and climate change - four shades of green. In: Kabisch N, Korn H, Stadler J, Bonn A (eds) Nature-based solutions to climate change adaptation in urban areas: linkages between science, policy, and practice. Springer International Publishing, Cham, pp 29–49
Randrup TB, Buijs A, Konijnendijk CC, Wild T (2020) Moving beyond the nature-based solutions discourse: introducing nature-based thinking. Urban Ecosyst 23:919–926. https://doi.org/10.1007/s11252-020-00964-w
Roberts D, Boon R, Diederichs N, Douwes E, Govender N, McInnes A, Spires M (2012) Exploring ecosystem-based adaptation in Durban, South Africa learning-by-doing at the local government coal face. Environ Urban 24(1):167–195
Roseland M (1997) Dimensions of the eco-city. Cities 14:197–202
Rosenzweig C, Solecki W (2018) Action pathways for transforming cities. Nat Clim Chang 8:756–759. https://doi.org/10.1038/s41558-018-0267-x
Secretariat of the Convention on Biological Diversity (2004) The ecosystem approach (CBD guidelines). Secretariat of the Convention on Biological Diversity, Montreal, 50 p

Sheppard S, Konijnendijk van den Bosch C, Croy O, Palamo AM, Barron S (2017) Urban forest governance and community engagement. In: Ferrini F, Konijnendijk van den Bosch C, Fini A (eds) Routledge handbook of urban forestry. Routledge, Abingdon, pp 205–221

Spano G, D'Este M, Giannico V, Elia M, Cassibba R, Lafortezza R, Sanesi G (2021) Association between indoor-outdoor green features and psychological health during the COVID-19 lockdown in Italy: a cross-sectional nationwide study. Urban Forestry Urban Greening 62:127156. https://doi.org/10.1016/j.ufug.2021.127156

Thomas C, Xing Y (2021) To what extent is biophilia implemented in the built environment to improve health and wellbeing? State-of-the-art review and a holistic biophilic design framework. In: Howlett RJ, Littlewood JR, Jain LC (eds) Emerging research in sustainable energy and buildings for a low-carbon future - advances in sustainability science and technology. Springer, Singapore, pp 227–247

UNEA (2022) Nature-based solutions for supporting sustainable development. https://www.unep.org/environmentassembly/unea-5.2/proceedings-report-ministerial-declaration-resolutions-and-decisions-unea-5.2

UNISDR (2012) Making Cities Resilient Report 2012. https://www.unisdr.org/files/28240_rcreport.pdf

UNISDR (2015) Sendai Framework for Disaster Risk Reduction 2015–2030. United Nations Office for Disaster Risk Reduction, Geneva, 53 p

United Nations (2017) New Urban Agenda. ISBN: 978-92-1-132731-1, Quito, Ecuador

Van den Berg AE, Jorgensen A, Wilson ER (2017) Evaluating restoration in urban green spaces: does setting type make a difference? Landsc Urban Plan 127:173–181. https://doi.org/10.1016/j.landurbplan.2014.04.012

van Ham C, Klimmek H (2017) Partnerships for nature-based solutions in urban areas: showcasing successful examples. In: Kabisch N, Korn H, Stadler J, Bonn A (eds) Nature-based solutions to climate change adaptation in urban areas: linkages between science, policy, and practice. Springer International Publishing, Cham, pp 275–289

Wilson EO (1984) Biophilia: the human bond with other species. Harvard University Press, Cambridge, MA

WMO (2021) Guidance on integrated urban Hydrometeorological, climate, and environment services. Volume II: Demonstration Cities. WMO-No. 1234. World Meteorological Organization, Geneva, pp 153

Biodiversity and Ecosystem Functions as Pillars of BioCities

Arne Sæbø, Hans Martin Hanslin, Bart Muys, David W. Shanafelt, Tommaso Sitzia, and Roberto Tognetti

1 Introduction

The BioCities concept builds on the integration of natural and human processes in urban design, with natural biotic and abiotic factors and processes integrated with the development of constructed features to provide for human well-being.

The diversity of plants, animals, and microorganisms, along with their genetic information and the ecosystems they form, make up the biological diversity that is central to dispensing nature's benefits to human society and to foster mitigation and adaptation to climate change. Cities may negatively impact biodiversity, however, either directly (e.g. soil destruction and degradation), or indirectly (e.g. changes in biogeochemical cycles; the introduction of non-native species (Pickett and Cadenasso 2009); changes in land use and landscape fragmentation (Szlavecz et al. 2011). For a long time, cities have disturbed, degraded, and even destroyed

A. Sæbø (✉) · H. M. Hanslin
Norwegian Institute of Bioeconomy Research (NIBIO), Ås, Norway
e-mail: Arne.Sabo@nibio.no

B. Muys
KU Leuven, Leuven, Belgium

D. W. Shanafelt
Institut National de Recherche pour l'Agriculture, l'Alimentation et l'Environnement (INRAE), Bureau d'Economie Théorique et Appliquée (BETA), Université de Lorraine, Université de Strasbourg, AgroParis Tech, Centre National de la Recherche Scientifique (CNRS), Lorraine, France

T. Sitzia
Department Land, Environment, Agriculture, and Forestry, Università degli Studi di Padova, Padua, Italy

R. Tognetti
Dipartimento di Agricoltura, Ambiente e Alimenti, University of Molise, Campobasso, Italy

G. E. Scarascia-Mugnozza et al. (eds.), *Transforming Biocities*, Future City 20,
https://doi.org/10.1007/978-3-031-29466-2_3

natural ecosystems, altering ecological processes, and species' habitats. These practices are in urgent need of change by incorporating the management of urban ecosystems for biodiversity conservation.

The way to transfer long-term benefits from nature to urban citizens in BioCities is to conserve, manage, and develop biodiversity, and to allow species to fulfill the functions needed to sustain urban ecosystem services. Understanding and using the concept of nature-based solutions (NBS), as defined by the International Union for Conservation of Nature and by the European Commission (IUCN 2016; Faivre et al. 2017), is central to integrating nature in BioCities. Nature-based solutions are increasingly being adopted in strategies and policies at the European level and applied in key innovative research projects (European Commission 2021), but its practical application is still under development (Dorst et al. 2019; Castellar et al. 2021). Urban forestry, biophysical green infrastructure, the design of green infrastructures onto buildings, and their related ecosystem services are all important in the NBS concept (Escobedo et al. 2019). Green solutions should be integrated with the development and maintenance of the urban ecosystem, including water, soil, and atmospheric factors and processes. When properly designed, several of the NBS yield co-benefits, such that ecological functions and integrity will contribute to building urban natural capital. However, there is a need to further develop NBS as tools to create high-quality and multifunctional solutions targeted to the needs and wants of the BioCity.

The abiotic factors inherent to urban biodiversity include, but are not limited to, climate and weather, soils (chemical, physical, and biological attributes), and water (subsurface, ponds, rivers, and lakes), and how these variables are spatially situated on the landscape. Microbial associations, plants, and animals form complex and dynamic biological communities that are dependent on these abiotic factors, and are always strongly affected by human activity, intentionally or not, as part of the urban metabolism (Faeth et al. 2011). Differences between urban and rural areas are many and are mainly related to the numerous human processes and constructed elements in urban areas. The transition between urban and rural areas can be described as a continuum, with the border between land uses being typically fuzzy and diffuse.

Climate change exerts growing pressure on urban environments, which in turn requires a new urban planning paradigm towards climate resilience. Another serious threat to sustainability is that urban expansion is continuing to accelerate biodiversity loss (Bullock et al. 2011; Zari 2018), further threatening functions and services delivered by biodiversity and ecosystems. Habitat fragmentation due to agriculture expansion is cited as the primary driving factor for species extinction (IUCN 2016), with growing world human population leading to increases in demand for agricultural products further exacerbating the problem. BioCities should contribute to breaking this negative circle of development.

This chapter reviews the current knowledge on how natural processes can influence the well-being of citizens and how novel approaches to nature-based urban design can complement existing green infrastructure. Urban ecosystems combine elements of natural systems and processes with those of constructed systems and management regimes. We will show that this platform can increase sustainability, resilience, and livability of cities. Hence, biodiversity will be the pillar that supports the green and sustainable development of the BioCity. Moreover, we suggest key areas that will be necessary for the sustainable development of the BioCities of the future.

2 State of the Art: Describing Processes and Possibilities

2.1 Essential Building Blocks of Biodiversity in Cities

Biodiversity, one of the main focuses of BioCities (as stated in chapter "Towards the Development of a Conceptual Framework of BioCities"), is essentially the result of abiotic environmental conditions with biotic life evolving to those conditions and perhaps further modifying its habitat over time. Because of the tremendous influence of humans building, managing, and continuously changing the city environment, the biotic and abiotic elements of the urban system are altered in ways that sustain components that might never have been present in a 'natural' environment.

The complex ecological interactions between abiotic resources (climate, geologic substrate, and topography), biotic resources (both species and genetic), and juxtaposition of habitats on the landscape, are the fundamental building blocks for their sustainable management, conservation, and restoration (Perring et al. 2013). By considering all of these components, we strengthen the ecological integrity of the urban landscape by providing resources and linkages to the surrounding areas and native organisms, supporting biodiversity in a wider context.

2.1.1 Abiotic and Biotic Factors

Fundamental resources are related to landscape, geomorphology, topography, and properties of the surface materials that are formed by geologic processes under certain climate regimes over long periods of time. The outcome of these processes has largely determined where most of the major world's cities have been established. Whilst many factors contributed to where cities are today, they tend to be on or adjacent to sites with nutrient rich and productive soils, game-rich forests that supply building materials and energy, and within easy access to large rivers, lakes, or the sea (Diamond 1997; Bosker 2022). We also find examples of cities, however, that have been developed with an imbalance in their natural resources. Phoenix, for example is the fifth largest city in the USA and exists in a desert, forcing major challenges to its water supply both now and into the future. Cities must be developed in ways that do not compromise natural resources, whilst still providing those resources needed to sustain people with food, water, and healthy environments. Cities that are fundamentally unsustainable can only be adapted to nature's boundaries if they are transformed, but also may be scaled back, balancing resource use to what is available at the site. Every city can be viewed as a dissipative, metabolic system (Giampietro

2019), creating entropy, both positive and negative, that is continuously compensated for by a flow of matter and energy from the wider surrounding environment. The solution is to find the balance, where a BioCity is designed to use only the resources that can be provided by the surrounding BioRegion (see chapter "From BioCities to BioRegions and Back: Transforming Urban-Rural Relationships") and that can be tolerated in the context of global sustainability. This is a major challenge for policymakers.

Water is increasingly a scarce resource, especially in cities. Paradoxically, climate change also increases problems of urban flooding, caused by exaggerated soil sealing in combination with extreme weather events as outlined in the IPCC's Sixth Assessment Report (Masson-Delmotte et al. 2021). To address these challenges, harvesting and reuse of stormwater and greywater have been developed as decentralised solutions (Campisano et al. 2017) in parallel with NBS to handle more extreme episodes of precipitation.

Large cities create their own climate, especially by increasing temperatures, which may have both positive and negative effects. A generally warmer environment within the city (the heat island effect) and increased heterogeneity in city temperatures can create microclimates that strongly affect both abiotic and biotic processes. Ossola and Lin (2021) warn against relying on NBS to mitigate climate change in cities with extremely high temperatures, however, as extreme temperature episodes may be detrimental to NBS themselves.

Soils, especially topsoils, are important for productive and sustainable land use. Anthropogenic activities, past and present, have resulted in grave soil degradation generally caused by lack of erosion protection or improperly constructed soil mixtures, sealing and compaction, and contamination from industry and traffic. The number of potentially contaminated sites has been estimated to total 2.5 million in Europe (Perez and Rodriguez 2018), with high costs for remediation. Brownfields appropriately treated and managed, however, may provide new opportunities for using previously unavailable areas for development of residential zones, new businesses, or urban green areas, thus saving other valuable areas for biodiversity preservation, forestry, or food production (Song et al. 2019). Urban soils, however, are often of extremely low physical, chemical, and biological quality and need improvement (Downing Day and Harris 2017).

Cities are home to thousands of species, which are all part of the tree of life. Urban-dwelling species are those that find suitable habitats within the city, and are flexible in adapting to the human-made environment. Often cities are considered dead, sterile environments with low diversity where only common, ruderal species occur (Concepción et al. 2015). This may be partially true, but we know, thanks to citizen science, that cities often contain a surprising number of species, including rare and red-listed species with high conservation value (Soanes et al. 2019). This can be explained by the presence of a relatively high diversity of habitats, hosts, and sources of food, and less use of harmful biocides and fertilisers when compared to the farmed countryside (Reichholf 2007).

Cities and their interconnections are hubs for non-native species to establish in new areas. City green elements and spaces can support urban ecological processes different from their local natural counterparts. Examples include ornamental plants escaping from gardens and contributing to spontaneous vegetation development of hybrid and novel urban ecosystems, often on brownfields or other areas with relaxed management. Naturalisation also occurs, such as captive parrots and turtles being released by their owners to become pseudo-native species in their new environments (Knowler and Barbier 2005). Some species have shown incredible flexibility when adapting to urban environments, like the common blackbird (*Turdus merula*) feeding on earthworms in lawns, or fox (*Vulpes vulpes*) populations in urban centre's surviving on food waste foraging. Many other species feel perfectly at home in the city, like the rock pigeon (*Columba livia*) and its natural predator the peregrine falcon (*Falco peregrinus*), who have substituted the cliff faces of their natural environment for the buildings of the city. The active use of non-indigenous species should be done carefully, however, acknowledging risks, unknowns, and potential consequences, yet taking advantage of the services provided by restored and rehabilitated novel and hybrid ecosystems (Klaus and Kiehl 2021). The accelerated pace of climate change will make the issue of the introduction of non-indigenous plant and animal species, and pests as well, a relevant issue for future BioCities. On the one hand, novel ecosystems can be self-regulating, without energy inputs from humans and possessing a lower carbon footprint than artificially maintained green spaces (Kowarik 2011). But on the other hand, they may also lack unique and specialised species, such as those that need the deep shade of a forest, the wetness of a swamp, or larger habitat ranges.

The urban forest, the sum of all the trees, woody shrubs, and associated habitats in a city along with created green infrastructures such as green roofs, green facades, infiltration zones, and other NBS, is increasingly used as a tool to moderate the climate of cities, especially to combat the urban heat island effect through shading and evapotranspiration (Ellison et al. 2017). Along the urban roads of Bangalore, India, afternoon ambient temperatures are on average 5.6 °C lower under the canopy of trees than on exposed roads, and surface temperatures are up to 27.5 °C lower (Vailshery et al. 2013). At the University of Melbourne, Berry et al. (2013) observed that temperatures of building walls might be reduced by as much as 9 °C under tree shade. These effects depend on the material, structure, geometry, and design of buildings, however, as well as on tree species, season, and orientation. Attention should be paid to choose the appropriate plant species that favour the stormwater management capabilities of green roofs and façades (Andenæs et al. 2021), reduce the risk of attracting urban pests like mosquitoes, causing unnecessary energy and water consumption, or other nuisances such as falling leaves, fruits, and limbs.

The design of these tools and the composition of species in such elements should reflect desired sustainable natural processes and functions, but also be adaptable to a changing climate. Drought-resistant Mediterranean rock plants, for example offer great promise as a species for urban green roofs (Van Mechelen et al. 2014). Selection and use of plants should reflect the present and expected future urban climates (Sjöman et al. 2016) to secure stability in the BioCity.

2.1.2 Providing Ecosystem Services

Urban nature provides multiple ecosystem functions, with supporting services like photosynthesis, respiration, transpiration, decomposition, infiltration, and nutrient cycling, which in turn support other ecosystem services useful for human beings (Gómez-Baggethun et al. 2013). It also provides provisioning, regulating, and cultural services. Provisioning services are delivered when urban people harvest homegrown strawberries or enjoy an autumnal walk in the urban forest collecting chestnuts or mushrooms. Regulating services are provided when riverine forests protect a city against inundation, or when trees absorb particulate matter pollution from the air. Cultural services are positive effects from nature on mental health, education, or by providing social meeting places in urban green areas. Figure 1 illustrates some of the connections between biodiversity, ecosystem services, and humans. The interactions are complex, however, and a complete picture with the BioCity is still unknown.

Systems with higher biodiversity tend to show greater performance in ecosystem functioning, and ultimately provide more benefits for human beings (Cardinale et al. 2012). This causal relationship between ecological structure (i.e. biodiversity) and ecosystem services is referred to as the ecosystem service cascade (La Notte et al. 2017). Individual species fulfill different functions in the ecosystem. More species generally means more overlap and redundancy in their functions and, due to differences between species (e.g. preferences for environmental conditions), greater ability of the ecosystem to maintain the same level of functioning in the face of natural or anthropogenic change (Loreau 2010). Increasing a city's biodiversity, including tree species richness of urban greenspaces (Wang et al. 2021), can support

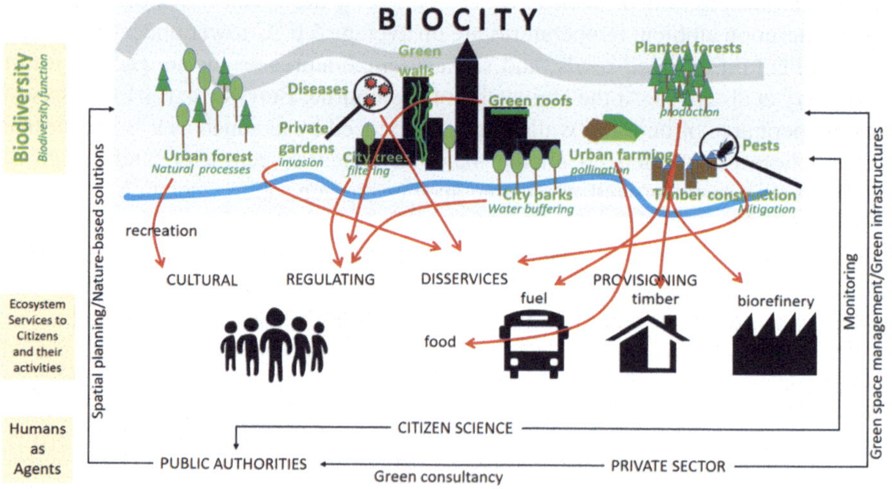

Fig. 1 Urban biodiversity as the essential base of the BioCity, with its underlying the ecosystem functions and ecosystem services to citizens. Human agencies can steer urban biodiversity by targeting and optimising species ecosystem service delivery and limiting disservices

a healthier, more resilient ecosystem, better performance in its service deliveries, and greater insurance against negative impacts of climate change. For this reason, urban management and planning that leaves more space and resources for restored nature will improve the vitality of the urban ecosystem and the health of its citizens (Aronson et al. 2017; Aerts et al. 2020).

Since green elements in cities tend to be scarce, they will often need to provide several ecosystem services at the same time, which assumes multifunctionality. In other words, the level of multifunctionality is given by how green space management targeting one ecosystem function improves other functions, contributing to multiple ecosystem services in a socio-ecological context (Hansen and Pauleit 2014). Multifunctionality can be measured at both the local and landscape levels, given that trade-offs between specific functions will limit the co-benefits within a given area.

2.1.3 Surveying Urban Green Spaces

Official international statistics (e.g. Urban Atlas by Eurostat and Green Capital Initiative by the European Commission) identify urban green spaces and green infrastructures in terms of surface area. A great variety of methods are used to estimate the urban green spaces. Recently, two-dimensional (2D) indicators derived from optical remote sensing (e.g. Landsat and Sentinel-2), such as Normalised Difference Vegetation Index (NDVI), have become one of the most widely used data sources to characterise and represent cities in official reports, as well as assessing exposure to green space in epidemiological studies. These 2D indicators could be considered as a good proxy for some urban structures, but they have limited capacity to take account of the differences in the type and quality of green spaces in the heterogenous structure of urban environments.

Some authors believe it is possible to characterise green spaces in a more adequate way (e.g. Giannico et al. 2016; Tan et al. 2016), according to the different levels of biomass in different types of urban green (e.g. trees in lawns), and using a ratio between 'green' and the surrounding environment, such as buildings and grey infrastructure (e.g. roads and parking lots). Green biomass is proven to provide a large variety of ecosystem services in scientific literature (Sanesi et al. 2009; Marziliano et al. 2013; Tan et al. 2016). In addition, an emerging body of evidence has highlighted the importance of the three-dimensional (3D) structure of green/grey areas for several health outcomes, such as mental fatigue, aggressive behaviour, and effectiveness in managing major life issues (Kuo and Sullivan 2001). With the increased availability of Light Detection and Ranging (LiDAR) point cloud data, the use of 3D indicators alongside 2D indicators has become possible (Giannico et al. 2016). A coupled 2D/3D approach in the description of urban green spaces can guarantee a characterisation that allows the definition of different characteristics and estimate the ecosystem services they could provide.

2.1.4 Trade-Offs and Disservices

The diversity of nature brought into cities through NBS will play a critical role in improving the adaptive capacity and resilience of BioCities in increasingly fragile urban contexts (Demuzere et al. 2014). In the context of climate change, examples of services that increase resilience and climate change adaptation are water retention and detention, and pollination services that depend on NBS, including biodiversity. Whilst research scientists and urban planners acknowledge the potential of NBS and green elements for the successful provisioning of climate-related ecosystem services (Matthews et al. 2015), experimental outcomes rarely draw explicit links to biodiversity in urban environments (Schwarz et al. 2017). Although biodiversity is often considered a co-benefit of NBS, quantifying the trade-offs and synergies between the conservation of urban biodiversity and the delivery of other benefits (e.g. climate resilience) needs to be addressed before implementing biodiversity assessments in the decision-making process.

Not all of the effects of NBS, however, are beneficial or socially acceptable. Examples of disservices of urban nature include pollen as an allergenic, pests, diseases, invasive species, vermin, and harmful insects. Plants can damage built infrastructure (above and below ground), generate dirt from shredded leaves and fruits, reduce visibility through tall tree crowns, and produce safety risks, which are not socially accepted (Heynen et al. 2006). The plane tree (*Platanus* sp.), for example is one of the most planted and maintained trees along city roads but is a source of allergens (Varela et al. 1997). People living in cities are on average more prone to allergies than rural people (Ehrenstein et al. 2000) since pollution exacerbates the effects of allergens (Molfino et al. 1991). Conversely, more urban green space reduces the allergenic effects (Stas et al. 2021). Research has shown that canopy cover, particularly along narrow urban roads, can produce local increases in gas pollutants and particulate matter if the tree canopy decreases air circulation (Sæbø et al. 2017). Moreover, in reaction to environmental stresses like excesses of light and temperature or not enough water, trees tend to emit biogenic volatile organic compounds (bVOC). The most abundant bVOC is isoprene (Seinfeld and Pandis 2016), which may contribute to ground-level ozone formation (da Silva et al. 2018). Urban forests with a high diversity of tree species could provide a refuge for introduced non-native forest insects, which come with many risks (Branco et al. 2019).

Some trade-offs are already well known, whilst others will certainly be detected and quantified during the implementation of the BioCity. For example, giving priority to vegetation in dry climates may have trade-offs regarding limited urban water supplies. If NBS should function as intended, planners must designate sufficiently large areas as green elements. Stakeholders may have different priorities, however, when it comes to the use of urban areas, where attention to profits and green solutions may be antagonistic, thus creating conflict. This is a major challenge for policymakers, who most often prioritise profits as the main incentive for the development of urban areas.

Understanding the benefits and costs of integrating nature in urban environments requires interdisciplinary knowledge of urban biodiversity and ecosystem services. Indeed, trade-offs may arise based on different designs of green elements, temporal scales of urban development, and spatial scales of urban elements. Landscape textures, city limits, ownership patterns, regulatory structures, and available technologies change over time. These changes may affect not only areas being inventoried and monitored, but also species distributions. Therefore, quantifying economic costs and measuring human and environmental benefits of green elements requires multi-dimensional indicators to assess the potential of biodiverse green elements to adapt ecological processes to societal needs, and mainstreaming these ideas into new policy frameworks. In this sense, policymakers and city managers need comprehensive multi-scale and cross-sectoral guidance and recommendations to effectively manage urban green elements in biodiversity friendly and climate-effective ways. This will undoubtedly be complex and demanding to manage in a comprehensive and holistic way.

3 A New Approach

A paradigm shift needs to happen when constructing or transitioning to a BioCity towards building the city by mimicking natural ecosystems, as a forest analogue instead of installing forests within the city (see chapters "Towards the Development of a Conceptual Framework of BioCities" and "Towards BioCities: The Pathway to Transition"). The primary goal should be to make the city a compartment of forest and nature. This expansion of the city, or rather of the forest, is what makes it realistic to develop BioCities built on NBS. People need nature, but nature can do perfectly well without humans. Only by acknowledging that we need to adapt to the nature that hosts us can we hope to save the organisms and natural systems that we rely on. There are obstacles of paramount importance, however, on our path to the holistic BioCity.

If biodiversity is to be integrated into the urban landscape to the degree where ecosystem services can be substantially supplied, it requires a shift in our attitudes towards land use allocation in cities. Locally, land sharing is often given priority over land sparing, to provide multifunctionality and multiple ecosystem services. Accepting that the city is an integrated compartment of the forest means that the connections between area types should be strengthened accordingly. Interfaces between the dense cityscape and rural areas will be of utmost importance for resource management of urban areas, and the exchange of biodiversity between the two should rely on effectively structured and functioning landscape and ecological connectivity. On this interface, exchanging sustainable food and fibre sources may be intensified, with the integration of recycling waste products building towards a more circular economy. However, this implies that urban development needs to include and embrace the strategic development of the surrounding rural areas (see chapter "From BioCities to BioRegions and Back: Transforming Urban-Rural

Relationships"). If divisions between policy, planning, and management of the two compartments continue, then the BioCity will remain a vision never realised.

From Designed to Novel Ecosystems
Green elements and social systems in cities together create a designed socio-ecological system (Macdonald and King 2018), which may eventually evolve into intermediate or hybrid systems between novel and current systems. Novel ecosystems retain characteristics of the historic system, such as basic ecological processes (primary production, biogeochemical cycling), but embody new characteristics driven by human wants and needs.

Cities can positively affect biodiversity or ecological processes, such as by restoring or mimicking natural structures. The use of vacant surfaces for vegetation will be more important in the future to promote biodiversity to provide desired ecosystem services. Although non-native species, habitat fragmentation, and disturbance regimes impact elements of native ecosystems in urban landscapes, novel urban ecosystems can spontaneously establish and contribute to ecological processes (Klaus and Kiehl 2021). These novel communities need to be considered as a reference point in setting management and restoration targets. The combination of natural and artificial structures (hybrid solutions) and combinations with smart (electronic) solutions have great potential but need to be developed in tandem with BioCities (Matasov et al. 2020; Torresan et al. 2021).

Transitioning to BioCities
Although urban ecosystems provide a plethora of services, ecosystem dynamics in the urban context are still poorly known (Pataki et al. 2011). According to the Resilient BioCity concept and panarchy theory, as stated in the BioCity Manifesto in the first chapter of this book, social-ecological systems are interlinked in adaptive cycles of growth, accumulation, restructuring, and renewal, whilst the equilibrium state is composed of a complexity of dynamic states of equilibria. This translates into a series of regeneration and reorganisation stages of blue and green nature in the urban area, but overall, nature-based systems should not be eroded or razed but rather conserved, improved, and increased in size. In cities, habitat loss and habitat fragmentation are interrelated, and both affect species richness and ecological processes (Liu et al. 2016). In the next decades, if built-up areas continue to increase with the urban population, it will be environmentally catastrophic. If, however, growth can be tailored to save or even increase biodiversity and green elements in cities, sustainability and resilience can be obtained. To this end, urban development should be planned in a holistic manner, viewing each urban and nature element as interconnected components in a designed ecosystem.

In transitioning to BioCities, we must use plants that are well adapted to the conditions today and those of future climate scenarios (Sjöman et al. 2016). Greater diversity in established or constructed vegetation will secure stability, thus contributing to continuous delivery of services even in the face of novel pests, diseases, and a changing climate. Water interacts with plants and soils, impacting the flow of materials, nutrients, and contaminants, connecting surface and ground waters. Water for urban greening should come from cleaned and reused water from human activities. The recycling, composition, and structure of urban green spaces can positively affect the quality and amount of water available to cities, but also its retention and detention capacity.

Understanding the complexity of future development entails an integrated analysis and assessment of landscape characteristics, disturbance gradients, and social issues for creating sustainable solutions in support of urban nature stewardship. It requires incorporating customised nature-driven urban designs into policy targets and guidance tools to enable adaptive governance (Elands et al. 2019). In this context, monitoring the outcomes of NBS in the face of continuously changing urban conditions is critical for determining the degree to which novel solutions depart from current or best practices. Monitoring depends on repeated measurements of data related to natural factors. Monitoring gives information on how ecosystems, ecosystem services, and resilience develop over time. Assessment frameworks designed to monitor the impacts of NBS include indicators on urban forest processes, biodiversity, and management. Monitoring of NBS in dynamic and complex urban systems requires technological solutions for data collection, processing, and utilisation at affordable costs (European Commission 2021). Networks of wireless, affordable, and multiparameter monitoring devices, based on the "Internet of Things" (IoT), represent opportunities to monitor ecosystem services offered by urban trees and forests, in the form of meaningful indicators for both human health and environmental policies. An example of these technologies allowing for real-time data transmission and numerous low-cost monitoring points is represented by the TreeTalker© system (Matasov et al. 2020), which create new opportunities also for a wider application of citizen science.

Morgenroth and Östberg (2017) emphasise the need for the standardisation of methods and indicators used in monitoring. Standardisation is a prerequisite for the comparison of data and record development over time, as well as for comparing data between cities.

3.1 Biodiversity, Digital Technology, and Environmental Awareness

Information and communication installed on tree stems and in urban soils provide the opportunity to monitor in real-time indicators of a wide array of ecosystem services (Matasov et al. 2020). For example, dendro-chemistry is a promising field of urban monitoring, which uses trees as an archive of historical events of air pollution (Alterio et al. 2020). Real time, diffused monitoring of urban biodiversity and ecosystems brings extensive opportunities for environmental education and citizen participation. Citizens can provide valuable data on urban green monitoring, complementing data collected from deployed devices and remote sensing (Heigl et al. 2019). Citizens may use open-source software (e.g. i-Tree), participatory apps (e.g. eBird, iNaturalist, pland@net, and observation.org), and web-based platforms.

As an example, indicators of biodiversity (e.g. proportion of green spaces and number of native species) and phenology (e.g. budburst and migration), or urban green management (e.g. distribution and accessibility), may be derived from citizen observations, at high frequencies and low costs (Torresan et al. 2021).

Methods for remote sensing are developing fast, with higher resolution and utilising more user-friendly software. Citizen science may help monitor urban forest spatial dynamics in space and time (Fraisl et al. 2020), conveying a combination of spatial data and on-site surveys in the proximity of NBS (e.g. mapping of the delivery of ecosystem services in locations where they are being provided) (De Vreese et al. 2016). Citizen involvement can generate a robust amount of data that can better support any analysis. But including citizens in the monitoring process is even more important in strengthening the dialogue between city governments, planners, management, and stakeholders, and to connect nature in BioCities to their citizens via the practical engagement with local communities (e.g. through citizen budgets, participatory management, citizen watering initiatives, and community-based fire management).

For advanced objectives, citizens may require specific training and capacity building (Anton et al. 2018). Lending or giving volunteers biodiversity recorders, which can be used during periodic surveys or targeted monitoring tasks, has enormous potential, especially when combined with conservation efforts in urban and peri-urban environments. Nevertheless, citizen commitment to undertake biodiversity monitoring raises questions on data quality and use, as well as legal questions and privacy issues (Ganzevoort et al. 2017). Citizens are not machines that mindlessly store data, but have specific motivations for collecting and sharing information. Therefore, involvement of volunteers in biodiversity-related citizen science requires creating proper incentives for collecting and reporting comparable monitoring data. Often the non-material rewards of aiding or engaging in nature conservation are sufficient, or participants could be directly compensated.

4 Priority Areas of Paramount Importance for the Realisation of BioCities

The fundamental reason for proposing the concept of BioCity is to address the crucial issue of climate change and health crises impacts on urban systems. They affect plant and animal species, as well as humans and the environment, in interlinked connections well exemplified by the One Health concept. Functional traits of BioCities already proposed in chapter "Towards the Development of a Conceptual Framework of BioCities", as the Self-Sufficient BioCity and the Urban-Rural Balanced BioCity, point to the crucial role of producing energy and bioresources whilst adapting to climate change and fostering interconnections with the surrounding BioRegion. There are tremendous challenges to be overcome when building natural processes and biodiversity as pillars of BioCities. Hence, there is a need to prioritising overcoming some of the biggest obstacles that require more immediate attention.

4.1 *Water Shortage and Flood Overflow Control*

The BioCity necessitates that we use water more efficiently. Strategies to harvest and reuse stormwater and reuse of greywater are approaches that can be implemented in the redesign of urban hydrology underpinning BioCities. Water that flows overland and subsurface bind the different physical and biological components of the watershed. Water may interact with biological organisms and urban structures, transforming and transporting materials, nutrients, and contaminants downstream. This flow depends on the topography and geomorphology of the urban watershed and may affect the composition, structure, and function of the urban ecosystem. Indeed, the hydrologic flows in urban riparian zones may influence and be affected by altered biophysical processes (e.g. rapid runoff and drought), which may impair the natural course of biogeochemical cycles (Pickett et al. 2020).

Reuse of water and increases in retention and infiltration of stormwater must be done without spreading pollutants or endangering water quality. More research is needed to find ways to safely reuse water for different purposes in cities. Disinfection may be required. However, the ability to store water of sufficient quantity and quality in periods of surplus, to be used in periods of drought, should be further developed. The trade-offs between the water demand from additional vegetation in BioCities and future water scarcity scenarios need to be addressed at the policy, planning, and management levels, and shifting towards better-adapted plant species should be considered.

4.2 *Food Production*

Food production in cities may come from traditional horticulture in open urban green spaces, and also from green roofs, although until now this has often been more important for social contexts (cohesion and education) than for food production. In the Mediterranean region of Europe, just 10–25% of urban areas would be needed for cultivation to meet the recommended consumption of vegetables by urban dwellers (Martellozzo et al. 2014). Yet even in Germany, the world leader in green roof technologies, only about 10% of all houses have installed green roofs (Huld et al. 2018). Whilst these roofs are environmentally beneficial, the majority of people do not go so far as to create food, energy, or social opportunities that rooftop greenhouses can provide. Therefore, substantial untapped potential still exists for expanding vegetable production in urban areas, especially on existing rooftops. In recent years, however, the emergence of community supported, often organic, agriculture (crop sharing or Community Supported Agriculture [CSA] model) in the urban fringe has paralleled the increasing interest of urban families to get access to healthy, locally produced, and sustainable fresh food. An example of urban food production is the Picasso Food Forest in Parma, Italy (Riolo 2019). The Picasso Food Forest represents a hotspot of biodiversity, hosting a plant nursery and wildlife shelter, whilst providing a genetic bank that conserves several heritage and local varieties of food plants.

The use of closed compartments has recently become an important trend, providing the basis for substantial production of fungi and microbial-based food in urban areas. The development of plant production in such systems is driven by the availability of less energy demanding light emitting diodes (LED), which supply light appropriate for photosynthesis at low costs. In such systems, mainly low-growing herbs and salad plants can be grown in vertical systems with high output, supporting the commercial production of food in urban areas. This food system and its related technology, however, are still in development. These innovations should embrace not only the technical issues of the production system, but also include how to make the production sustainable by recycling and reutilising primary resources such as water, nutrients, and biomass. Can there be a coupling between the products of human living and activity with food production? In such systems, both production factors, like nutrients and soils, and the food products will travel very short distances from farm to table, contributing to a sustainable production.

4.3 Landscape Scale Management of Urban Greening and NBS

The success of BioCities will depend on how services and functions are provided in a larger landscape, including optimising co-benefits through landscape connectivity and the juxtaposition of land areas allocated for specific functions. NBS designed for one specified purpose often have impacts (co-benefits) far beyond those targeted. Whilst the impact on a single NBS may be small, the total impact summed across all factors can be large. The grand network of NBS and their interactions will need to be documented and monitored, however, as well as their optimisation in a holistic system.

The dynamics of spatial patterns in the urban setting may influence biogeophysical processes, including cycles and fluxes of key ecosystem resources (e.g. energy, nutrients, and materials) that underlie changes in land use and land cover. Avoiding ecological homogenisation of biogeochemical processes and ecological functions across cities requires context-tailored urban planning, which may operate at the landscape scale as a mosaic of heterogeneous and interrelated ecosystems (Pickett et al. 2010).

Spatial heterogeneity of the urban landscape depends upon urban morphology, vegetation type, building structure, and paving material. Maintaining ecological processes and biodiversity at the landscape scale requires understanding dynamic variables including edge effects, patch size, habitat quality for organisms, and structural fragmentation (Loreau 2010). The inclusion of spatial heterogeneity as a design principle for providing multiple functions and services merits further exploration. This heterogeneity has to be balanced, however, by approaches to defragment the urban green space that support flow of individuals their genes in the landscape, to reduce local extinctions caused by fragmentation.

Designing the BioCity requires both top-down and bottom-up approaches. It is a great challenge to formulate all the important connections, interactions (positive and negative), and the (sometimes surprising) feedback loops. Pocket parks, larger parks, and the urban forest can contribute to maintaining connectivity of natural areas. In the core of cities, small parks can be connected by treed boulevards, for example creating corridors between habitat patches and mitigating for fragmentation of the BioCity ecosystem. The role of vegetated buildings in landscape connectivity and creating wildlife corridors has also been recognised but is yet to be fully articulated (Mayrand and Clergeau 2018).

The theory of island biogeography (MacArthur and Wilson 1967) is a key concept in landscape ecology and for its application to land use and urban planning. As highlighted for parks and woodlots in general (Alvey 2006), even in small spontaneous patches, plant species richness tends to increase with patch size. For example, in the small city of Padua, Italy, the diversity of woody species on different patches of urban forests was related to the size of the patch (Sitzia et al. 2016). Larger patches are also generally more accessible and used by people in greater frequency (Cambria et al. 2021). This calls for new planning approaches for BioCities, which would take advantage of the relationships between biodiversity and spatial properties of urban greenspaces, such as patch size, shape, and connectivity.

Spontaneous development of vegetative communities in urban areas is acknowledged as a potential NBS, providing that it is integrated with societal demand for ecosystem services. For example, woodland patches that spontaneously develop into wild woodlands can play an important role in urban biodiversity by forming novel ecosystems that did not exist in the past, and refuges for dispersing or migrating native species. Wilderness in cities, however, has been commonly interpreted as wasteland, a sign of abandoned or derelict places, and lack proper management from private landowners or proper land use allocation from public institutions. On the contrary, abandoned human spaces represent opportunities for the recovery of natural processes in growing and shrinking cities. Unfortunately, before their potential is comprehended, exploited, or realised, they are often subjected to aggressive land use transformation, with a reduction of ecosystem services for citizens (Foster 2014; Zipperer 2002). Instead, transient measures, allowing the use by people, could be applied until the foreseen building development is realised (Kattwinkel et al. 2011). In other cases, when the biodiversity of these places becomes relevant for sustaining the BioCity, they should be protected like many of the semi-natural habitats in rural areas, such as those within the European Union Habitats Directive. These sites are important in BioCities because they do not require energy inputs and have a lower carbon footprint than artificially established greenspaces (Kowarik 2021).

4.3.1 Soil Quality Management

Soil functions and soil quality are critical for the sustainable development of BioCities. Soils are a limited resource and should be considered non-renewable, since soil formation is a very slow organic process. Consequently, soils should be subject to a strong protection regime. Living soils are the most organism- and function-diverse habitat that exists on our planet, and multiple services can be expected from soil environments (see van Elsas et al. 2019 for references). Such services can be to decrease plant diseases, use substances that promote plant growth, and bioremediation of organic and inorganic pollutants. Growing knowledge of soil functions will only increase their value for ecological services, boosting the impacts and co-benefits of good soil management.

Low-quality soils (texture, structure, and nutrients) affect the growth, functionality, and longevity of urban forests. The main problems are related to low water storage capacity, compaction, contamination, and other suboptimal soil factors. Knowledge on how to establish and manage soils for urban trees, with respect to soil quality and need for soil volume, even under harsh city conditions, is available (Grabosky and Bassuk 2017) but would benefit from further research. A vital focus is to remove soil sealing and increase the rooting substrate volume and soil quality to decrease plant stress (Godefroid and Koedam 2007). When looking for how to establish and improve conditions in urban forests, site assessments, including soil evaluations, are necessary (Bassuk 2017).

Manufacturing artificial soils is an alternative to importing soil from rural areas. Further on, the reuse of building materials and organic matter produced in the BioCity could be building blocks for the manufactured soils. The reuse of soils from urban development sites should be implemented as routine when striving to increase circularity, as well as improving conditions in urban habitats, provided that soil quality is carefully monitored.

The physical properties of various soil types determine their infiltration capabilities as mitigation for stormwater runoff and play a large role in determining which plants will grow at a given site. After an appropriate soil has been developed or installed (with sufficient depth and volume), the planning of above-ground vegetation and structures can be conducted. A large body of knowledge already exists on soils as biofilters (Beryani et al. 2021; Fang et al. 2021), but it should be further researched, developed, and exploited for innovative uses and adaptations. Ecological engineering related to water and soils needs to be strengthened. Overall, this body of work relates to roof gardens, green roofs, green walls, flower beds, and infiltration zones, optimising the linkage between vegetation and soils to maximise function, filtration, and biodiversity.

5 Case Studies: Good Practices

1. *Urban Forest Patches—Berlin, Germany*
   Nature-based projects and solutions are currently applied in cities around the world. One of the most iconic examples is the integration of spontaneous woodlands, green infrastructure, and restoration in the Natur-Park Südgelände in Berlin (Fig. 2). Here, undisturbed forest patches, occasionally grazed clearings, and designed physical features are all present in the same area on former freight yards.
2. *Abandoned Industrial Areas or Urban Oasis?—Rome, Italy*
   The former SNIA Lake in Rome is another example (Fig. 3). It is a former industrial site, a mosaic of semi-natural habitat and industrial ruins which developed after illegal activities by the former private owner. It is currently designated as a natural monument. The ruins of the buildings, though never finished, were not demolished. On the contrary, they are embedded in the surrounding wild nature, inspiring architects, citizens, and artists.

   Regardless of their size and shape, nature-based solutions can be implemented in urban areas and contribute to biogeographical processes (Sitzia et al. 2016). This can include the spontaneous development of vegetated lines in streets, fallow patches in gardens, or vegetation retaking infrequently maintained walls. The establishment of artworks that contrast wild nature can be seen as a cultural NBS, to integrate remnants of a possibly painful or unwanted past into new green spaces, and combining modern design approaches with novel urban wilderness (Kowarik 2021).

Fig. 2 The Natur-Park Südgelände in Berlin. Here, undisturbed forest patches, occasionally grazed clearings, and designed physical features are all present in the same area on former freight yards (Photo by Tommaso Sitzia)

Fig. 3 The former SNIA Lake. A mosaic of semi-natural habitat and industrial ruins that followed illegal activities by the former private owner. The ruins of the buildings, though never finished, were not demolished. On the contrary, they are embedded in the surrounding wild nature, inspiring architects, citizens, and artists (Photo by Forum Territoriale Parco delle Energie)

Before nature-based projects reach their final completion, they experience various stages of succession, including transitions in land cover and land use, along with changes in the local plant and animal communities. For example, small patches of agricultural land may still be present in the built matrix of expanding cities, neglected military sites can be rewilded, or low-profitable mining and construction sites may be abandoned. This heterogeneity provides a range of possibilities from which to choose novel and experimental designs.

In contrast to wild spaces in cities, public opinion is anchored to the artificial establishment of greenspaces, through tree planting and construction of green spaces, without acknowledging the economic and ecological costs of the related intensive maintenance. By combining nature and culture, and offering positive signs of active management, BioCity managers should seek to remove the negative connotation of wild spaces in cities, allowing people with different values to find common identities, including the removal of vegetation that damages historical monuments.

3. *Xeriscaping in Urban Landscape Design—Phoenix, Arizona*

Depending on the region and season, between 30 and 60% of household water consumption in the United States is dedicated to outdoor water use, much of which is spent on watering grass and trees (EPA 2020). To combat this, desert

Fig. 4 Examples of xeriscaping from Phoenix, Arizona (Photo courtesy of Courtney M. Currier, with permission)

communities have promoted the use of xeriscaping in urban landscape design (Fig. 4). Xeriscaping can be defined as 'quality landscaping that conserves water and protects the environment' (EPA 1993, 2002), which includes the implementation of native xeric flora as an alternative to nonnative grasses and trees that require more water. This has been shown to reduce water consumption whilst promoting native biodiversity (EPA 1993, 2017; Sovocool and Morgan 2005).

4. *Green Buildings—Paris, France*

   Incorporating gardens into the walls and roofs of buildings is gaining popularity in the architectural design of many cities, with notable architects such as Patrick Blanc, Phillippe Samyn, Stefano Boeri, and Jean Frizzi practicing this approach. This type of green infrastructure brings aesthetic beauty, sound reduction, temperature regulation, and the potential for other ecosystem services such as carbon storage and biodiversity. The city of Paris has adopted this principle with the installation of 'City Trees', the construction of the Jardin BioPark (Fig. 5), and the forthcoming '50 Montaigne' development, amongst others.

Fig. 5 Jardin BioPark in Paris, is an example of green buildings increasing community aesthetics and ecosystem services (Photos by David W. Shanafelt, with permission)

6 Outcomes and Concluding Remarks

To maximise the contribution of biodiversity to BioCities, the authors propose special attention be given to four areas as part of the transition to BioCities: (1) water management in BioCities; (2) food production; (3) landscape scale management of urban green infrastructure, urban forestry, and NBS; and (4) soil quality management.

However, to encourage the implementation of these green elements as pillars of the BioCity, we also need to (1) illustrate the social and economic value of all green elements, (2) specify what services are provided by which NBS, (3) determine which valuation terms (quantitative or qualitative) are used by diverging stakeholders, and (4) improve communication with stakeholders. Disservices must be identified to properly weigh the benefits and trade-offs of NBS in design, planning, and management.

Sufficiently increasing nature in cities and building the BioCity will not be easy, considering the high ambitions. Nevertheless, some key topics and opportunities exist for the BioCity concept:

1. Map and secure water resources of the present and future BioCities.
2. Establish and strengthen connectivity between the city and rural areas, including the relationship between humans and nature.
3. Develop urban biophysical green infrastructure in tandem with the planning of other infrastructures including:
   - The fundamental building blocks for biodiversity in cities include soil, water, climate, and species, each of which requires proper planning, management, and monitoring.
   - Utilise NBS to improve the long-term integrity of the BioCity and the production of ecosystem services.
4. Designate enough areas in the BioCity for ecosystem service provisioning of a sufficient scale to affect sustainability and quality of life of urban dwellers.
5. Increase soil quality and soil functions to more fully realise their potential for ecosystem service provisioning.
6. Compromising natural resources must be avoided by:
   - Managing resources to sustainably provide ecosystem benefits for the BioCity, in a joint consideration of the respective needs of humans and nature.
   - Work towards a realised circular bioeconomy.

References

Aerts R, Stas M, Vanlessen N, Hendrickx M, Bruffaerts N, Hoebeke L, Dendoncker N, Dujardin S, Saenen ND, Nieuwenhuyse A, Aerts JM, Orshoven J, Nawrot TS, Somers B (2020) Residential green space and seasonal distress in a cohort of tree pollen allergy patients. Int J Hyg Environ Health 223(1):71–79

Alterio E, Cocozza C, Chirici G, Rizzi A, Sitzia T (2020) Preserving air pollution forest archives accessible through dendrochemistry. J Environ Manag 264:110462

Alvey AA (2006) Promoting and preserving biodiversity in the urban forest. Urban Forestry Urban Greening 5:195–201

Andenæs E, Time B, Muthanna T, Asphaug S, Kvande T (2021) Risk reduction framework for blue-green roofs. Buildings 11:185. https://doi.org/10.3390/buildings11050185

Anton V, Hartley S, Geldenhuis A, Wittmer HU (2018) Monitoring the mammalian fauna of urban areas using remote cameras and citizen science. J Urban Ecol 4(1):1–9

Aronson MF, Lepczyk CA, Evans KL, Goddard MA, Lerman SB, MacIvor JS, Nilon CH, Vargo T (2017) Biodiversity in the city: key challenges for urban green space management. Front Ecol Environ 15(4):189–196

Bassuk N (2017) Site assessment: the key to sustainable urban landscape establishment. In: Ferrini F, Konijnendijk van den Bosch CC, Fini A (eds) Routledge handbook of urban forestry. ISBN: 978-1-138-64728-2

Berry R, Livesley SJ, Aye L (2013) Tree canopy shade impacts on solar irradiance received by building walls and their surface temperature. Build Environ 69:91–100

Beryani A, Goldstein A, Al-Rubaei AM, Viklander M, Hunt WF III, Blecken GT (2021) Survey of the operational status of twenty-six urban stormwater biofilter facilities in Sweden. J Environ Manag 297:113375

Bosker M (2022) City origins. Reg Sci Urban Econ 94:103677

Branco M, Nunes P, Roques A, Fernandes MR, Orazio C, Jactel H (2019) Urban trees facilitate the establishment of non-native forest insects. NeoBiota 52:25–46. https://doi.org/10.3897/neobiota.52.36358

Bullock JM, Aronson J, Newton AC, Pywell RF, Rey-Benayas JM (2011) Restoration of ecosystem services and biodiversity: conflicts and opportunities. Trends Ecol Evol 26:541–549

Cambria VE, Campagnaro T, Trentanovi G, Testolin R, Attorre F, Sitzia T (2021) Citizen science data to measure human use of green areas and forests in European cities. Forests 12:779. https://doi.org/10.3390/f12060779

Campisano A, Butler D, Ward S, Burns MJ, Friedler E, DeBusk K, Fisher-Jeffes LN, Ghisi E, Rahman A, Furumai H, Han M (2017) Urban rainwater harvesting systems: research, implementation and future perspectives. Water Res 115:195–209. https://doi.org/10.1016/j.watres.2017.02.056

Cardinale BJ, Duffy JE, Gonzalez A, Hooper DU, Perrings C, Venail P, Narwani A, Mace GM, Tilman D, Wardle DA, Kinzig AP, Daily GC, Loreau M, Grace JB, Larigauderie A, Srivastava DS, Naeem S (2012) Biodiversity loss and its impact on humanity. Nature 486:59–67. https://doi.org/10.1038/nature11148

Castellar JAC, Popartan LA, Pueyo-Ros J, Atanasova N, Langergraber G, Säumel I, Corominas L, Comas J, Acuña V (2021) Nature-based solutions in the urban context: terminology, classification and scoring for urban challenges and ecosystem services. Sci Total Environ 779:146237

Concepción ED, Moretti M, Altermatt F, Nobis MP, Obrist MK (2015) Impacts of urbanisation on biodiversity: the role of species mobility, degree of specialisation and spatial scale. Oikos 124(12):1571–1582

da Silva CM, Corrêa SM, Arbilla G (2018) Isoprene emissions and ozone formation in urban conditions: a case study in the city of Rio de Janeiro. Bull Environ Contam Toxicol 100(1): 184–188

De Vreese R, Leys M, Fontaine CM, Dendoncker N (2016) Social mapping of perceived ecosystem services supply: the role of social landscape metrics and social hotspots for integrated ecosystem services assessment, landscape planning and management. Ecol Indic 73:517–533

Demuzere M, Orru K, Heidrich O, Olazabal E, Geneletti D, Orru H, Bhave AG, Mittal N, Feliu E, Faehnle M (2014) Mitigating and adapting to climate change: multi-functional and multi-scale assessment of green urban infrastructure. J Environ Manag 146:107–115

Diamond J (1997) Guns, germs, and steel. W. W. Norton, New York. ISBN: 9780099302780

Dorst H, van der Jagt A, Raven R, Runhaar H (2019) Urban greening through nature-based solutions: key characteristics of an emerging concept. Sustain Cities Soc 49:101620

Downing Day S, Harris JR (2017). Improving soil quality for urban forests. In: Ferrini F, Konijnendijk van den Bosch CC, Fini A (eds) Routledge handbook of urban forestry. Routledge, London

Ehrenstein V, Mutius V, Kries V (2000) Reduced risk of hay fever and asthma among children of farmers. Clin Exp Allergy 30(2):187–193

Elands BHM, Vierikko K, Andersson E, Fischer LK, Gonçalves P, Haase D, Kowarik I, Luz AC, Niemelä J, Santos-Reis M, Wiersum KF (2019) Biocultural diversity: a novel concept to assess human-nature interrelations, nature conservation and stewardship in cities. Urban Forestry Urban Greening 40:29–34. https://doi.org/10.1016/j.ufug.2018.04.006

Ellison D, Morris CE, Locatelli B, Sheil D, Cohen J, Murdiyarso D, Gutierrez V, van Noordwijk M, Creed IF, Pokorny J, Gaveau D, Spracklen DV, Bargués Tobella A, Ilstedt U, Teuling AJ, Gebreyohannis Gebrehiwot S, Sands DC, Muys B, Verbist B, Springgay E, Sugandi Y, Sullivan CA (2017) Trees, forests and water: cool insights for a hot world. Glob Environ Chang 43:51–61

Environmental Protection Agency (EPA) (1993) Xeriscape landscaping: preventing pollution and using resources efficiently. Report EPA-840-B-93-001

Environmental Protection Agency (EPA) (2002) Water-efficient landscaping: preventing pollution and using resources wisely. Report EPA832-F-02-002

Environmental Protection Agency (EPA) (2017) Water efficiency management guide: landscaping and irrigation. Report EPA 832-F-17-016b

Environmental Protection Agency (EPA) (2020) WaterSense. Retrieved from www.epa.gov/watersense/outdoor

Escobedo FJ, Giannico V, Jim CY, Sanesi G, Lafortezza R (2019) Urban forests, ecosystem services, green infrastructure and nature-based solutions: nexus or evolving metaphors? Urban Forestry Urban Greening 37:3–12

European Commission, Directorate-General for Research and Innovation (2021) Evaluating the impact of nature-based solutions: a handbook for practitioners. Publications Office of the European Union. https://doi.org/10.2777/244577

Faeth SH, Bang C, Saari S (2011) Urban biodiversity: patterns and mechanisms. Ann N Y Acad Sci 1223(1):69–81

Faivre N, Fritz M, Freitas T, de Boissezon B, Vandewoestijne S (2017) Nature-based solutions in the EU: innovating with nature to address social, economic and environmental challenges. Environ Res 159:509–518

Fang H, Jamali B, Deletic A, Zhang K (2021) Machine learning approaches for predicting the performance of stormwater biofilters in heavy metal removal and risk mitigation. Water Res 200:117273

Foster J (2014) Hiding in plain view: vacancy and prospect in Paris' Petite Ceinture. Cities 40:124–132

Fraisl D, Campbell J, See L, When U, Wardlaw J, Gold M, Moorthy I, Arias R, Piera J, Oliver JL, Masó J, Penker M, Fritz S (2020) Mapping citizen science contributions to the UN sustainable development goals. Sustain Sci 15:1735–1751

Ganzevoort W, van den Born RJG, Halffman W, Turnhout S (2017) Sharing biodiversity data: citizen scientists' concerns and motivations. Biodivers Conserv 26:2821–2837. https://doi.org/10.1007/s10531-017-1391-z

Giampietro M (2019) On the circular bioeconomy and decoupling: implications for sustainable growth. Ecol Econ 162:143–156

Giannico V, Lafortezza R, John R, Sanesi G, Pesola L, Chen J (2016) Estimating stand volume and above-ground biomass of urban forests using LiDAR. Remote Sens 8(4):339

Godefroid S, Koedam N (2007) Urban plant species patterns are highly driven by density and function of built-up areas. Landsc Ecol 22(8):1227–1239

Gómez-Baggethun E, Gren Å, Barton DN, Langemeyer J, McPhearson T, O'Farrell P, Andersson E, Hamstead Z, Kremer P (2013) Urban ecosystem services. In: Elmqvist T (eds) Urbanisation, biodiversity and ecosystem services: challenges and opportunities. Springer, Dordrecht, pp 175–251. ISBN 978-94-007-7088-1 (eBook)

Grabosky J, Bassuk N (2017) Design options to integrate urban tree root zones and pavement support within a shared soil volume. In: Ferrini F, Konijnendiijk van den Bosch CC, Fini A (eds) Routledge Handbook of Urban Forestry. ISBN: 978-1-138-64728-2

Hansen R, Pauleit S (2014) From multifunctionality to multiple ecosystem services? A conceptual framework for multifunctionality in green infrastructure planning for urban areas. Ambio 43: 516–529. https://doi.org/10.1007/s13280-014-0510-2

Heigl F, Kieslinger B, Paul KT, Uhlik J, Dörler D (2019) Toward an international definition of citizen science. Proc Natl Acad Sci USA 116(17):8089–8092. https://doi.org/10.1073/pnas.1903393116

Heynen N, Perkins HA, Roy P (2006) The political ecology of uneven urban green space: the impact of political economy on race and ethnicity in producing environmental inequality in Milwaukee. Urban Aff Rev 42(1):3–25

Huld T, Bódis K, Pinedo Pascua I, Dunlop E, Taylor N, Jäger-Waldau A (2018) The rooftop potential for PV systems in the European Union to deliver the Paris Agreement. Eur Energy Innov 12

IUCN (2016) Defining nature-based solutions. WCC-2016-Res-069-EN. www.iucn.org/sites/dev/files/content/documents/wcc_2016_res_069_en.pdf

Kattwinkel M, Biedermann R, Kleyer M (2011) Temporary conservation for urban biodiversity. Biol Conserv 144(9):2335–2343

Klaus VH, Kiehl K (2021) A conceptual framework for urban ecological restoration and rehabilitation. Basic Appl Ecol 52:82–94. https://doi.org/10.1016/j.baae.2021.02.010

Knowler D, Barbier E (2005) Importing non-native plants and the risk of invasion: are market-based instruments adequate? Ecol Econ 52(3):341–354

Kowarik I (2011) Novel urban ecosystems, biodiversity, and conservation. Environ Pollut 159(8–9):1974–1983

Kowarik I (2021) Working with wilderness: a promising direction for urban green spaces. Landscape Architect Front/Views Criticism 9:92–103

Kuo FE, Sullivan WC (2001) Aggression and violence in the inner city: effects of environment via mental fatigue. Environ Behav 33(4):543–571

La Notte A, D'Amato D, Mäkinen H, Paracchini ML, Liquete C, Egoh B, Geneletti D, Crossman ND (2017) Ecosystem services classification: a systems ecology perspective of the cascade framework. Ecol Indic 74:392–402

Liu Z, He C, Wu J (2016) The relationship between habitat loss and fragmentation during urbanisation: an empirical evaluation from 16 world cities. PLoS One 11(4):e0154613. https://doi.org/10.1371/journal.pone.0154613

Loreau M (2010) From populations to ecosystems: theoretical foundations for a new ecological synthesis. Monographs in Population Biology. Princeton, Princeton University Press. www.jstor.org/stable/j.ctt7s78j

Macdonald E, King EG (2018) Novel ecosystems: a bridging concept for the consilience of cultural landscape conservation and ecological restoration. Landsc Urban Plan 177:148–159

Martellozzo F, Landry JS, Plouffe D, Seufert V, Rowhani P, Ramankutty N (2014) Urban agriculture: a global analysis of the space constraint to meet urban vegetable demand. Environ Res Lett 9(6):064025. https://doi.org/10.1088/1748-9326/9/6/064025

Marziliano PA, Lafortezza R, Colangelo G, Davies C, Sanesi G (2013) Structural diversity and height growth models in urban forest plantations: a case-study in northern Italy. Urban Forestry Urban Greening 12(2):246–254

Masson-Delmotte V, Zhai P, Pirani A, Connors SL, Péan C, Berger S, Caud N, Chen Y, Goldfarb L, Gomis MI, Huang M, Leitzell K, Lonnoy E, Matthews JBR, Maycock TK, Waterfield T, Yelekçi O, Yu R, Zhou B (eds) (2021) Climate change 2021: the physical science basis. Contribution of working group I to the sixth assessment report of the intergovernmental panel on climate change. Cambridge University Press, Cambridge, pp 33–144. https://doi.org/10.1017/9781009157896.002

Matasov V, Belelli Marchesini L, Yaroslavtsev A, Sala G, Fareeva O, Seregin I, Castaldi S, Vasenev V, Valentini R, Io T (2020) Monitoring of urban tree ecosystem services: possibilities and challenges. Forests 11(7):775. https://doi.org/10.3390/f11070775

Matthews T, Lo AY, Byrne JA (2015) Reconceptualising green infrastructure for climate change adaptation: barriers to adoption and drivers for uptake by spatial planners. Landsc Urban Plan 138:155–163

Mayrand F, Clergeau P (2018) Green roofs and green walls for biodiversity conservation: a contribution to urban connectivity? Sustainability 10:985. https://doi.org/10.3390/su10040985

Molfino NA, Wright S, Katz I, Tarlo S, Silverman F, McClean PA, Slutsky AS, Zamel N, Szalai JP, Raizenne M (1991) Effect of low concentrations of ozone on inhaled allergen responses in asthmatic subjects. Lancet 338(8761):199–203

Morgenroth J, Östberg J (2017) Measuring and monitoring urban trees and urban forests. In: Ferrini F, Konijnendiijk van den Bosch CC, Fini A (eds) Routledge handbook of urban forestry. ISBN: 978-1-138-64728-2

Ossola A, Lin BB (2021) Making nature-based solutions climate-ready for the 50°C world. Environ Sci Policy 123:151–159

Pataki DE, Carreiro MM, Cherrier J, Grulke NE, Jennings V, Pincetl S, Pouyat RV, Whitlow TH, Zipperer WC (2011) Coupling biogeochemical cycles in urban environments: ecosystem services, green solutions, and misconceptions. Front Ecol Environ 9:27–36. https://doi.org/10.1890/090220

Perez PA, Rodriguez EN (2018) Status of local soil contamination in Europe: revision of the indicator "Progress in the management contaminated sites in Europe". EUR 29124 EN, Publications Office of the European Union, Luxembourg, 2018. ISBN 978-92-79-80073-3 (print),978-92-79-80072-6 (pdf). https://doi.org/10.2760/093804 (online), https://doi.org/10.2760/503827 (print), JRC107508

Perring MP, Manning P, Hobbs RJ, Lugo AE, Ramalho CE, Standish RJ (2013) Novel urban ecosystems and ecosystem services. In: Hobbs RJ, Higgs ES, Hall C (eds) Novel ecosystems: intervening in the new ecological world order, pp 310–325. ISBN: 978-1-118-35422-3

Pickett STA, Cadenasso ML (2009) Altered resources, disturbance, and heterogeneity: a framework for comparing urban and non-urban soils. Urban Ecosyst 12:23–44

Pickett STA, Cadenasso ML, Grove JM, Boone CG, Groffman PM, Irwin EG, Kaushal S, Marshall V, McGrath BP, Nilon CH, Pouyat RV, Szlavecz K, Trot A, Warren P (2010) Urban ecological systems: scientific foundations and a decade of progress. J Environ Manag 92(3):331–362. https://doi.org/10.1016/j.jenvman.2010.08.022

Pickett STA, Cadenasso ML, Baker ME, Band LE, Boone CG, Buckley GL, Groffman PM, Grove JG, Irwin EG, Kaushal S, LaDeau SL, Miller AJ, Nilon CH, Romolini M, Rosi MJ, Swan CM, Szlavecz K (2020) Theoretical perspectives of the Baltimore ecosystem study: conceptual evolution in a social–ecological research project. Bioscience 70(4):297–314. https://doi.org/10.1093/biosci/biz166

Reichholf JH (2007) Stadtnatur – Eine neue Heimat für Tiere und Planzen. Oekom Verlag, München. ISBN: 978-3-86581-042-7

Riolo F (2019) The social and environmental value of public urban food forests: the case study of the Picasso Food Forest in Parma, Italy. Urban Forestry and Urban Greening 45:126225

Sæbø A, Janhäll S, Gawronski SW, Hanslin HM (2017) Urban forestry and pollution mitigation. In: Ferrini F, Konijnendijk van den Bosch C, Fini A (eds) Routledge handbook of urban forestry. Taylor & Francis Group, New York. ISBN 978-1-138-64728

Sanesi G, Padoa-Schioppa E, Lorusso L, Bottoni L, Lafortezza R (2009) Avian ecological diversity as an indicator of urban forest functionality: results from two case studies in northern and southern Italy. J Arboric 35(2):80

Schwarz N, Moretti M, Bughalo MN, Davies ZG, Haase D, Hack J, Hof A, Melero Y, Pett TJ, Knapp S (2017) Understanding biodiversity-ecosystem service relationships in urban areas: a comprehensive literature review. Ecosyst Serv 27:161–171

Seinfeld JH, Pandis SN (2016) Atmospheric chemistry and physics: from air pollution to climate change. Wiley. ISBN: 978-1-118-94740-1, 1152 pp

Sitzia T, Campagnaro T, Weir RG (2016) Novel woodland patches in a small historical Mediterranean city: Padova, Northern Italy. Urban Ecosyst 19(1):475–487

Sjöman H, Morgenroth J, Deakin Sjöman J, Sæbø A, Kowarik I (2016) Diversification of the urban forest - can we afford to exclude non-native tree species? Short communication. Urban Forestry Urban Greening 18:237–241

Soanes K, Sievers M, Chee YE, Williams NS, Bhardwaj M, Marshall AJ, Parris KM (2019) Correcting common misconceptions to inspire conservation action in urban environments. Conserv Biol 33(2):300–306

Song Y, Kirkwood N, Maksimović C, Zheng X, O'Connor D, Jin Y, Hou D (2019) Nature based solutions for contaminated land remediation and brownfield redevelopment in cities: a review. Sci Total Environ 663:568–579

Sovocool KA, Morgan M (2005) Xeriscape conversion study. Southern Nevada Water Authority

Stas M, Aerts R, Hendrickx M, Delcloo A, Dendoncker N, Dujardin S, Somers B et al (2021) Exposure to green space and pollen allergy symptom severity: a case-crossover study in Belgium. Sci Total Environ 781:146682

Szlavecz K, Warren PS, Pickett STA (2011) Biodiversity in the urban landscape. In: Cincotta RP, Gorenflo LJ (eds) Human population: its influences on biological diversity, ecological studies, vol 214. Springer, Berlin, pp 75–101. https://doi.org/10.1007/978-3-642-16707-2_6

Tan Z, Lau KKL, Ng E (2016) Urban tree design approaches for mitigating daytime urban heat Island effects in a high-density urban environment. Energ Buildings 114:265–274

Torresan C, Garzón MB, O'Grady M, Robson TM, Picchi G, Panzacchi P, Tomelleri E, Smith M, Marshall JD, Wingate L, Tognetti R, Rustad L, Kneeshaw DD (2021) A new generation of sensors and monitoring tools to support climate-smart forestry practices. Can J For Res 51:1751. https://doi.org/10.1139/cjfr-2020-0295

Vailshery LS, Jaganmohan M, Nagendra H (2013) Effect of street trees on microclimate and air pollution in a tropical city. Urban Forestry Urban Greening 12(3):408–415

van Elsas JD, Trevors JT, Rosado AS, Nannipieri P (2019) Modern soil microbiology, 472 pp, 3rd edn. CRC Press, Taylor and Francis Group. ISBN 978-1-4987-6353-0

Van Mechelen C, Dutoit T, Hermy M (2014) Mediterranean open habitat vegetation offers great potential for extensive green roof design. Landsc Urban Plan 121:81–91

Varela MD, Subiza J, Subiza JL, Rodriguez R, Garcia B, Jerez M, Jimenez JA, Panzani R (1997) Platanus pollen as an important cause of pollinosis. J Allergy Clin Immunol 749(100), 6, part 1

Wang X, Dallimer M, Scott CE, Shi W, Gaoe J (2021) Tree species richness and diversity predicts the magnitude of urban heat Island mitigation effects of greenspaces. Sci Total Environ 770: 145211

Zari MP (2018) The importance of urban biodiversity – an ecosystem services approach. Int J Biodivers 2:357–360

Zipperer WC (2002) Species composition and structure of regenerated and remnant forest patches within an urban landscape. Urban Ecosyst 6(4):271–290

Green Infrastructure and Urban Forests for BioCities: Strategic and Adaptive Management

Thomas B. Randrup, Märit Jansson, Johanna Deak Sjöman, Koenraad Van Meerbeek, Marie-Reine Fleisch, David W. Shanafelt, Andreas Bernasconi, and Evelyn Coleman Brantschen

But this expenditure [. . .], has nothing excessive, compared to the services rendered by the plantations. They are indispensable to renew the stale air [. . .]. They provide shade, so necessary to the many public [. . .]. Finally, they contribute greatly to the decoration of the city.
—Adolphe Alphand, Director of Public Roads and Promenades in Paris (1873)

1 Introduction

Urban nature in the form of trees and parks has played an important part in European cities, and in many other cities of the world, at least since industrialisation in the eighteenth century. As such, the practices of planning, designing and managing urban nature have a long history, with an evolution of the associated roles and

T. B. Randrup (✉) · M. Jansson · J. D. Sjöman
Department of Landscape Architecture, Planning and Management, Swedish University of Agricultural Sciences, Alnarp, Sweden
e-mail: thomas.randrup@slu.se

K. Van Meerbeek
Department of Earth and Environmental Sciences, KU Leuven, Leuven, Belgium

M.-R. Fleisch
AgroParisTech, UMR Silva, Nancy, France

D. W. Shanafelt
Centre national de la Recherche Scientifique (CNRS), Institut National de la Recherche Agronomique (INRA), Bureau d'Économie Théorique et Appliquée (BETA), Université de Lorraine, Université de Strasbourg, AgroParis Tech, Strasbourg, France

A. Bernasconi
Pan Bern AG, Bern, Switzerland

E. Coleman Brantschen
School of Agricultural, Forest and Food Sciences HAFL, Bern University of Applied Sciences, Bern, Switzerland

G. E. Scarascia-Mugnozza et al. (eds.), *Transforming Biocities*, Future City 20,
https://doi.org/10.1007/978-3-031-29466-2_4

responsibilities. In many instances, the formal planning and designing of urban nature began via local government institutions (Persson et al. 2020). They have seen the value of developing nature in cities for an increasing urban population in need of contemplation and recreation outside of working hours in what were then dense, filthy, and unhealthy urban agglomerations. Public management developed along with developing these spaces and was seen as a public service. Planning, designing, and management are often seen as separate and unique practices in the development of urban nature, with implications for how successful and long-term the development can become (Jansson et al. 2019). However, instead of separate practices, they should be turned into holistic processes which allow nature to prevail, adapt and change over time in the development of urban nature as a central aspect of future BioCities, as envisioned by the Resilient BioCity concept, in chapter "Towards the Development of a Conceptual Framework of BioCities".

Incorporating nature into urban areas, for a true transition to BioCities, requires firstly that careful consideration be given to citizen perceptions and relationship with green spaces. Thus, urban planners, and green space managers should take into consideration the fact that people may perceive and use different types of urban green space (UGS) in various ways. A network of accessible good quality green space comprising various types and sizes of UGS should be considered instead of favouring only certain types of UGS.

Ideally, results of human–nature relationship studies should be used more regularly in urban and green space planning and management, but this is not always the case. For the best/most comprehensive results, whenever possible a representative sample of citizens should be included covering different ages, gender and ethnicities.

2 Public Urban Nature Management

Although the primary responsibilities for planning, designing and management of nature in urban areas lie with local governments in most parts of the world, it is generally not a mandatory obligation. Whilst funding for management was not a problem in large parts of the world throughout much of the twentieth century, with local taxation providing necessary resources and democratic organisations to coordinate management activities (Persson et al. 2020), this financial model has later proven to be fragile as local governments saw an increasing demand for meeting other statutory responsibilities such as education, health, and social care.

During the 1980s a new movement formed in an effort to reform public administrations, referred to as new public management (NPM) (Hood 1995). NPM introduced new steering mechanisms for management and outsourcing of service tasks to the private sector, which in many instances included operational maintenance. This reflects a more decentralised and fragmented approach to management, requiring new service delivery models to achieve efficiency. With NPM, the public was seen as a provider of public goods for the users, who may be seen as customers of those goods, in this case 'urban nature'. The NPM approach to public service delivery was further intensified by the financial crash in 2008, which saw public finances decrease substantially. More recently, local government management models have started to show greater diversity in the organisation of urban nature management. In line with an increasing interest in public engagement, numerous initiatives have, in many countries, gained traction to engage and even transfer responsibilities to not only private companies and charitable trusts but also to local citizen groups (Buijs et al. 2016). Such initiatives may come from the government itself but are also sometimes driven by an increasing demand from the public to participate.

2.1 Green Infrastructure

Urban nature has been defined and described in various ways, and often synonymously with the term green infrastructure (GI). GI can broadly be defined as a network of vegetation, water and permeable surfaces and may include parks, street trees, sports areas, schoolyards, private gardens, townscapes, vertical gardens, community gardens, peri-urban agricultural landscapes, housing environments, cemeteries, wetlands and urban forest. Typologies may be interconnected or overlapping, such as a city park encompassing a lake. GI thus constitutes an integrated part of and contributes to the 'urban matrix', consisting of green (e.g. parks, gardens and allotments), blue (e.g. ponds and lakes), brown (e.g. abandoned harbour or industrial areas) and grey spaces (e.g. squares and plazas). In most cases, urban areas contain mixtures of each (Haase et al. 2020).

The European Commission (2013a) defined GI as '*a strategically planned network of natural and semi-natural areas, including green and blue spaces and other ecosystems, designed and managed to deliver a wide range of ecosystem services at various scales*'. According to this definition, GI is planned, designed and managed, and in line with Davies et al. (2015), involves at least four detailed and interlinked goals: (1) securing a connection between individual spaces, (2) securing multifunctionality, (3) integrating with other infrastructures and (4) operating on multiple scales. Thus, planning, designing and managing GI are ongoing processes that operate on different scales, both geographical and temporal. This rather idealistic definition does not seem to include or even regard unplanned typologies, e.g. brownfields which are not necessarily part of a formal planning, designing and management regime, but still provides several important ecological and social values. As a concept, GI has been aligned with other important urban infrastructures, such as traffic or electricity. However, in comparison, it is not perceived as 'equal' in the government planning process (Hislop et al. 2019), as local governments are often not obliged to develop GI, neither theoretically nor in practice (de Magalhães and Carmona 2009). Yet, planning for green infrastructure should be done in tandem with planning for other infrastructures. Roads and paved and impermeable areas must be optimally placed and designed to give space for NBS. The role of vegetated buildings in landscape connectivity and wildlife corridors to strengthen biodiversity should be better investigated (Mayrand and Clergeau 2018). This presents a great challenge for policymakers, planners and urban foresters.

2.2 Urban Forests

Urban forests are the backbone of GI, bridging rural and urban areas and ameliorating a city's environmental footprint (FAO 2016). Urban forests (UF) can be defined as networks of systems comprising all woodlands, groups of trees, and individual trees located in urban and peri-urban areas. They are therefore regarded as an integral and significant part of GI, 'representing' urban trees, whether they be grown in woodlands or forests, along streets, or in parks or private gardens (Randrup et al. 2005). The Society of American Foresters defined the planning and managing urban forests as urban forestry: '*the art, science and technology of managing trees and forest resources in and around urban community ecosystems for the physiological, sociological, economic, and aesthetic benefits trees provide society*' (Helms 1998, p. 193). The tradition of carrying out forestry in urban areas is not new either, as several cities of the world, particularly in Europe, have owned and managed forests for centuries (Konijnendijk 1999). Whilst urban forestry has focused on managing both individual trees and tree stands in urban areas, it also maintains a strong social perspective—being urban. This requires planning, designing and management activities related to people's needs and preferences.

Whilst the traditions of planning, designing and managing GI and UF have a long history in local European governments, the challenges facing GI and UF, including the physical spaces they represent, the services they provide and the related management organisations, are dramatically changing. Green infrastructure is threatened by increased urbanisation (unpopulation.org 2018), often leading to densification in urban areas (FAO 2017), which results in increased land use change. Likewise, contemporary urban challenges such as climate change and pollution affect the processes around GI and UF.

2.3 The Need for New Approaches

A number of concepts and approaches have been suggested over past decades to address the challenges related to GI and UF, affecting planning, designing and management. Relevant public agendas and action plans developed key methodologies and concepts such as sustainable urban development (World Commission on Environment and Development 1987), Local Agenda 21 (UN 1992) and green infrastructure (European Commission 2013b); whilst new approaches were elaborated by research initiatives as ecosystem-based adaptation (Colls et al. 2009), ecosystem services evaluation (TEEB 2010) and Nature's Contribution to People (Diaz et al. 2018). Recently, nature-based solutions (European Commission 2015), and the 17 Sustainable Development Goals (UN 2015) have continued to focus on how to address human needs and actions, whilst respecting and even using nature to restore and develop urban areas. The concepts and approaches launched have yet to result in definite changes leading towards healthier and more livable cities. Hence, there remains an urgent need to make cities an attractive environment for both humans and nature/biodiversity.

The benefits of GI and UF for humans have been expressed as ecosystem services (ESs), which for GI and UF are well-known and well-described (MEA 2005; WHO 2016). However, the provisioning of ESs is challenged in current planning, designing and management practices due to organisational regimes such as NPM, and reduced public funding (Dempsey et al. 2014; Jansson and Randrup 2020). Governance-related dimensions such as a lack of leadership, responsibilities, funding, standards and institutional capacity, including fragmented organisational structures leading to disconnections between planning, designing and management, are often described as limiting factors for the improvement of GI and UF (Qiao et al. 2018; Ordonez et al. 2019). Thus, there is a need to address planning, designing and management as holistic processes, which allow nature to prevail, adapt and change over time, instead of considering those different steps as separate practices. This will require that conventional government structures are re-defined to include a more cyclical approach, to planning, designing, construction and management—as well as allowing GI and UF to develop *with* people, rather than primarily *for* people. We propose a strategic and adaptive management approach connected to new governance arrangements to achieve this.

3 Processes of Planning, Designing and Management to Develop BioCities

In local government and similar organisations that deliver GI and UF and associated services, planning, designing and management are all performed by different actors or different divisions within one or more organisations, such as a local government and/or its departments, consultants, and private companies. Each phase in the logic has its own expertise based on separate and specialised educational backgrounds, its own organisational residence and specific institutional logics and traditions (Jansson et al. 2019). As a consequence, the processes of GI and UF are not well connected, e.g. planning and designing not being sufficiently coupled to management (Dempsey et al. 2014; Jansson et al. 2019; Jansson and Randrup 2020), which constitutes challenges to the possibility of delivering GI and UF in a sustainable, long-sighted and adaptive way (Fig. 1).

Within local government, the expertise related to GI and UF may be complex as the formal responsibilities for different parts of the GI are located in different departments. Sports fields may be the responsibility of the cultural department, whilst green spaces in relation to retirement homes may be the responsibility of

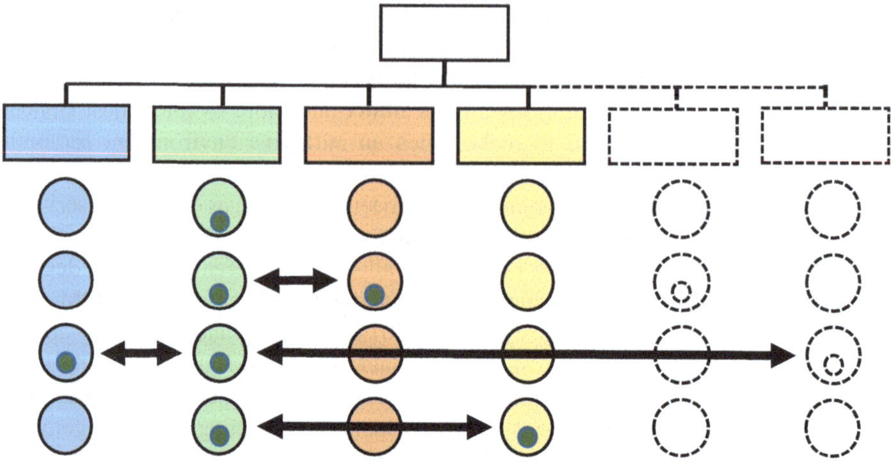

Fig. 1 Conceptual model of how a traditional organisation (e.g. a local government) is divided into specialised departments—or silos—focusing on social aspects, cultural aspects and technical aspects. Often GI and UF matters lie within a technical department, requiring that relations to human health and well-being are performed across departments (silos). Also, GI and UF planning, designing and management may be divided into sub-departments, which will require further cooperation across sub, and main departments. Green dots illustrate green expertise located within different departments

the social department. Parks and roadside trees are commonly the responsibility of the technical or highways departments, but may even here be subdivided into specialised 'parks' and 'roads' sections. Hence, in many local governments, it is difficult to create a full overview of the GI resource (Persson et al. 2020), just as the entire GI resource is lacking a clear responsible agent—a BioCity 'champion'.

On a practical level, the organisational division between processes can be illustrated via the process of planting trees. If planning and designing do not take into consideration the local context by providing sufficient room for root growth or planting species suitable for local environmental conditions, then future management of those trees will experience problems. In turn, this causes limitations in addressing contemporary challenges such as climate change or pollution. Ideally, the long-term management aspects should be incorporated as an integral part of both planning and designing.

In general, a long-term approach is needed towards long-term development of ecological processes of GI and UF. For example, Franch (2018) describes designing on site in existing GI settings as a basis for interventions and a 'differentiated management'. In such a site-specific approach, designing, construction and management intermingle (Franch 2018), and might allow various actors, including urban citizens, to engage in the combined process. Such engagement towards nature connectedness can relate to, such as perceptions of biodiversity values and restorative qualities, but also vary quite differently between different citizens and stakeholder groups depending on age, gender, culture and to which extent an already existing relation to nature exists or habits of spending time in nature are established (Hoyle et al. 2018).

Whilst a long-term approach is needed towards development of ecological processes, like the succession of species composition and structure, this can also lead to instant aesthetic effects if the designing has thorough recognition to the time dimension (Sjöman et al. 2017), and to how humans engage with nature at a different scale from that of ecological processes and environmental phenomena (Gobster et al. 2007).

Communication with different stakeholder groups as part of the management process is crucial for reaching a general acceptance and appreciation of naturalistic designing and ecological processes (Hoyle et al. 2018). Different mechanisms can be used for this purpose where readily available tools may include mobile connections and smart applications thus using technological innovations as a means to reach citizens not primarily interested or used to spending time in naturalistic settings (Nitoslawski et al. 2019). New ways of thinking and handling GI and UF include recognition of the integrated parts in socio-ecological-technical system (SETS) (McPhearson et al. 2016). Nature, citizens and technical infrastructures are not separate entities but rather an interconnected whole.

These examples emphasise the need to not only look across departments within an organisation, but also to develop new governance approaches beyond them. It also accommodates for long-sighted cyclical approaches in the management of ecological processes reaching beyond traditional and conventional time scales of public management, and how to contribute to a biophilic relationship between urban citizens and GI. We suggest that although the individual entities within an organisation play an important role towards the planning, designing and management of GI and UF we rather foresee the processes of interconnectivity and trans-disciplinarily between different sectors and stakeholder groups to be decisive mechanisms. We describe these approaches within the context of strategic management, adaptive management, and governance.

3.1 Strategic Management

Strategic management embraces a holistic and long-term perspective in the development of urban nature, being it GI, UF or other green-blue elements. The concept acknowledges that once a space is planned, designed and constructed, it may continue to develop over centuries (Randrup and Persson 2009). During this time, continuous maintenance is required, but adaptation to a changing context is needed too. This includes changes in local demographics and social needs, in relation to available resources, as well as to global challenges such as climate change and urbanisation. Thus, whilst maintaining in the short term, there is also a constant need for adaptive re-planning, re-designing, and re-construction. Hence planning, designing, construction and maintenance are realised as a long-term cyclic process. Strategic management is long term in the sense that it embraces the continuous development and upkeep of the environment, relating to various actors and governance perspectives (Jansson et al. 2019). Strategic management breaks the traditional logic of planning, designing, construction and maintenance by exemplifying a non-linear process where decisions and adaptation in relation to changed user patterns, demography, climate etc. may occur at any point. The progressive and long-term process of strategic management thus needs to be adaptive in order to be relevant for future implementation into planning and management.

The organisational prerequisites and support of a strategic management approach are numerous. One is to recognise that often a certain responsibility or task will be related to its context and thus, in its nature, be cross-sectoral and trans-disciplinary. Another is to recognise that management is performed at least at three different organisational levels, most notably policy, tactical, and operational (Randrup and Jansson 2020).

3.2 Adaptive Management

Adaptive management as a concept has a long history and stems from the forestry profession, as a means to transfer and implement policy decisions into management. FEMAT (1993) described adaptive management as '*[the] process of implementing policy decisions as scientifically driven management experiments that test predictions and assumptions in management plans and using the resulting information to improve the plans*' (FEMAT 1993). Later, adaptive management has been described as '*a systematic process for continuously improving management policies and practices by learning from the outcomes of previously employed policies and practices*' where '*management is treated as a deliberate experiment for purposes of learning*' (MEA 2005). Whilst strategic management can be viewed as a cross-departmental approach to planning, designing, construction and maintenance, adaptive management is an inter-departmental approach coupling visionary policy making with operational maintenance.

Where strategic management emphasises the need for a long-sighted and cyclic process in developing nature, adaptive management also includes new governance arrangements, monitoring and evaluation processes. Adaptive management implies an iterative, collective decision-making and learning with knowledge co-production that integrates not only scientists and managers but also other stakeholders (Kingsford et al. 2017). In addition to its trans-disciplinary character, another fundamental point of this approach is the existence of dynamic feedback loops, something that fits well with the non-linear process of strategic management. Monitoring is critical to check if targets are met to redefine them if necessary, or to correct planning practices or change their implementation (Ahern et al. 2014), like through re-planning, re-designing, and re-construction (Jansson et al. 2019).

Adaptive management and "safe-to-fail" designing imply much closer collaboration between urban planners, designers, managers, environmentalists and other stakeholders than currently practiced, as also called for in strategic management. A holistic focus on GI and UF can form a united goal for strategic and adaptive management, allowing both for cross-sectorial and for organisational multi-level approaches. Further, different stakeholder groups should be engaged to contribute to the innovation and decision-making processes. This, in total, calls for new governance modes and approaches.

3.3 New Governance Approaches

Another change in the processes and mindsets that can further bridge planning, designing and management into strategic and adaptive processes for BioCities is the development 'from government to governance'. That phrase describes a shift within public management where cross-dimensional networks of governance replace traditional hierarchical modes of government (Lo 2018). Governance is often related to processes and relations when several actors, public and private, governmental and non-governmental, are involved in the steering of a public good (Arts and Visseren-Hamakers 2012; Smith et al. 2014). Governance is thus highly relevant in the perspective of GI and UF and similar types of urban nature.

Governance structures can be described via a policy arrangement, and any given policy arrangement can be described by four overall dimensions: (1) actors (institutions, organisations, groups, users) and how they are affected by (2) rules of the game, (3) power and resources as well as (4) discourses (Arts et al. 2006). The interrelations between the four dimensions can be relatively stable over time, but often, a policy arrangement will be unstable. Discourses may change and affect funding or citizen interest, and new rules and regulations may change power relations between stakeholders. These relationships also depend on how much the governing organisation keeps a steering role, the level of hierarchy in the organisation, and if initiatives are undertaken from top-down or bottom-up. Thus, governance approaches range from closed co-governance arrangements where the governing organisation keeps much of the steering, to open co-governance or even un-hierarchic self-governance with a larger degree of power given to private citizens (Jansson et al. 2019).

Often, there will be a need to work with a variety of governance approaches simultaneously, allowing different arrangements to appear depending on local contexts. A number of stakeholders may have important roles to play; citizens at large, active citizens, entrepreneurs, NGOs and others. One way of diversifying the governance perspective is through so-called 'mosaic governance approaches' (Buijs et al. 2016). This name implies developing a palette of governance approaches that can involve various stakeholders and support initiatives that are initiated bottom-up as well as top-down, specific to local contexts. Such complex and multiple approaches show similarities with the polycentric systems of governance (Ostrom 1990; Carlisle and Gruby 2017) and can form parts of strategic and adaptive management approaches (Kingsford et al. 2017).

Through a mosaic governance approach, a multitude of arrangements can co-exist and lead to new links between stakeholders, where new processes and expertise can develop. This can be used to develop BioCities in a more place-based, participatory way, allowing various governance arrangements depending on the local needs and possibilities. Such a mosaic of governance structures creates complex interrelations between stakeholders, but can, if successful, allow for an inclusive and holistic approach to planning, designing and management of GI and UF. Amongst the possible challenges is the risk of excluding certain users or interest groups. New governance approaches thus require new mindsets, competences and roles for local governments and other governing organisations. This needs to extend to their associated stakeholders too, such as decision makers, planners, designers, managers and maintenance personnel, being able to not only work with GI and UF directly, but also to work with people, through processes and facilitation.

Case Study #1: A City Scale Approach

The Akerselva River project in Oslo illustrates how long-term visions of re-integrating one of the largest rivers running through the city is possible through strategic management and collaboration across governmental departments. Lost to culverts and pollution from nineteenth century industrialisation, calls to clean the 10-km long river and re-establish its recreational qualities began in 1915. The twentieth century was marked by efforts to improve water quality whilst exploring the potential for hydroelectricity on the river. During the past 40 years, efforts were made to remove culverts constructed in the industrialised era. The development of parks and inclusive green space along the river have resulted in a recreational blue greenway throughout the city, embedded with new housing and commercial properties that pay homage to the cultural heritage of the riverbank. The initiatives for the Akerselva River have resulted in similar endeavours for other rivers in Oslo and future visions for the Akerselva, include means of addressing climate adaptation, providing clean water, and incorporating ecological values such as increasing biodiversity in tandem with recreation and their subsequent effects on peoples' well-being. The success of Akerselva River is the result of a continuous political will where planning, designing and management have occurred non-linearly and adapted to challenges over time. Current efforts jointly consider past and futures contexts whilst incorporating different expertise along the way to learn during the process. The future path for strategically and adaptively managing the Akerselva River is to continue exploring a longitudinal and experimental approach involving technological innovation, 'learning by doing', providing open data for and from the public realm, and embracing a new mindset of making space for nature and time for ecological processes within new and adaptive policy frameworks (Photo 1).

Photo 1 Akerselve River running through previous industrial and current cultural urban areas. (credit: Thomas B. Randrup)

Case Study #2: A Project-Level Approach

The redevelopment of Garibaldi Street in Lyon, France, is an example of strategic and adaptive management in favour of creating a better quality of life in urban areas by the means of a transformation of a single streetscape. This street was designed in the 1960s as an 'urban highway' to facilitate traffic in the city centre and formed an urban barrier between the western and eastern parts of the city. The metropolis of Lyon wished to launch a vast transformation of this street for it to become a living space for transport as well as recreation. The long-term objective was to create a new green infrastructure of nearly three kilometres between two parks.

The realisation of the street redevelopment was carried out in several phases to take into account traffic reorganisation, the amount of work to be done, and financial constraints but also to learn from the experience of the first phase. It began with a preliminary consultation with local citizens through several workshops. The views of different stakeholders made it possible to restore safe spaces for pedestrians and bicycles, and to integrate a planted 'promenade' whose watering is ensured by a former hopper converted into a storage space of the rainwater. These improvements were fully in phase with the 'climate plan' adopted by the metropolis of Lyon in 2017, which includes actions to combat the urban heat island effect. The two main tools proposed are; (1) a project to facilitate the infiltration and evapotranspiration of water in soils and (2) the 'canopy plan', which concerns the protection of existing trees and the development of new plantations throughout the territory, in order to increase the canopy cover from 27% to 30% by 2030.

Garibaldi Street has served as an experimental site for managers and scientists to test and learn with an adaptive approach. The site also tested how to increase the refreshing efficiency of the trees during heatwaves by carrying out short-time irrigation of plants to increase evapotranspiration, and cooling of the local area. Thanks to evapotranspiration, the trees of Garibaldi Street have contributed to refreshing their environment. The temperature difference compared to the nearest weather station was about −2 °C, with a thermal comfort gain of 10°UTI (Universal Thermal Climate Index). This gain will tend to increase when trees are further watered. The Garibaldi Street scheme has been used to refine watering decision-making to be replicated and extended to other sites in the city. In accordance with the plan, the metropolis of Lyon has been experimenting and developing vegetative solutions throughout the city to mitigate the effects of global warming. The still ongoing redevelopment of this street is a test bed or 'urban laboratory' for experimenting with governance approaches through social involvement, incorporating different expertise and including close collaborations with researchers to find sustainable and replicable solutions to improve city life.

Case Study #3: Citizen Science as a Means Towards a BioCity

Leuven 2030 is a non-profit organisation founded by local inhabitants, companies, civil society organisations, knowledge institutions and public authorities in the Belgian university city of Leuven. This bottom-up partnership aims to move towards a healthy, liveable and climate-neutral city by combining science (scientific framework), social power (bringing people together) and storytelling (inspiring stories about the steps taken). Together with a large group of experts from various disciplines, Leuven 2030 identified challenges and actions to take towards a climate-neutral city, which are structured into 13 programmes and drawn up in the 'Roadmap 2025–2035–2050'.

One of these programmes focuses on strengthening the local green and blue spaces as an important pillar of a healthy, vibrant and climate resilient city. To ensure optimal planning, designing and management of green and blue spaces aiming at mitigating the urban heat island (UHI) effect, the University of Leuven (KU Leuven) started the Leuven. Cool citizen science project in collaboration with Leuven 2030. The UHI effect is measured by more than 100 weather stations in gardens and public places across Leuven and is linked to human heat stress and the spatial arrangement of green and blue spaces. Real-time measurements of wind, temperature and rainfall are available on an online dashboard to inform and engage citizens (https://leuven.cool/) (Photo 2).

4 Promoting BioCities: From Silos to Synergies

Developing BioCities requires a new way of thinking and acting whilst addressing some of the major contemporary challenges facing society. So, how do we mimic adaptive approaches—which in fact mirror how society and nature work under good conditions—as complex adaptive systems? How do we allow for natural complexity and self-organisation within governance arrangements? And how do we support resilience within governance structures, that is, by recognising governance as a complex adaptive system in itself?

A strategic and adaptive approach to management, including a diverse (mosaic) governance structure may be a way forward. Whilst the many concepts addressing the decline of biodiversity, climate change and urbanisation all have a strong anthropogenic perspective, it has been claimed that this human-centred focus is not sufficient to create real transformations in addressing nature's role to solve these contemporary societal challenges (Randrup et al. 2020). Likewise, the European Commission's definition of green infrastructures may be seen as an ideal situation in planning, designing and managing urban nature, but not necessarily pursuing a holistic approach as it tends to have focus on the already planned, designed and managed green and blue elements of the urban matrix, and not on the ongoing processes that can form BioCities over time. Too often, hierarchical, chronological

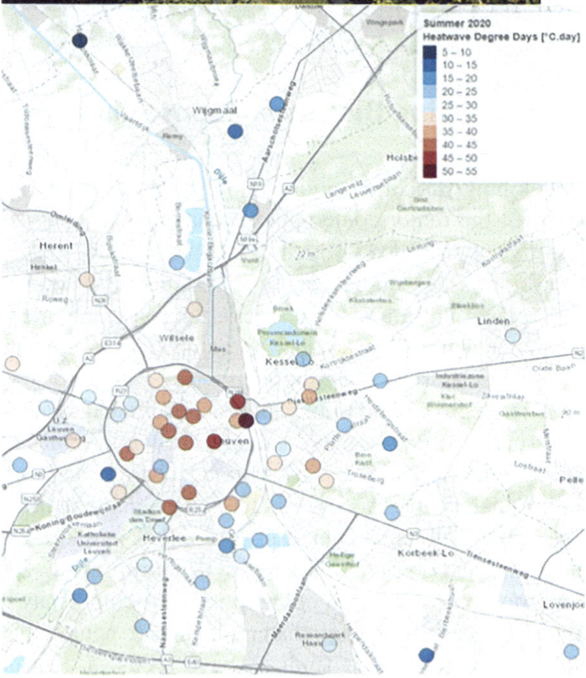

Photo 2 and Map 1: A citizen science project in Leuven, Belgium, seeks to evaluate the impact spatial planning, design, and management of green infrastructure on urban heat islands. The photos show weather stations in public and private space, whilst the map displays the number of heatwave degree-days (sum of the number of degrees above a certain threshold during heat waves) in 2020 recorded by the weather stations (Map and photo credit: Eva Beele)

and single-disciplinary approaches prevail when GI and UF are developed. New approaches and new ways of thinking are needed for involving a variation of practices and tools that help make the processes of developing GI and UF intelligible.

4.1 A Network Approach

The current decentralised approach to management is a reaction to years of organisational development trying to find the optimum degree of efficiency and service delivery. This approach may be seen as a constant reliance on rational decision-making in the development of society. The traditional organisational development of 'silos' has generated results in forms of improvements in welfare systems, education and housing. But it also comes with a potential collapse in rules and control mechanisms (Andersson 2020), as well as a fragmentation of expertise related to planning, designing and management decisions in separated practices.

There is a need to break the hierarchical, linear distribution of actors both within and across the formal organisations. It is generally accepted that there are two necessary parts to ensure efficient functioning of an organisation: the structural elements (as expressed by an organisational chart) and the process (Thommen 2016). Figure 1 illustrates an organisation chart, and in Fig. 2, we illustrate how the individual actors may assemble across silos in various governance structures.

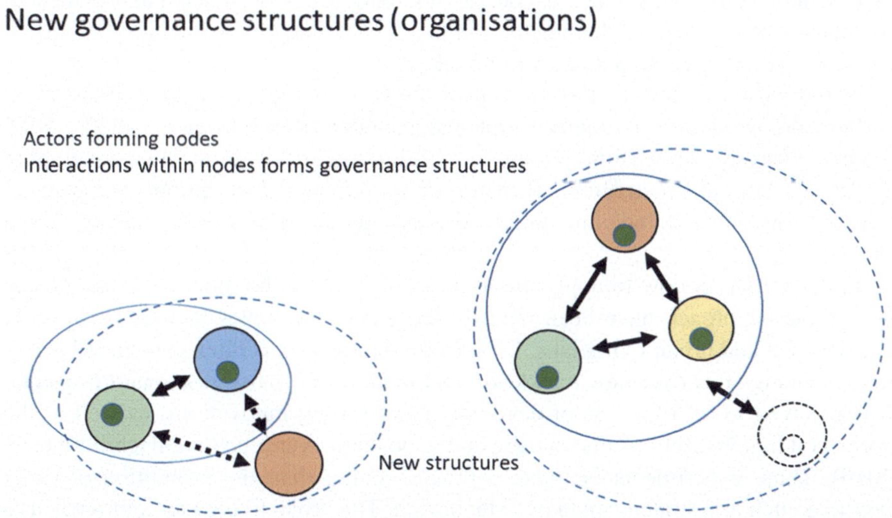

Fig. 2 Conceptual model of how actors related to departments of public and private institutions, policy makers, local stakeholders, citizens at large etc.—form the nodes of the governance structure, and the interactions between them formalise them. Communication is facilitated by memory banks (smart technologies). When new structures occur, via interactions between actors, BioCity Champions develops

As an actor is brought into the planning, designing or management process, a node is formed; as it *interacts* with another actor, a *structure* is formed. The creation and removal of nodes and structures can change over time and vary across space, allowing feedbacks between and amongst nodes. Such nodes may in some cases also be denoted as Life-Labs or Living Labs (McCormick and Hartmann 2017). However, there will be no 'true' organisation or governance structure, but a layering of multiple structures or a mosaic of structures (Buijs et al. 2016). The mosaic governance structures or the physical urban matrix, based on its spaces (patches) and the related connections, resembles ecological networks.

Continuous learning and communication should occur, based on transdisciplinary approaches, both within, but especially between the organisational structures. Focus should be on how the overall coordination, and the related knowledge transfer between planners, designers and managers, can be better solved. In this perspective, institutional memory and experience become assets for knowledge transfer—not only between different professionals and stakeholder groups but also in regard to longitudinal perspectives and how governance towards nature-based approaches can find support in the past and present to adapt to future uncertainties. Andersson and Barthel (2016) described the creating of 'memory banks' to delineate different ecological and social memory carriers, e.g. using smart technology. Interactive logbooks with records of geographical data, ecological information, citizen participation and of management and operational failures and successes could be an example of a supporting tool. However, knowledge transfer is new for the many stakeholders and institutions involved in current planning, designing and management of GI and UF, and although knowledge transfer exists in some cases, often there is little or no coordinated exchange of experience. Coordinated and evidence-based tests and associated documentation in a local context is particularly important for new and unconventional paths to be taken.

Nature-based Thinking (NBT) was recently presented as a new mindset to address urban nature governance and management (Randrup et al. 2020). NBT requires that three dimensions are always taken into consideration when developing GI and UF: the green structures themselves, the citizens, and the formal organisations owning the land, including politicians, and (public) planners and managers, see Fig. 3.

Frameworks on how to apply the mindset of NBT or the more concrete mosaic governance approach have been readily developed. Examples include urban field experiments and urban living labs. Urban experimentation is often considered a way to seed change that over time may lead to a fundamental transformation of a system. It is believed to be a process of developing and testing innovations, as well as the corresponding institutions that nurture and scale them over time (Fuenfschilling et al. 2019). Thus, experiments facilitate a process rather than the promotion of exact products such as a strategic plan or a site design. The urban living labs approach, as a form for urban experiments, offers a way to foster new collaborative, trans-disciplinary ways of thinking in urban planning, designing and management, and provides a real-world testing ground for urban innovation and transformation (McCormick and Hartmann 2017). For those involved, it is particularly important

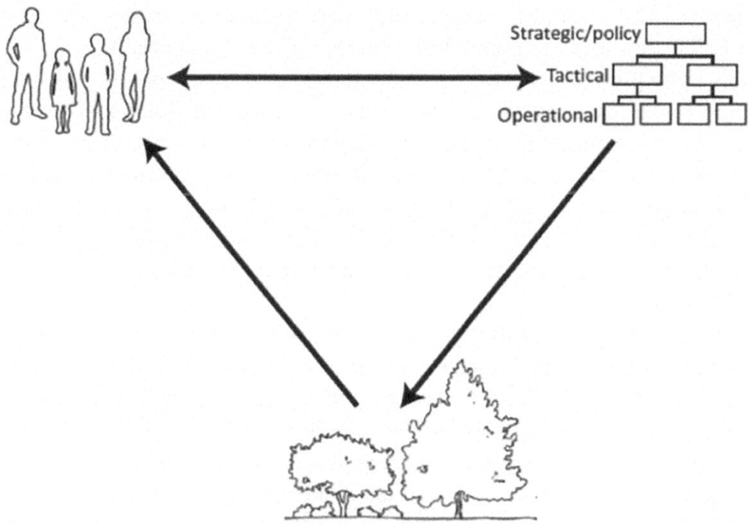

Fig. 3 Conceptual model of the three dimensions related to Nature-based Thinking: the green structures themselves, the citizens, and the formal organisations owning the land, including politicians, and (public) planners and managers. The arrows indicate primary relationships where the relation between green structures and users can be denoted as a Community–Ecological nexus; the relation between the organisation and the green structures can be denoted as an Ecological–Governance nexus, and the relation between the users and the organisation can be denoted as a Community–Governance nexus. Adapted from Randrup et al. (2020)

that the experiences are systematically collected, documented and new knowledge is transferred back into practice. Such experimental arrangements can be launched on a small scale—for example at the neighbourhood level—or on a large scale—for example at the city level. The irrepressible will learn and understand the city and biosphere as unity is a special characteristic of BioCities. Urban experiments and living labs are thus ideal instruments for future BioCities.

4.2 *Co-responsibility*

Multiple and complex interactions are needed in line with a simultaneous need for agility of the many actors involved. These interactions require new forms of cooperation. The main challenge lies in the fact that as the actors in each process are different, separate logics, languages and traditions are likely to appear. In addition, there need to be interactions between the sub-networks, such as when a local government manages the public trees of an urban forest but has little influence on the management of the private part of the same urban forest. An important prerequisite for effective interaction between the various sub-networks is functioning interfaces and the associated feedback loops (as a central part of adaptive management). The inter-connectivity between institutions, networks as well as between authorities and the population (citizen science) is central here.

For this to be possible, a different understanding of 'responsibility' is needed. In an analogy to co-learning, we could speak of *co-responsibilities*. Key tasks are distributed amongst different sectors and departments, and only through the interaction of the actors do sustainable solutions emerge. An important prerequisite for this is that the individual experts are given the competence and time to get directly involved at their own level and to act and decide in a solution-oriented way. In this way, the most diverse experts from the most diverse institutions each become 'BioCity Champions'.

Thus, the future of BioCities is closely related to user involvement and user engagement, and is directly connected to the idea of citizen science as shown in the case study from Leuven. Citizens can become a part of research and development, being able to actively contribute to the creation of knowledge. This can be in planning and monitoring or also in operational management (maintenance) of specific places. Further on, citizens can also take on public tasks, as for example, a possible deployment of citizens as BioCity Champions, or non-professionals who take over tasks in the public spaces as partners of the public authorities.

4.3 The Potential of Brown Spaces

Within the 'urban matrix', the untouched and unplanned areas, the so-called brown spaces or brownfields, take on a special significance. They are places of discovery and innovation, pioneering sites free from formal governance constraints. When managing existing spaces there is indeed an increasing trend to simply 'let go', and let nature prevail (Randrup et al. 2021). In that sense there are two societal discourses at play; the general lack of funding and the loss of biodiversity. Lack of funding may lead to an approach of not planning, not designing and not maintaining, which could be logically aligned with the argument for creating more local biodiversity. Gradually nature will come back—simply by letting go (see chapter "Biodiversity and Ecosystem Functions as Pillars of BioCities"). However, in reality is not as simple as that. Social-ecological landscapes are not promoted by doing nothing (Dunnett and Hitchmough 2004), just as there are numerous issues related to this too. Such issues include aesthetic expressions, inequality matters and fire risks, just to mention some.

Within the existing governance frameworks, there is the possibility of planning for brown spaces or existing green spaces not to be detailed designed and managed. In contracts to the ongoing urban densification trend, there is a need to reconsider brown spaces as construction sites for densification, or to just leave them as potentials to become spaces of innovation, or 'recovery fields'. In other words, planning can also be the decision 'not to control', but to, e.g. flexibly manage for long-term natural development. This in turn relies on ecological knowledge of both place and of vegetation, recognising the potential of designing and aesthetics in relation to pioneer and late successional species and how captivating places can be made in either young and immature plant communities or maturing vegetation. It also stimulates the approach of addressing different functions depending on future demands. Brownfields can thus not only be addressed for recovery but also as timeservers for future uncertainties.

4.4 *Benefits and Trade-offs*

When planning, designing and managing BioCities, it is imperative to consider the benefits and trade-offs associated with each strategy. Empirical evidence suggests that GI is not integrated at the same level and importance as grey infrastructure in the planning, designing and management processes (Hislop et al. 2019; de Magalhães and Carmona 2009). It is only when considering GI and UF at the same level that trade-offs can be discussed. We do not wish to claim that, for example, adaptive or strategic management should be unilaterally practiced without exception, but rather that these strategies are viable ways to increase the health, well-being, and livelihood of the BioCity and its citizens. Indeed, there are benefits and trade-offs to each strategy and situation. For instance, the costs of implementation, monitoring and management might make one strategy preferable to another, or the public might not value (or even devalue) the natural components of the BioCity. Similarly, efficiency requires effective coordination within and between actors in the organisational structure, with open and clear communication a key.

The aesthetics of 'naturalness' triggers further attention to whether characteristics of GI and UF need to be chaotic or ordered, uncontrolled or managed. It pinpoints the classical quest of the relationship between nature and humankind and to which extent human involvement is either beneficial or ominous to ecological processes and natural qualities. However, many of our cultural landscapes define and represent how human activities have been part of supporting and shaping landscapes that often are perceived as 'natural beauty' (Gordon 2018). Thus, the processes of strategic or adaptive management do not exclude human-controlled management or aesthetics deriving from symbiotic relationships between nature and humankind. Indeed, nature itself carries illustrative examples of order and symmetry, with most notable examples the patterns revealed in the Fibonacci sequence and in fractals—complex structures that exist in various systems at various scales and delineate a repetitive arrangement (Thwaites and Simkins 2007).

The case studies from Oslo and Lyon show how green and blue spaces are incorporated into the urban matrix and as part of a city's GI are valued and prioritised in line with other important urban infrastructures, such as the grey infrastructures of roads and traffic corridors. Here, the trade-offs of nature values expressed as cultural ecosystem services have proven to outweigh the previous conditions, in Lyon now documented via regulatory monitoring, and in Olso via Akerselva's proven success in terms of attracting businesses, social life and in increasing biodiversity—all the way from the outskirts of the city to the very city centre. Both projects demonstrate how planning, designing and long-term management, coupled with innovative and integrative governance structures, have generated transformed city areas using GI and UF.

Restructuring city landscape should also pay attention to land property rights as their use could strongly affect direction and outcome of city transformation and the transition to BioCity. Legal mechanisms for spatial reorganisation can have a potent impact at the urban or territorial scale, for instance, transfer of development rights (TDR) programmes, which allow 'sending area' landowners with holdings identified by local authorities as in need of preservation or protection to sever and sell their transferable development rights, or TDRs, to developers in less sensitive, or even strategically prioritised, 'receiving areas' where growth is desired and can accommodate the increased density. In exchange for the TDRs, local governments typically require that the property owner record a restrictive covenant or easement, limiting further development. Increased residential density is the most popular incentive for a developer to buy TDRs, yet other development thresholds, such as floor area or building height limits, can also be leveraged (Nelson et al. 2011).

5 Outcomes and Concluding Remarks

Green infrastructure is not 'the icing on the cake', as something which may create a green solution to an already established project. It should be dealt with as a whole, taking into account ecological processes. Cyclical processes to plan, design, and manage GI and UF should be developed across organisational departments and expertise. The time dimension should be acknowledged, balancing strategic and visionary approaches with incremental and adaptive planning processes. Thus, the development of BioCities is long-sighted in allowing nature to prevail and change, and to allow involved stakeholders and institutions to learn and adapt.

GI and its many stepping stones (e.g. UF), are important elements that must be incorporated into urban plans from the very beginning of the planning processes. In particular, the implementation of NBT and NBS means that urban planning processes are linked to the knowledge and requirements of designing, constructing and maintaining, and thus ensuring GI and UF. There is a need to see the preservation of building culture in line with the preservation and safeguarding of the ecosystem services provided by GI and UF to ensure BioCities of the future. This includes considering which ecosystem services should be provided by GI in a given area, and to integrate the necessary requirements into planning and maintenance procedures. Such thinking contributes to the integration of the three dimensions of NBT, as the concept of ecosystem services is related to green structures (GI and UF), the use of green structures which should be the result of a continued dialogue and trade-off process between the involved users and formal organisations over time.

Future BioCities will not only have different qualities and appearances compared to today, but will also need to be realised, implemented and developed over time through a new set of adaptive processes that are strategic, circular, multifunctional, research-based and inclusive. However, context is key; some BioCities may be organised in a rather classic way, whilst other may test new forms and processes. In any case, the involved actors should move towards co-responsibility and trans-disciplinarity in the likelihood that this is more important than the surrounding organisational structures.

References

Ahern J, Ciliers S, Niemelä J (2014) The concept of ecosystem services in adaptive urban planning and design: a framework for supporting innovation. Landsc Urban Plan 125:254–259. https://doi.org/10.1016/j.landurbplan.2014.01.020

Andersson SL (2020) Like an oarsman - moving forward, looking backwards towards the past. In: Nilsson K, Weber R (eds) Green visions : greenspace planning and design in Nordic cities. Arvinius + Orfeus Publishing, Stockholm, pp 36–55

Andersson E, Barthel S (2016) Memory carriers and stewardship of metropolitan landscapes. Ecol Indic 70:606–614

Arts B, Visseren-Hamakers IJ (2012) Forest governance: mainstream and critical views. ETFRN News 53:3–10

Arts B, Leroy P, van Tatenhove J (2006) Political modernisation and policy arrangements: a framework for understanding environmental policy change. Public Org Rev 6(2):93–106

Buijs AE, Mattijssen TJM, van der Jagt APN, Ambrose-Oji B, Andersson E, Elands BHM, Møller MS (2016) Active citizenship for urban green infrastructure: fostering the diversity and dynamics of citizen contributions through mosaic governance. Curr Opin Environ Sustain 22:1–6

Carlisle K, Gruby RL (2017) Polycentric systems of governance: a theoretical model for the commons. Policy Stud J. https://doi.org/10.1111/psj.12212

Colls A, Ash N, Nyman NI (2009) Ecosystem-based adaptation: a natural response to climate change. IUCN commission on ecosystem management (CEM), nature conservancy, US. IUCN, 2009, Gland, p 16. ISBN: 978-2-8317-1215-4

Davies C, Hansen R, Rall E, Pauleit S, Lafortezza R, De Bellis Y, Santos A, Tosics I (2015) Green infrastructure planning and implementation. The status of European green space planning and implementation based on an analysis of selected European City-regions. Green surge report 5.1. Brussels. European publication-detail/-/publication/fb117980-d5aa-46df-8edcaf367cddc202/

de Magalhães C, Carmona M (2009) Dimensions and models of contemporary public space management in England. J Environ Plan Manag 52(1):111–129

Dempsey N, Smith H, Burton M (eds) (2014) Place-keeping: open space management in practice. Routledge, London

Diaz S, Pascual U, Stenseke M, Martin-Lopez B, Watson RT, Molnar Z, Hill R, Chan KMA, Baste IA, Brauman KA, Polasky S, Church A, Lonsdale M, Larigauderie A, Leadley PW, Eadley, APE VO, Van Der Plaat F, Schroter M, Lavorel S, Aumeeruddy-Thomas Y, Bukvareva E, Davies K, Demissew S, Erpul G, Failler P, Guerra CA, Hewitt CL, Keune H, Lindley S, Shirayama Y (2018) Assessing nature's contributions to people. Science 359(6373):270–272. https://doi.org/10.1126/science.aap8826

Dunnett N, Hitchmough J (eds) (2004) The dynamic landscape: design, ecology and management of naturalistic urban planting, 1st edn. Taylor & Francis. https://doi.org/10.4324/9780203402870

European Commission (2013a) Ecosystem services and green infrastructure. ec.europa.eu/environment/nature/ecosystems/index_en.htm

European Commission (2013b) Building a green infrastructure for Europe. Luxembourg: Publications Office of the European Union 2013, pp. 24. ISBN: 978–92–79-33428-3. https://doi.org/10.2779/54125

European Commission (2015) Nature-based solutions and re-naturing cities: final report. Directorate-General for Research and Innovation–Climate Action, Environment, Resource Efficiency and Raw Materials. op.europa.eu/en/

FAO (2016) FAO forestry paper no. 178: guidelines on urban and peri-urban forestry. (Salbitano F, Borelli S, Conigliaro M, ChenY) Food and Agriculture Organisation of the United Nations, Rome. www.fao.org/3/i6210e/i6210e.pdf

FAO (2017) The state of food and agriculture: leveraging food systems for inclusive rural transformation. Food and Agriculture Organisation of the United Nations (FAO), Rome, p 2017

FEMAT (1993) Forest ecosystem management: an ecological, economic, and social assessment. Forest Ecosystem Management Assessment Team, US Government Printing Office, Washington, DC

Franch M (2018) Drawing on site: Girona's shores. J Landsc Architect 13(2):56–73

Fuenfschilling L, Frantzeskaki N, Coenen L (2019) Urban experimentation and sustainability transitions. Eur Plan Stud 27(2):219–228. https://doi.org/10.1080/09654313.2018.1532977

Gobster PH, Nassauer JI, Daniel TC, Fry G (2007) The shared landscape: what does aesthetics have to do with ecology? Landsc Ecol 22:959–972

Gordon JE (2018) Geoheritage, geotourism and the cultural landscape: enhancing the visitor experience and promoting geoconservation. Geosciences 8(4):136

Haase D, Pauleit S, Randrup TB (2020) Urban open spaces and the urban matrix: elements, form and functions. In: Jansson M, Randrup TB (eds) Urban open space governance and management. Routledge, London, pp 30–50. ISBN: 9780367173036

Helms J (ed) (1998) The dictionary of forestry. Society of American Foresters, Bethesda

Hislop M, Scott AJ, Corbett A (2019) What does good green infrastructure planning policy look like: developing and testing a policy assessment tool within Central Scotland UK. Plan Theory Pract 20(5):633–655. https://doi.org/10.1080/14649357.2019.1678667

Hood C (1995) The "new public management" in the 1980's: variations on a theme. Account Org Soc 20(2/3):93–109

Hoyle H, Jorgensen A, Hitchmough JD (2018) What determines how we see nature: perceptions of naturalness in designed urban green spaces. People Nat 1:167–180

Jansson M, Randrup TB (2020) Urban open space governance and management. Routledge, London, 224 p. ISBN: 9780367173036

Jansson M, Vogel N, Fors H, Randrup TB (2019) The governance of landscape management: new approaches to urban space development. Landsc Res 44(8):952–965. https://doi.org/10.1080/01426397.2018.1536199

Kingsford RT, Roux DJ, McLoughlin CA, Conallin J, Norris V (2017) Strategic adaptive management (SAM) of intermittent rivers and ephemeral streams. In: Datry T, Bonada N, Boulton A (eds) Intermittent rivers and ephemeral streams: ecology and management, pp 535–562

Konijnendijk CC (1999) Urban forestry in Europe: a comparative study of concepts, policies and planning for forest conservation, management and development in and around major European cities. Doctoral dissertation. Research Notes 90. Faculty of Forestry, University of Juensuu, Finland

Lo C (2018) Going from government to governance. In: Farazmand A (ed) Global encyclopedia of public administration, public policy, and governance. Springer, Cham. https://doi.org/10.1007/978-3-319-31816-5_3282-1

Mayrand F, Clergeau P (2018) Green roofs and green walls for biodiversity conservation: a contribution to urban connectivity? Sustainability 10:985. https://doi.org/10.3390/su10040985

McCormick K, Hartmann C (2017) The emerging landscape of urban living labs: characteristics, practices and examples. The International Institute for Industrial Environmental Economics, Lunds University. lup.lub.lu.se/record/77262ed5-1219-4798-89d9-872286efdb7b

McPhearson T, Haase D, Kabisch N, Gren Å (2016) Advancing understanding of the complex nature of urban systems. Ecol Indic 70:566–573. https://doi.org/10.1016/j.ecolind.2016.03.054

MEA (2005) Millennium ecosystem assessment report - ecosystems and human well-being: synthesis. Island Press, Washington, DC

Nelson A, Pruetz R, Woodruff D (2011) The TDR handbook. Island Press. islandpress.org/books/tdr-handbook

Nitoslawski SA, Galle N, Konijnendijk Van Den Bosch C, Steenberg JWN (2019) Smarter ecosystems for smarter cities: a review of trends, technologies, and turning points for smart urban forestry. Sustain Cities Soc 51:101770

Ordonez C, Threlfall CG, Kendal D, Hochuli DF, Davern M, Fuller RA, van der Ree R, Livsley SJ (2019) Urban forest governance and decision-making: a systematic review and synthesis of the perspectives of municipal managers. Landsc Urban Plan 189:166–180

Ostrom E (1990) Governing the commons: the evolution of instituitions for collective action. Indiana University. https://doi.org/10.1017/CBO9780511807763

Persson B, Neal P, Steidle A, Randrup TB (2020) Organisations related to urban open spaces. In: Jansson M, Randrup TB (eds) Urban open space governance and management. Routledge, London, pp 51–66

Qiao X, Kristoffersson A, Randrup TB (2018) Challenges to implementing sustainable stormwater management from a governance perspective: a literature review. J Clean Prod 196:943. https://doi.org/10.1016/j.jclepro.2018.06.049

Randrup TB, Jansson M (2020) Strategic management of urban open spaces. In: Jansson M, Randrup TB (eds) *Urban open space governance and management*, pp 190–203

Randrup TB, Persson B (2009) Management of public green space in the Nordic countries: development of a new strategic green space management regime. Urban Forestry Urban Greening 8:31–40

Randrup TB, Konijnendijk C, Dobbertin MK, Prüller R (2005) The concept of urban forestry in Europe. In: Konijnendijk CC, Nilsson K, Randrup TB, Schipperijn J (eds) Urban forests and trees: a reference book. Springer, Berlin, pp 9–21

Randrup TB, Buijs A, Konijnendijk C, Wild T (2020) Moving beyond the nature-based solutions discourse: introducing nature-based thinking. Urban Ecosyst 23:919–926. https://doi.org/10.1007/s11252-020-00964-w

Randrup TB, Sunding A, Svännel J, Jansson M, Sang ÅO (2021) Urban open space management in the Nordic countries: identification of current challenges based on managers' perceptions. Cities 115:103225. https://doi.org/10.1016/j.cities.2021.103225

Sjöman H, Hirons A, Deak Sjöman J (2017) Criteria in the selection of urban trees for temperate urban environments. In: Ferrini F, Konijnendijk van den Bosch C, Fini A (eds) Routledge handbook of urban forestry, pp 339–362

Smith H, Pereira M, Hull A, Konijnendijk van den Bosch C (2014) The governance of open space decision-making around place-keeping. In: Dempsey N, Smith H, Burton M (eds) Place-keeping: open space management in practice. Routledge, London, pp 52–75

TEEB (2010) The economics of ecosystems and biodiversity: mainstreaming the economics of nature: a synthesis of the approach, Conclusions and Recommendations of TEEB

Thommen JP (2016) Betriebswirtschaft und Management: Eine Managementorientierte Betriebswirtschaftslehre, Teil 9: Organisation. Versus, Zurich

Thwaites K, Simkins I (2007) Experiential landscape. An approach to people, place and space. Routledge, Oxon

UN (1992) United Nations Conference on Environment & Development. Rio de Janeiro, Brazil, 3 to 14 June 1992. AGENDA 21. sustainabledevelopment.un.org/content/documents/Agenda21.pdf

UN (2015) The 17 sustainable development goals. www.un.org/sustainabledevelopment/

UN Population.org (2018)

WHO (2016) Urban green spaces and health. World Health Organisation, Regional Office for Europe, Copenhagen

World Commission on Environment and Development (1987) Our common future. Oxford University Press, Oxford

Mitigation and Adaptation for Climate Change: The Role of BioCities and Nature-Based Solutions

Silvano Fares, Teodoro Georgiadis, Arne Sæbø, Ben Somers, Koenraad Van Meerbeek, Eva Beele, Roberto Tognetti, and Giuseppe E. Scarascia-Mugnozza

1 Introduction

Cities of the world, hosting more than half of the world's population, are at the centre of the climate change mitigation agenda. Cities account for 60–80% of overall energy usage consumption (UN DESA 2022) and up to 70% of global GHG emissions in both industrialised countries and emerging economies (Henninger 2008; Ramachandra et al. 2015). This is primarily due to transportation demands, industrial emissions, resource consumption, and energy infrastructures (Park et al. 2013). Accordingly, a fundamental role of BioCities will be to aim at a zero net-emission target by applying one of the key functional traits described in the *BioCities Manifesto* in chapter "Towards the Development of a Conceptual Framework of BioCities", namely *the BioCity as a net Carbon sink.*

S. Fares (✉)
National Research Council of Italy, Institute for Agriculture and Forestry Systems in the Mediterranean, Naples, Italy
e-mail: silvano.fares@cnr.it

T. Georgiadis
National Research Council of Italy, Institute of BioEconomy, Bologna, Italy

A. Sæbø
Norwegian Institute of Bioeconomy Research (NIBIO), Ås, Norway

B. Somers · K. Van Meerbeek · E. Beele
Department Earth & Environmental Sciences, Katholieke Universiteit Leuven, Leuven, Belgium

R. Tognetti
Dipartimento di Agricoltura, Ambiente e Alimenti, University of Molise, Campobasso, Italy

G. E. Scarascia-Mugnozza
University of Tuscia, Viterbo, Italy

G. E. Scarascia-Mugnozza et al. (eds.), *Transforming Biocities*, Future City 20,
https://doi.org/10.1007/978-3-031-29466-2_5

A largely diversified set of transformative actions should be adopted by urban areas in the world (ASLA 2022) to greatly contain their carbon footprint: improving building efficiency, adopting more sustainable transportation systems, and advancing landscape design with seamless interconnections of streets, parks, and plazas to encourage alternative mobility that makes life healthier (see chapter "BioCities as Promotors of Health and Wellbeing"). Also, *urban trees and green space may reduce cities' overall carbon footprint* by absorbing carbon dioxide whilst decreasing energy use, thanks to their shading and air-cooling effects. Cities may also reduce emissions through wiser waste management involving materials recycling and food waste composting. It would be extremely useful, however, to document the comparative characteristics of these interventions, including their relative efficacy, potentials, and emissions reductions.

A systematic scoping of these actions, describing a spectrum of urban solutions, available instruments, targeted sectors, expected or verified mitigation potential, and social outcomes, has been conducted by Sethi et al. (2020), which provides insights into mitigation and adaptation prospects of urban areas from all over the world. Via a detailed mapping of more than 800 case studies on sectors such as buildings, energy, transport, waste, industry, agriculture, forestry, and other land uses, the authors were able to identify 41 different urban solutions with an average GHG abatement potential ranging from 5.2 to 105%, with most of them clustered in the building, energy, waste materials, and transport sectors (Sethi et al. 2020). The greatest GHG abatement potential lies in technology-oriented interventions in carbon-neutral buildings, energy efficiency, waste material recycling and treatment, more sustainable transport, and green roofs and facades (prominently at lower latitudes). Whilst system-wide interventions, such as urban form and landscape planning, energy from biomass, parks, afforestation, and other green spaces, have lower abatement potential but wider scope.

In light of that framework, this chapter will describe the contribution of Nature-Based Solutions (NBS) for climate mitigation and adaptation. On the other hand, chapter "Innovative Design, Materials, and Construction Models for BioCities" will deal mostly with C-neutral buildings, material recycling, and energy management; whilst chapter "Forests, Forest Products and Services to Activate a Circular Bioeconomy for City Transformation" will examine the role of a circular bioeconomy.

Trees, forests, and other green infrastructures, in and around cities, provide a wide spectrum of solutions to combat the effects of climate change through their normal biotic activity. Research has commenced in discerning the many options that trees, forests, and green spaces may offer to the future of cities. In this chapter, we analyse the current state of the art on how green infrastructures mitigate and adapt to climate changes and pollution, how they may improve urban air quality, increase green mobility, and can promote other important ecosystem benefits such as water cycle regulation and supply. Relevant case studies will be also described, as gaps and future perspectives will be analysed towards reaching the full potential of urban forests and other green spaces, for BioCities in Europe and beyond.

2 Urban Forest and Other NBS Contributions to Climate Change Mitigation

2.1 Climate Change Issues and Trends

Trees, forests, and other green infrastructures contribute to mitigating climate change and do so in several ways (Fig. 1): directly storing carbon in the biomass and in the soil, and indirectly by substituting wood material for other construction material (e.g. cement and steel) that have a larger carbon footprint than wood. They can also be used as a renewable substitute for fossil fuels, such as using wood boilers in power plants. Even a small stand of mature forest (i.e. a 0.1-ha beech forest) can sequester all the CO_2 that a car emits in 1 year (driving 15,000 km emits two tons of CO_2) (Scarascia-Mugnozza and Matteucci 2014). Urban forests and trees are even more valuable since they do much more than this—they are intrinsically multi-functional and aligned to the classic pollution model of source (fossil fuel and other), pathway (air and earth), receptor (people) and cycling (i.e. water cycle). Urban forests can act on all of these components.

Tree foliage canopies also control irradiance by interception and absorption, whilst cooling by evapotranspiration ensures the release of energy and heat. The reduced 'heat island effect' and corresponding increase in thermal comfort helps in reducing energy costs for heating and cooling buildings, by up to 50%, and helps mitigating the temperature and drought extremes in urban areas (Wang 2016). Pollution is often associated with climate change because high temperatures promote photochemical smog. Climate change, therefore, exacerbates pollution. Urban green infrastructures improve urban air quality since they clean the air by removing up to

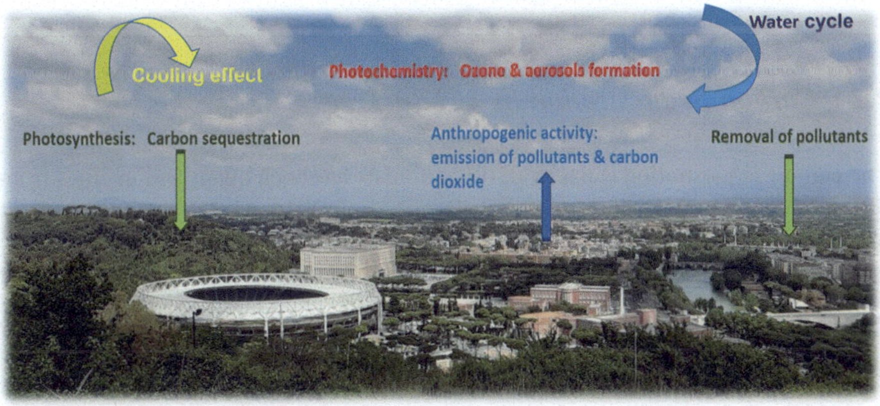

Fig. 1 Ecological processes simultaneously occurring at the urban–atmosphere interface: primary emission of pollutants from anthropogenic activities, photochemical processes affecting air quality, phytoremediation with removal of carbon and pollutants from the atmosphere, water cycle, and cooling effect by urban forests and other green spaces

20% of ozone and particulate matter emitted by transport and burning of fossil fuels (Fares et al. 2016).

Trees act as a sink for CO_2 by fixing carbon during photosynthesis and storing it as biomass, in both above- and below-ground structures (Fares et al. 2015). This is tightly correlated to urban soils, which also offer carbon storage and act as a valuable organic matter biome. Roots account for 20–26% of tree biomass (Liberloo et al. 2009). Nevertheless, the tree canopy releases CO_2 during the respiration processes, which is required for metabolising carbohydrates for tree energy production. This adds to soil CO_2 fluxes through microbial decomposition of organic matter and respiration from roots and mycorrhizae (Godbold et al. 2006). On a daily and seasonal basis, CO_2 fluxes vary as a function of atmospheric circulation. A decrease in convective movement of the air in winter generally leads to an increase in atmospheric CO_2 concentration. Vegetation activity is influenced by daily and annual variations in photosynthetic CO_2 consumption, and varies along with citizen habits such as traffic density reduction during weekends and holidays (Gratani and Varone 2005).

Warmer climates, longer growing seasons, and elevated atmospheric CO_2 concentrations may accelerate tree growth (Piao et al. 2013), but a harsh urban environment can lead to effects that may reduce tree carbon sequestration capacity (e.g. water shortage, high temperature, and air pollution). When urban greenspace is properly managed in arid landscapes, urban forests can store more carbon than adjacent suburban and rural areas (McHale et al. 2015), but water stress remains a major constraint for tree growth. Heat stress and soil sealing (e.g. pavement) may induce water stress in urban trees (Haase and Hellwig 2022). This warrants careful selection and diversification of plant material to enhance the resilience of urban forests. In this sense, botanical gardens and forest nurseries may have a renewed role in understanding the response of trees to changing the environment of cities, and to produce suitable genetic material (Hirons et al. 2020).

2.2 *Carbon Sequestration by Urban Trees*

The carbon sequestration capacity of urban forests has been quantified in cities from different continents (Zhao et al. 2013; Raciti et al. 2014). Trees may account for more than 95% of the carbon stored in above-ground vegetation (Davies et al. 2011). Table 1 shows the carbon sequestration rate in urban forests of different cities of the world (Tang et al. 2016; Liu and Li 2012; Jim and Chen 2009; Nowak and Crane 2002; Fares et al. 2020).

Table 1 C-sequestration by urban forests in different cities of the world (Mg C ha^{-1} $year^{-1}$)

China				USA		EU	
Beijing	Shenyang	Guangzhou	Hangzhou	Atlanta	Jersey City	Roma	Torino
1.3	2.84	4.0	1.66	1.23	0.23	1.01	0.5

Nowak et al. (2013) found that whole-tree carbon storage densities average 76.9 Mg C ha^{-1} in urban forest areas in the USA, with an average annual sequestration rate of 2.8 Mg C ha^{-1} per year. This is equal to a total annual sequestration of 25.6 Tg C. Wilkes et al. (2018), employing multi-scale LIDAR technology, estimated a median aboveground carbon density of urban forests in London to be as high as 24.3 Mg C ha^{-1}, values that are comparable to temperate and tropical forests. In terms of monetary benefits, total tree carbon storage in US urban areas is estimated at $50.5 billion, whilst annual carbon sequestration is estimated at $2.0 billion (Nowak et al. 2013). These results suggest that urban areas will become even more important as carbon sinks, and effective tools to assess carbon densities in these areas, including the soil components, are therefore vital.

2.3 Substitution Effect of Wood Use

There is an on-going debate over the role that wood-based products play in climate mitigation. In addition to carbon storage in trees, soils, and wood products; using wood to substitute for greenhouse gas-intensive materials (chemical compounds, construction elements, textile fibres) and fossil fuels (energy services) may have climate benefits (Sathre and O'Connor 2010). Incentivising both wood-based products and increasing urban tree density might benefit carbon sequestration and boost citizen awareness, especially amongst city dwellers.

A substitution effect typically describes how much GHG emissions would be avoided if a wood-based product is used instead of another product to provide the same function—be it a chemical compound, a construction element, an energy service, or a textile fibre. Leskinen et al. (2018) computed that for each kilogram of carbon in wood products that substitute non-wood products (i.e. substitution or displacement factor), there occurs an average emission reduction of approximately 1.2 kg of carbon. The use of wood and wood-based products is associated with lower fossil and process-based emissions when compared to non-wood products. Substitution benefits are largely gained due to reduced emissions during the initial production and the end-of-life product stages, particularly when post-use wood is recovered for energy. The use of forests for biomass and energy production is controversial, however, considering the growing role of forests as efficient carbon sinks for climate change mitigation. A recent study by Favero et al. (2020) suggests that the simultaneous uses of forests for biomass and for carbon sequestration may be reconciled if an economic analysis, accounting for the interactions between demand and supply and forest management, is considered. In fact, results of this study show that an expanded use of wood for biomass production will result in net carbon benefits, but an efficient policy also needs to regulate forest carbon sequestration.

Planning and managing of urban forests by municipalities can act as important leverage to carbon sequestration and influence urban citizen lifestyles, promoting a shift towards greener building demands and greener mobility strategies. Please note, however, that poor management of urban greenspaces, such as poor tree pruning, tree bole and root wounds, leaf litter removal, soil compaction, and root constraints, may strongly impact carbon fluxes and impair storage capacities. On the other hand, residuals from urban tree pruning and tree plantations in peri-urban areas may provide wood for bioenergy, as well as raw materials for construction and furniture, hence contributing towards a circular bioeconomy (see also chapters "Forests, Forest Products and Services to Activate a Circular Bioeconomy for City Transformation" and "Innovative Design, Materials, and Construction Models for BioCities"). Managing an effective trajectory for integrating wood-based products and bioenergy in urban NBS management will require a mix of research and policy that encourage appropriate land-use policy and technology innovation.

2.4 Need to Reconcile Mitigation and Adaptation

Although urban trees can sequester atmospheric CO_2 and serve as long-term carbon sinks, in general, rarely do urban local authorities incorporate forest carbon storage and sequestration policies into their planning. Examining current urban forestry plans for effective carbon mitigation could reveal several ways more efficient carbon sequestration. Interventions highlighting the spatial juxtaposition of green infrastructure at the municipal scale, for instance, could offset increasing atmospheric CO_2 concentration, as would policy efficiencies at broader spatial scales (regional or national) (Baró et al. 2014). Urban areas could increase canopy cover through new tree planting or adopting appropriate management strategies for existing canopy cover. Linking these actions to interconnected green infrastructure planning will also deliver additional benefits in terms of reduction in urban heat island and storm water runoff effects, thus reconciling mitigation and adaptation strategies to address challenges posed by climate change. Nevertheless, urban areas of all sizes need support from the governance and research sectors, and society at large, to meet mitigation and adaptation targets. Without this support, it is difficult to envision how change can occur to the scale required.

3 Urban Green Infrastructures Contribute to Climate Change Adaptation

3.1 Climate Change and Urban Heat Islands: A Dangerous Mix

Climate change will continue to raise average air temperatures leading to an increased frequency and severity of heat waves. The consequences of these heat waves are most acute in urban areas where the urban heat island (UHI) effect further increases air temperatures and forms a direct threat to human health and well-being

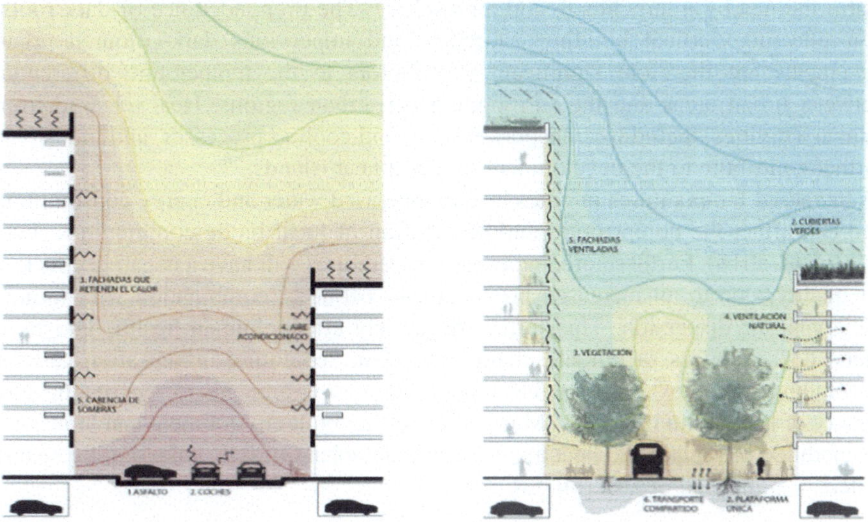

Fig. 2 Urban Heat Island and the role of urban vegetation as mitigation. ©IAAC

(Campbell et al. 2018). Already, this causes a yearly average of 12,000 deaths only in the USA and has led to the UHI effect and its associated urban heat events to become known as a silent killer (WHO 2018). The 2003 heat wave in Europe, for example caused the deaths of over 70,000 people (Robine et al. 2008). The mechanism for this mortality is that exposure to high temperatures increases heat strokes and exhaustion, as well as aggravating already existing cardiovascular, pulmonary, and renal diseases (Shindell et al. 2020). Additionally, high temperatures increase human heat stress, or the uncomfortable feeling people experience when the human body fails to regulate its internal temperature (Fig. 2).

The enhancement of UHIs during the last few decades is a direct consequence of worldwide urbanisation and related urban sprawl. An UHI develops through processes that impact the absorption of solar radiation during the day and the subsequent release during nighttime. Anthropogenic activity and the urban environment greatly influence the intensity of UHIs (Piracha and Chaudhary 2022). In cities, 50%–70% of all surfaces are impervious pavements (roads, parking lots, squares), buildings' vertical surfaces, and roofs (Kuang et al. 2019). These surfaces contribute to the UHI and far exceed the surfaces that reduce the UHI, such as parks, trees and urban forests, gardens, and water bodies. Urbanisation has altered the urban morphology resulting in narrow canyon-like streets in which both short and longwave radiation are trapped, hereby reducing the amount of long-wave radiation loss and thus cooling during the night. Also, surface modifications in which bare soil and vegetation are replaced by impervious human structures (i.e. buildings and paved streets) generally result in a lower albedo, higher thermal emissivity, and higher heat capacity. Materials used in urban surfaces, such as roads, pavements, roofs, and walls, have commonly low solar reflectance. These urban materials heat up and warm the nearby microclimate and atmosphere. The preponderance of dark roofs, non-reflecting vertical buildings' façades, and impervious dark-coloured urban pavements are therefore significant contributors to the temperature differential between urban areas and the surrounding peri-urban regions. Heat released from human activities, including transport, heating and cooling processes, and industries, further contribute to the development of urban heat islands.

Increased temperatures in cities lead to increased water and energy consumption and can affect the composition and distribution of local biotic communities (Leal Filho et al. 2018). Furthermore, increased temperatures will have a negative effect on air quality, due to an increased production of ozone in combination with limited horizontal air dispersion. This further negatively affects human health. The combined consequences of climate change and expected future urbanisation call for developing strategies and resources to cool urban environments to secure a healthy quality of life within our cities. Many of the urban built environment can, however, be modified or replaced to become greener and cooler surfaces, able to mitigate and reflect heat, instead of absorbing it.

3.2 The Role of Urban Green Spaces to Mitigate the Urban Heat Island (UHI) Effect

Urban greenspaces, and especially trees, have been identified as important regulators of air temperature. The mitigation of heat at the city level has a crucial effect on the thermal comfort of citizens and can also induce energy savings, thus indirectly reducing CO_2 emissions. The cooling effect is significant due to evapotranspiration, which is clearly much higher from vegetation than from sealed surfaces (Saaroni et al. 2018). Transpiration of water through their leaves is driven by the absorption of ambient heat, resulting in a cooling of the environment. Evapotranspiration additionally increases the air humidity, which helps relieve the UHI effect. A large tree, for example, can transpire up to 600 l of water per day (Purcell 2021).

The other fundamental mechanism by which trees can improve urban microclimate and thermal comfort is through shading. Leaves and branches intercept incoming shortwave solar radiation, hereby reducing the amount of radiation that reaches the underlying built-up surface. The amount of sunlight transmitted through the canopy varies based on tree species, but in the summertime generally only 10–30% of the sun's energy reaches the ground below a tree (Hardy et al. 2004). The remainder is absorbed by leaves and used for photosynthesis or for sensible and latent heat within the tree canopy, or reflected to the atmosphere.

Due to their lower specific heat capacity, trees reflect more of the incoming solar radiation (higher albedo) in comparison with their impervious surroundings, hence they provide further cooling during summertime. In a study in California, USA, Scott et al. (1999) demonstrated that tree shading during summertime reduces the temperatures inside parked cars by up to 25 °C. Coutts et al. (2016) found maximum daytime cooling by street trees in shallow canyons up to 1.5 °C, and a detailed analysis in several urban areas throughout Europe showed that compared to continuous urban fabric, LSTs observed under urban trees are on average 0–4 °C lower in Southern European regions and 8–12 °C lower in Central Europe (Schwaab et al. 2021). This provides evidence that urban trees are effective natural air conditioning systems, as long as urban forest and trees are healthy and do not suffer from water stress during drought.

3.3 Green Infrastructures Reduce Energy Costs

Summer energy consumption in the city is mitigated by green infrastructures that provide natural air conditioning that can significantly reduce energy usage. Climate change will cause an increase in global cooling energy demand both in temperate and in the world's warmest regions, therefore building sector adaptation measures are urgently needed (Gamero-Salinas et al. 2021). Appropriate passive cooling design of buildings, including roof insulation, natural ventilation, glazing properties, and envelope solar absorptance, could help lower the indoor overheating risk. Proper building design strategies, however, have rarely been adopted in many temperate and warm climatic regions. People worldwide are facing cooling access risk that could result in an increase in cooling energy demand and increased world carbon emissions by 2100 (Santamouris et al. 2015), especially if we consider that presently using air conditioners accounts for nearly 20% of the total electricity used in buildings around the world (IEA 2018). A fulcrum point was reached earlier this century when the energy demand during the summer, in major cities of Central and Mediterranean Europe, overtook that of winter (Georgiadis 2019). A wide array of urban green infrastructures could be utilised to contribute to reducing the risk of building overheating, such as green roofs, green walls, street trees, and trees in parks and urban forests.

Urban vegetation, when not in a condition of water stress, maintains its own physiological temperature that is much lower than that of a sunny wall. Vegetation therefore acts as a cool-screen for long-wave radiation exchange, allowing the establishment of a radiative flow from the buildings towards the urban forest, and thereby reduces the energy demand for cooling (Young et al. 2015). Furthermore, the shading of building walls produces an additional cooling effect depending on the time of day (Pandit and Laband 2010). Together with the direct cooling effects produced by the shadow or by the radiative exchange with the buildings, the urban forest when configured as a public park or as a tree-line road, produces a 'cold well' effect (Zupancic et al. 2015). This 'cold well' is of great importance from the point of view of urban air mass movements.

Studies conducted on green infrastructures demonstrate that radiative and thermal loads on buildings, as well as wall and roof temperature, were significantly lower in the shade of trees compared to unshaded areas (Wang 2016). Thanks to the shading effect of trees, houses surface temperature can be reduced by 7.0–25.0 °C in different cities of the world (Akure, Nigeria; Melbourne, Australia; Sacramento, United States). Moreover, the indoor air temperature was reported to be lowered by up to 3.0 °C (Morakinyo et al. 2013). The cooling energy required for maintaining a constant indoor temperature can also be reduced by 10–30% in the shaded building compared to the unshaded building (Wang 2016). This not only refers to the immediate vegetated area but also the surrounding regions too. In some cases, a cooling effect can stretch hundreds of metres with a reduction in temperatures from 2.3 to 4.8 °C (Aram et al. 2019).

The cooling effect of tree canopies depends on many factors, such as the width, shape, type, and amount of vegetation cover and, of course, the climatic region in which the 'green infrastructure' is placed. All this scientific information should be integrated into an urban planning approach combining different types of green infrastructures in the urban ecosystem, including blue infrastructures such as lakes, canals, pools, falls, and fountains, to alleviate urban heat islands and enhance thermal comfort in different functional areas of a city. Cao et al. (2022) recently combined field data and spatial modelling at a city scale to provide scientific basis for urban designers and planners to formulate optimal greening schemes to improve urban microclimate. The results showed the cooling effect was best when the water bodies were added dispersedly with the vegetation. Vertical greening was also considered for urban design and planning, where green walls and green roofs have real cooling and humidifying effects, particularly in combination with dispersed tree and shrub vegetation around the buildings (Cao et al. 2022). The positive effects of green and blue infrastructures on improving the microclimates in cities produces not only a reduction in the energy cost [i.e. New York City's urban forest is estimated to reduce annual residential energy costs by $17.1 million per year (USDA-FS 2022)] but it also can increase in the value of the property due to both the reduced energy costs and the increased demand for the buildings in the neighbourhood. Moreover, urban trees also contribute to maintain a comfortable air temperature by creating a wind shield that reduces warm wind from blowing in the summer, similarly to evergreen trees reducing cold wind blowing in the winter. The evapotranspiration of vegetation adds to ambient moisture, which could also raise the outdoor and indoor humidity (Wang 2016).

Finally, it is crucial that research should provide scientific guidance for adapting urban vegetation planning to different climate zones on a larger scale. Wang et al. (2022) utilised Landsat-8 data of land surface temperatures for 30 cities worldwide to show that the larger the green space in a city, no matter the climate zone, the better the cooling effect. Additionally, complex shapes of green surfaces were found to have a greater cooling intensity, especially in tropical and temperate zones. The study also showed that the maximum temperature drop caused by a specific patch of green space, increased with latitude. The lower cooling effect occurred in arid zones of the Tropics.

4 Urban Green Infrastructures Improve Urban Air Quality

4.1 Air Pollution: A Threat to European Cities

Ambient air quality in cities may contain high levels of pollutants that cause human health problems. Over 80% of citizens are exposed to air quality levels that exceed WHO thresholds (WHO 2021). Ground-level concentrations of ozone and particulate matter, which have increased since pre-industrial times in urban and rural regions, are associated with cardiovascular and respiratory mortality and have a significant impact on human and ecosystem health (WHO 2021). Outdoor air pollution kills approximately 8 million people across the world every year (WHO 2018), with a global cost estimated at 1.7 trillion dollars (OECD 2014). The recent European Environment Agency's report on air quality (EEA 2021) shows that almost all Europeans living in cities are exposed to air pollution levels that exceed the health-based air quality guidelines set by the WHO. An analysis based on the latest official air quality data from more than 4000 monitoring stations across Europe states that exposure to air pollution (mainly particulate matter [PM], ozone [O_3], and nitrogen oxides [NO_x]) caused about 300,000 premature deaths in the European Union (EU) in 2019 (EEA 2021). Compared with the EU limit values, fine particulate matter concentrations were too high in seven EU Member States in 2017 (Bulgaria, Croatia, Czechia, Italy, Poland, Romania, and Slovakia). In addition, four EU Member States, (Bulgaria, Hungary, Poland, and Slovakia) have not yet met the EU's 2015 target for the 3-year average exposure for fine particulate matter. Air pollution, associated with urban lifestyle is one of the major causes of non-communicable diseases in the world (Schraufnagel et al. 2019). In urban societies, the new emerging health problem and causes of death are directly linked to pollution and lifestyles. The role of green infrastructure including urban forests is key in approaching and solving these problems.

4.2 The Role of Green Infrastructure

Urban trees and forests, including other types of vegetation and vertical greening, have received increasing attention for their potential contribution for reducing urban pollution and promoting citizens health and well-being. Studies have highlighted how green spaces can play an important role in improving air quality through the removal of air pollutants such as PM, O_3, NO_x, sulphur dioxide (SO_2), and polycyclic aromatic hydrocarbons (PAHs) (Tiwari et al. 2020).

Dry deposition represents the main receptor pathway in plant ecosystems for ozone and other pollutants (Clifton et al. 2020). This 'sink' capacity of plants results from interactions between meteorology, chemical, and physical characteristics of the pollutants and the properties of the canopy. Whilst the photosynthetic process and carbon assimilation has been widely investigated, plants' leaves can additionally

Fig. 3 A view of the experimental site in Castelporziano, Rome, Italy. Top—Holm oak urban forest taken with a drone; bottom right—experimental tower hosting sensors; bottom left—tridimensional sonic anemometer and closed-path sensors used to measure fluxes of greenhouse gases and pollutants are measured with micrometeorological techniques

absorb pollutants when they penetrate through stomata (Dusart et al. 2019). Particles are also intercepted by vegetation and retained on the surface of leaves, trunks bark, or can be absorbed into plant tissues (Han et al. 2020). Ozone deposition has been described in several agricultural and forest ecosystems, displaying two separate sinks: leaf stomata, and plants/soil surfaces (Clifton et al. 2020).

Green surfaces have an added benefit when compared to non-vegetation surfaces. Not only does the stomatal sink represents on average 45% of total sequestration, with peaks up to 70%, but leaves also have the capacity to detoxify absorbed pollutants once they penetrate inside intercellular spaces (Dusart et al. 2019).

Healthy urban forests may help clean the air of cities by removing ozone and particulate matter emitted by transport and burning of fossil fuels (Fares et al. 2020; Fig. 3). As leaves and tree crowns are the active interface between plant and atmosphere, canopy attributes like leaf area index (LAI) strongly influence plant ability to intercept atmospheric particles and gases (García de Jalón et al. 2019). LAI, hairiness, and wax content affect deposition, but also meteorological variables (precipitation, solar radiation, humidity, wind speed, temperature, and turbulence) have an impact on the magnitude of deposition velocity and thus the capacity of plants to ameliorate air quality (Xing and Brimblecombe 2020; Barwise and Kumar 2020). Research shows that the choice of species is also important. Some species can absorb particulate matter more than 10 times that of less efficient species (Sæbø et al. 2012). The increase in leaf area in cities is an important aim for increasing air purification capacity, as well as for increasing the other ecosystem services of plants. The fate of the deposited PM, however, will differ in different environments. Rain washes deposited matter off surfaces and the runoff should be channelled to appropriate drainage or to soils that can absorb, inactivate, and digest pollutants.

In a UK study, large scale vegetation (not just the trees within the city) was found to significantly reduce pollutant concentrations by 10% for PM, 30% for SO_2, 24% for NH_3, and 15% for O_3, when compared with areas without forests and shrub land (Nemitz et al. 2020). The urban forest of Florence (Italy) can remove up to 15% of the emissions generated from the business-as-usual activities of the city (Bottalico et al. 2017). Research modelling air pollution removal based on i-Tree (https://www.itreetools.org/about) has been extensively tested for US urban forests and widely applied globally. For example, trees in New York City currently remove roughly 1100 tons of air pollution per year, for and economic value estimated at $78 million per year.

5 Urban Green Infrastructures and the Water Cycle

Urban forests and other greenspaces offer also other environmental benefits that are relevant for climate change adaptation and mitigation. Green infrastructures can regulate the water cycle by reducing the impacts of rainstorm, runoff, and soil erosion, particularly in densely inhabited areas like cities.

Climate change is already affecting the water cycle both locally and globally (Masson-Delmotte et al. 2021), and its impact is likely to increase in the future. It is estimated the atmosphere can hold 7% more water vapour for every 1 °C increase in air temperature, thereby increasing the risk of flash flooding. According to the IPCC Report, the number of heavy precipitation events over land has increased globally because of human-influenced climate change, with confidence highest for North America, Europe, and Asia. If global warming continues to increase, short-duration extreme rainfall events, such as thunderstorms, and more intense individual showers are most likely to occur, even though the total annual local rainfall may increase or decrease depending on regional differences throughout of the world.

These alarming scenarios call for urgently embracing adaptive strategies, especially in urban areas where impervious surfaces, such as roofs, parking lots, and roads, convert precipitation to stormwater runoff, which cause water quality and quantity problems. Research has shown that trees can play a substantial role in reducing stormwater runoff via canopy interception loss, transpiration, facilitating infiltration into the soil, and by coupling trees with other green infrastructure technologies such as raingardens, bioswales, and permeable pavements (Berland et al. 2017). Urban forest patches can infiltrate on average 68% of rainfall events, mostly by improving soil infiltration due to the expansion of roots, which generate small channels into the soil (Phillips et al. 2019). However, the amount of precipitation infiltrated into the soil below urban trees, however, can vary from 40% to 90% depending on soil texture, degree of soil compaction, and rainfall characteristics (Phillips et al. 2019). Trees can also improve the eco-hydrological performance of other green infrastructures, such as bioswales and raingardens, by adding adequate control of soil moisture thanks to their evapotranspiration, as well as contributing with other environmental benefits such as providing shade, noise control, and mitigating pollution of soil and groundwater by phytoremediation (Berland et al. 2017).

It is possible to regulate the evapotranspiration of urban trees, namely their water requirements, by selecting the best adapted tree species for the respective site environment. This also benefits microclimate cooling and thermal comfort, as was shown in previous sections. As reported by Berland et al. (2017), daily water consumption by urban trees is highly variable according to the species, ranging from 3 kg $tree^{-1}day^{-1}$ in *Pinus canariensis*, a drought tolerant conifer, to 177 kg $tree^{-1}day^{-1}$ in hybrid *Platanus x acerifolia*, a water-consuming deciduous tree. Remarkably, large intra-specific variation in transpiration has been observed in urban tree and shrub species, providing scope for genetic breeding of urban vegetation for adaptation to climate change and environmental stresses.

The vegetation covering the building's roof may also help to mitigate the risk of urban flooding by increasing detention (delayed runoff) and retention (water released from vegetation by evapotranspiration) of stormwater. Blue-green roofs have been developed in Nordic countries and in other parts of the world as layered substrate infrastructure used for rainwater detention, with live vegetation overtop as part of a stormwater management strategy (Andenæs et al. 2021). Vegetated roofs can play an important role in warmer seasons, and in hot climatic regions, in mitigating the risk of building overheating. Andenæs et al. (2021) report from a large survey conducted in Nothern Europe, however, that there are special requirements for constructing and maintaining this type of green infrastructure. Amongst the main construction and maintenance challenges associated with blue-green roofs is water intrusion into the roof structure, drainage and drains functioning, fire protection, structural loads, and wind. Although technical risks associated with blue-green roofs are numerous, they are manageable provided that building projects comply with a framework of recommendations drafted according to scientific investigations (Andenæs et al. 2021).

The wide array of ecosystem services offered by green infrastructures to urban areas for better adaptation to climate change are, conversely, jeopardised by global warming impacts on functioning vegetation. Under extreme drought, as has occurred in the last few years in Central Europe and in other parts of the world, C-sequestration can be reduced up to 50%, and microclimate cooling can decrease 50–70%, depending on the drought tolerance and water use efficiency of various urban tree species (Rötzer et al. 2021). Extensive damage to urban trees and shrubs are likely to occur, however, during heat waves and drought spells, as illustrated in Germany in 2020 or in Australian during the Millennium Drought from 2001 to 2009. Extensive analyses of tree damages in cities in these countries indicated that moderate to extreme damage, with crown defoliation and branch dieback, could affect 30–75% of a city's trees depending on the tree species (Haase and Hellwig 2022). In Canberra (Australia), similarly, it was observed that the percentage of healthy trees in the city's streets and parks decreased from 80% to 37%, over

400,000 public urban trees, in the last 20 years compared to 1980–1999 (Zhang and Brack 2021). It is crucial to select the appropriate species of trees, shrubs, and other plants, for the urban green spaces, based on present city environmental conditions, but also taking into account the changing climate. Research is currently very active in this field to identify a wide array of plant material, mostly native but also exotic species, for the various needs of ecosystem services in urban areas at global scale, such as in China (Liu et al. 2021), Latin America (Guillen-Cruz et al. 2021), Australia (Marchin et al. 2022), and Europe (Schütt et al. 2022).

Considering climate change impacts, research on urban vegetation is increasingly focusing on the need to reduce the consumption of blue water for irrigating urban trees, parks, and lawns, which presently can account for 50–60% of the overall water consumption in temperate climate cities (Nouri et al. 2019). In the future there will be increasing need for implementing strategies for conservation of water resources in urban areas by promoting water harvest with rain gardens, bioswales and water pools, whilst reducing water consumption with appropriate selection of planting material and informed management of vegetation and urban forests.

Forest fires linked to climate change are also an increasing risk to cities, when hot temperatures, severe drought, excessive fuel load in green spaces, and the extension of interfaces between urban vegetation and residential areas cause a dramatic increase in the vulnerability of infrastructure, buildings, and people (Price and Bradstock 2014). Careful urban planning, appropriate infrastructure design, and active management of urban forests and other green spaces, will highlight intense management cooperation and community involvement to help mitigate these risks.

6 Green Mobility and Greener Urban Landscapes

Transportation accounts for 23% of all carbon emissions worldwide, making a large contribution to climate change and its impacts (Cepeliauskaite et al. 2021). With the European Green Deal the EU committed to a 90% reduction in transport-related GHG emissions by 2050 compared with 1990 levels, it will require a major transformative change of transportation modes towards more sustainable approaches (Tsavachidis and Le Petit 2022). Urban mobility makes a large contribution to overall transport energy consumption, as it accounts for 40% of all CO_2 emissions and is responsible for 70% of pollutants produced from transportation (Cepeliauskaite et al. 2021). It also represents one of the major causes of health problems for society, as described in the section on air pollutants in this chapter, and in chapter "BioCities as Promotors of Health and Wellbeing".

A series of studies conducted worldwide demonstrate that it is possible to adopt policy measures, city planning approaches, and innovative technological solutions, to tackle the challenge of reducing the impact of transport on climate change. In the case of Auckland (New Zealand), policies that promote public transport over private vehicles through road pricing have been found to reduce total emissions by 40% (Cepeliauskaite et al. 2021). Providing multiple travel options through multimodality, however, is the key choice to foster urban mobility transition (Tsavachidis and Le Petit 2022). A wide study conducted in seven major European cities clearly showed that moving from the private mobility model, mainly based on private car transport, towards green sustainable mobility approaches, based on walking, cycling, e-biking and public transport, would greatly reduce the C-footprint of urban transportation. The per capita emissions in the EU would drop from 1.2 tons CO_2 per year to almost zero, corresponding to about 50% of the overall GHG emissions reduction planned for transport by 2050 (Brand et al. 2021). Green mobility would also benefit greatly from applications of Information Technology to enhance integrated mobility, to improve infrastructure and services for transport system operators and users, and to provide data generation and sharing for decision-making (Cepeliauskaite et al. 2021). The transition towards green and active mobility is a fundamental aspect of urban area transformation into BioCities, and will require new approaches for urban planning for enhancing links between travel, land use, and urban form (Banister 2011). Special attention should be paid to integrate thoughtful landscape planning and design with the greening of infrastructures for active and sustainable human mobility. Urban trees, forests, meadows, and raingardens represent an attractive opportunity to implement ecological connectivity whilst improving environmental quality in our cities (Dall'Ara et al. 2019).

7 Outcomes and Concluding Remarks

The capacity of green infrastructures to provide a wide array of ecosystem services, from C-sequestration to thermal comfort to runoff regulation, has been widely recognised. Benefits can be highly variable, however, depending on plant species, canopy cover, geometry of green spaces, and on prevailing meteorological conditions in city environments, including exposure to sun and wind. Furthermore, the urban landscape matrix is often neglected in existing studies (Yu et al. 2020). Whilst urban forests are the most powerful air conditioning systems that cities can provide for their present and the future, it is doubtless that more research is needed to provide effective guidance to optimise landscape design and maximise the benefits delivered by green infrastructures.

Urban trees are vital for their capacity to mitigate the UHI effect, yet they too are impacted by a warming climate and UHI. Climate change, along with the global increase of trade and transport, is also leading to a spread of pests and plant diseases, probably at a rate faster than natural selection and adaptation in vegetation can cope. Adapting tree and shrub species composition of urban forests to focus on climate urban goals is important to ensure future functioning, but it is often ignored by urban planning. Therefore, the species selection process should consider the needs, structure and function of the modern BioCity at the time those trees will mature (e.g. 30–60 years) rather than for present conditions.

Most studies focus on the effect of urban forests on land surface temperature (LST) that are directly derived from thermal remote sensing satellite images, whilst only a few pay attention to urban air temperatures. Arnfield (2003) explains that there is no consistent relationship between the two. Under calm weather conditions, LST is more dependent on microscale site characterisations. Urban air temperatures, however, are more closely related to human health and comfort, and thus deserve at least an equal amount of attention. In the BioCity, a more effective and intensive monitoring of the urban ecosystem should require the development of networks of low-cost sensors (i.e. Internet-of-Things), which offer an important opportunity for adaptive design and management of urban landscape. Environmental modelling experiments are able to indicate to urban planners the optimal designs to be implemented to achieve the overall resilience objectives of the city. Finally, the focus of many studies concentrates on the cooling effect of urban forests, but only a few studies have made the link with heat-related health benefits whose importance for society is rapidly growing.

Roadside trees and urban forests are able to mitigate pollution at both local and regional levels. Whilst some tree species may be ideal candidates for reforestation projects in peri-urban areas, they may not be suitable for urban parks and street trees due to their species-specific environmental limitations, and they could generate disservices, such as releasing pollen or Volatile Organic Compounds. Matching the right species for the right place, therefore, is an important issue that requires expert evaluation.

The selection of the most appropriate plant species for urban green spaces requires evaluating tree species' tolerance to pollution. Recent work by Fares et al. (2019) suggests that ozone alone is responsible for a decrease of up to 5% of carbon assimilation of peri-urban Holm oak forest in Rome. More research is needed to disentangle the effects of pollutants from other environmental stressors, such as drought stress, which is often coupled to high levels of pollutants in Mediterranean cities.

An underestimated factor affecting urban air quality is soils. Most city soils are underneath pavement or buildings (e.g. sealed), which keeps them from providing natural ecosystem services. Making sealed soils more permeable, combined with creating more vegetative cover, could contribute to increased absorption of pollutants and their degradation in urban soils.

Finally, more studies are needed to develop a more realistic model ensemble to predict ecosystem services provided by urban trees, shrubs, and green infrastructures, in general. The development of such models requires a joint and intense collaboration of interdisciplinary research groups spanning plant ecophysiology, social sciences, and landscape architecture. Indeed, there is a need to development of 3-D spatially explicit models with an open access approach, and the capacity to utilise species-specific parameters. Ideally, these models will be embedded into a user-friendly Decision Support System for stakeholders from various backgrounds. Policymakers and practitioners need to know how to design vegetation projects under the different pollution scenarios and where to locate filtering vegetation in the cityscape, to maximise PM absorption and to divert pollution-carrying air streams away from where people are habituating.

References

Andenæs E, Time B, Muthanna T, Asphaug S, Kvande T (2021) Risk reduction framework for blue-green roofs. Buildings 11:185. https://doi.org/10.3390/buildings11050185

Aram F, Higueras García E, Solgi E, Mansournia S (2019) Urban green space cooling effect in cities. Heliyon 5(4):01339

Arnfield AJ (2003) Two decades of urban climate research: a review of turbulence, exchanges of energy and matter, and the urban heat Island. Int J Climatol 23:1–26

ASLA (2022) Climate change mitigation: cities. The American Society of Landscape Architects. www.asla.org/mitigationurban.aspx

Banister D (2011) Cities, mobility and climate change. Concepts and policy framework

Baró F, Chaparro L, Gomez-Baggethun E, Langemeyer J, David J, Terradas J (2014) Contribution of ecosystem services to air quality and climate change mitigation policies: the case of urban forests in Barcelona, Spain. AMBIO 43:466–479

Barwise Y, Kumar P (2020) Designing vegetation barriers for urban air pollution abatement: a practical review for appropriate plant species selection. Clim Atmos Sci 3:1–19

Berland A, Shiflett SA, Shuster WD, Garmestani AS, Goddard HC, Herrmann DL, Hopton ME (2017) The role of trees in urban stormwater management. Landsc Urban Plan 162:167–177

Bottalico F, Travaglini D, Chirici G, Garfì V, Giannetti F, De Marco A, Fares S, Marchetti M, Nocentini S, Paoletti E, Salbitano F, Sanesi G (2017) A spatially-explicit method to assess the dry deposition of air pollution by urban forests in the city of Florence, Italy. Urban Forestry Urban Greening 27:221. https://doi.org/10.1016/j.ufug.2017.08.013

Brand C, Dons E, Anaya-Boig E, Avila-Palencia I, Clark A, de Nazelle A, Gascon M, Gaupp-Berghausen M, Gerike R, Götschi T, Iacorossi F, Kahlmeier S, Laeremans M, Nieuwenhuijsen MJ, Ojuela JP, Racioppi F, Raser E, Rojas-Rueda D, Standaert A, Stigell E, Sulikova S, Wegener S, Panis LI (2021) The climate change mitigation effects of daily active travel in cities. Transp Res D 93:102764

Campbell S, Remenyi TA, White CJ, Johnston FH (2018) Heatwave and health impact research: a global review. Health Place 53:210–218

Cao S, Wang Y, Ni Z, Xia B (2022) Effects of blue-green infrastructures on the microclimate in an urban residential area under hot weather. Front Sustain Cities 4:824779. https://doi.org/10.3389/frsc.2022.824779

Cepeliauskaite G, Keppner B, Simkute Z, Stasiskiene Z, Leuser L, Kalnina I, Kotovica N, Andin SJ, Muiste M (2021) Smart-mobility services for climate mitigation in urban areas: case studies of Baltic countries and Germany. Sustainability 13:4127

Clifton OE, Fiore AM, Massman WJ, Baublitz CB, Coyle M, Emberson L, Fares S, Farmer DK, Gentine P, Gerosa G, Guenther AB, Helmig D, Lombardozzi DL, Munger JW, Patton EG, Pusede SE, Schwede DB, Silva SJ, Sörgel M, Steiner AL, Tai APK (2020) Dry deposition of ozone over land: processes, measurement, and modeling. Rev Geophys 58:e2019RG000670

Coutts AM, White EC, Tapper NJ, Beringer J, Livesley SJ (2016) Temperature and human thermal comfort effects of street trees across three contrasting street canyon environments. Theor Appl Climatol 124:55–68

Dall'Ara E, Maino E, Gatta G, Torreggiani D, Tassinari P (2019) Green mobility infrastructures: a landscape approach for roundabouts' gardens applied to an Italian case study. Urban Forestry Urban Greening 37:109–125

Davies ZG, Edmondson JL, Heinemeyer A, Leake JR, Gaston KJ (2011) Mapping an urban ecosystem service: quantifying above-ground carbon at a city-wide scale. J Appl Ecol 48: 1125–1134. https://doi.org/10.1111/j.1365-2664.2011.02021.x

Dusart N, Gérard J, Le Thiec D, Collignon C, Jolivet Y, Vaultier MN (2019) Integrated analysis of the detoxification responses of two Euramerican poplar genotypes exposed to ozone and water deficit: focus on the ascorbate-glutathione cycle. Sci Total Environ 651:2365–2379

EEA (2019) The European environment—state and outlook 2020: Knowledge for transition to a sustainable Europe, pp 499

EEA (2021) Air quality in Europe 2021 report. European Environmental Agency, Copenhagen
Fares S, Scarascia-Mugnozza G, Corona P, Palahí M (2015) Five steps for managing Europe's forests. Nature 519:407–409
Fares S, Savi F, Fusaro L, Conte A, Salvatori E, Aromolo R, Manes F (2016) Particle deposition in a peri-urban Mediterranean forest. Environ Pollut 1278–1286. https://doi.org/10.1016/j.envpol.2016.08.086
Fares S, Alivernini A, Conte A, Maggi F (2019) Ozone and particle fluxes in a Mediterranean forest predicted by the AIRTREE model. Sci Total Environ 682:494–504. https://doi.org/10.1016/j.scitotenv.2019.05.109
Fares S, Conte A, Alivernini A, Chianucci F, Grotti M, Zappitelli I, Petrella F, Corona P (2020) Testing removal of carbon dioxide, ozone, and atmospheric particles by urban parks in Italy. Environ Sci Technol 54:14910–14922. https://doi.org/10.1021/acs.est.0c04740
Favero A, Daigneault A, Sohngenet B (2020) Forests: carbon sequestration, biomass energy, or both? Sci Adv 6:eaay6792
Gamero-Salinas J, Monge-Barrio A, Kishnani N, López-Fidalgo J, Sánchez-Ostiz A (2021) Passive cooling design strategies as adaptation measures for lowering the indoor overheating risk in tropical climates. Energy Buildings 252:111417
García de Jalón S, Burgess PJ, Curiel Yuste J, Moreno G, Graves A, Palma JHN, Crous-Duran J, Kay S, Chiabai A (2019) Dry deposition of air pollutants on trees at regional scale: a case study in the Basque Country. Agric For Meteorol 278:107648
Georgiadis T (2019) Role of climate and city pattern (Chapter 3). In: Fabbri K (ed) Urban fuel poverty. Academic Press, pp 41–62
Godbold DL, Hoosbeek MR, Lukac M, Cotrufo MF, Janssens IA, Ceulemans R, Polle A, Velthorst EJ, Scarascia-Mugnozza G, De Angelis P, Miglietta F, Peressotti A (2006) Mycorrhizal hyphal turnover as a dominant process for carbon input into soil organic matter. Plant Soil 281:15–24
Gratani L, Varone L (2005) Daily and seasonal variation of CO_2 in the city of Rome in relationship with the traffic volume. Atmos Environ 39:2619–2624
Guillen-Cruz G, Rodríguez-Sanchez AL, Fernandez-Luqueño F, Flores-Rentería D (2021) Influence of vegetation type on the ecosystem services provided by urban green areas in an arid zone of northern Mexico. Urban Forestry Urban Greening 62:127135
Haase D, Hellwig R (2022) Effects of heat and drought stress on the health status of six urban street tree species in Leipzig, Germany. Trees, Forests People 8:100252
Han D, Shen H, Duan W, Chen L (2020) A review on particulate matter removal capacity by urban forests at different scales. Urban Forestry Urban Greening 48:126565
Hardy JP, Melloh R, Koenig G, Marks D, Winstral A, Pomeroy JW, Link T (2004) Solar radiation transmission through conifer canopies. Agric For Meteorol 126:257–270
Henninger S (2008) Analysis of near surface CO_2 variability within the urban area of Essen, Germany. Meteorol Z 17:19–27
Hirons M, Beauchamp E, Whitfield S, Conway D, Asare R, Malhi Y (2020) Resilience to climate shocks in the tropics. Environ Res Lett 15:100203
IEA (2018) The future of cooling: opportunities for energy-efficient air conditioning, pp 98. moien.lu/wp-content/uploads/2018/08/The_Future_of_Cooling.pdf. Accessed 5 Mar 2021
Jim CY, Chen WY (2009) Ecosystem services and valuation of urban forests in China. Cities 26: 187–194
Kuang W, Liu A, Dou Y, Li G, Lu D (2019) Examining the impacts of urbanisation on surface radiation using Landsat imagery. GIScience Remote Sensing 56(3):462–484. https://doi.org/10.1080/15481603.2018.1508931
Leal Filho W, Echevarria Icaza L, Neht A, Klavins M, Morgan EA (2018) Coping with the impacts of urban heat islands: a literature based study on understanding urban heat vulnerability and the need for resilience in cities in a global climate change context. J Clean Prod 171:1140e1149
Leskinen P, Cardellini G, González-García S, Hurmekoski E, Sathre R, Seppälä J, Smyth C, Stern T, Verkerk PJ (2018) Substitution effects of wood-based products in climate change mitigation. From Science to Policy 7. European Forest Institute. https://doi.org/10.36333/fs07

Liberloo M, Lukac M, Calfapietra C, Hoosbeek MR, Gielen B, Miglietta F, Scarascia-Mugnozza GE, Ceulemans R (2009) Coppicing shifts CO_2 stimulation of poplar productivity to above-ground pools: a synthesis of leaf to stand level results from the POP/EUROFACE experiment. New Phytol 182:331–346. https://doi.org/10.1111/j.1469-8137.2008.02754.x

Liu C, Li X (2012) Carbon storage and sequestration by urban forests in Shenyang, China. Urban Forestry Urban Greening 11:121–128. https://doi.org/10.1016/j.ufug.2011.03.002

Liu M, Zhang D, Pietzarka U, Roloff A (2021) Assessing the adaptability of urban tree species to climate change impacts: a case study in Shanghai. Urban Forestry Urban Greening 62:127186

Marchin RM, Esperon-Rodriguez M, Tjoelker MGG, Ellsworth DS (2022) Crown dieback and mortality of urban trees linked to heatwaves during extreme drought. Sci Total Environ 850: 157915

Masson-Delmotte V, Zhai P, Pirani A, Connors SL, Péan C, Berger S, Caud N, Chen Y, Goldfarb L, Gomis MI, Huang M, Leitzell K, Lonnoy E, Matthews JBR, Maycock TK, Waterfield T, Yelekçi O, Yu R, Zhou B (eds) (2021) Climate change 2021: the physical science basis. Contribution of Working Group I to the Sixth Assessment Report of the Intergovernmental Panel on Climate Change. Cambridge University Press, Cambridge, pp 33–144. https://doi.org/10.1017/9781009157896.002

McHale M, Pickett S, Barbosa O, Bunn DM, Cadenasso ML, Childers DL, Gartin M, Hess GR, Iwaniec DM, McPhearson T, Peterson MN, Poole AK, Rivers L, Shutters ST, Zhou W (2015) The new global urban realm: complex, connected, diffuse, and diverse social-ecological systems. Sustainability 7:5211–5240. https://doi.org/10.3390/su7055211

Morakinyo TE, Balogun AA, Adegun OB (2013) Comparing the effect of trees on thermal conditions of two typical urban buildings. Urban Clim 3:76–93

Nemitz E, Vieno M, Carnell E, Fitch A, Steadman C, Cryle P, Holland M, Morton RD, Hall J, Mills G, Hayes F, Dickie I, Carruthers D, Fowler D, Reis S, Jones L (2020) Potential and limitation of air pollution mitigation by vegetation and uncertainties of deposition-based evaluations: air pollution mitigation by vegetation. Philos Trans R Soc A Math Phys Eng Sci 378:20190320. https://doi.org/10.1098/rsta.2019.0320

Nouri H, Chavoshi Borujeni S, Hoekstra AY (2019) The blue water footprint of urban green spaces: an example for Adelaide, Australia. Landsc Urban Plann 190:103613

Nowak DJ, Crane DE (2002) Carbon storage and sequestration by urban trees in the USA. Environ Pollut 116:381–389

Nowak DJ, Greenfield EJ, Hoehn RE, Lapoint E (2013) Carbon storage and sequestration by trees in urban and community areas of the United States. Environ Pollut 178:229–236. https://doi.org/10.1016/j.envpol.2013.03.019

OECD (Organisation for economic co-operation and development) (2014) The cost of air pollution health impacts of road transport

Pandit R, Laband DN (2010) Energy savings from tree shade. Ecol Econ 69:1324–1329

Park M, Joo SJ, Lee CS (2013) Effects of an urban park and residential area on the atmospheric CO_2 concentration and flux in Seoul, Korea. Adv Atmos Sci 30:503–514

Phillips TH, Baker ME, Lautar K, Yesilonis I, Pavao-Zuckerman MA (2019) The capacity of urban forest patches to infiltrate stormwater is influenced by soil physical properties and soil moisture. J Environ Manag 246:11–18

Piao S, Sitch S, Ciais P, Friedlingstein P, Cong NAN, Huntingford C, Jung M (2013) Evaluation of terrestrial carbon cycle models for their response to climate variability and to CO_2 trends. Glob Chang Biol 19:1–16. https://doi.org/10.1111/gcb.12187

Piracha A, Chaudhary MT (2022) Urban air pollution, urban heat Island and human health: a review of the literature. Sustainability 14:9234. https://doi.org/10.3390/su14159234

Price O, Bradstock R (2014) Countervailing effects of urbanisation and vegetation extent on fire frequency on the wildland urban interface: disentangling fuel and ignition effects. Landsc Urban Plan 130:81–88

Purcell L (2021) Purdue landscape report: how do trees use water? Extension Forestry & Natural Resources, Purdue University. https://www.purdue.edu/fnr/extension/purdue-landscape-report-how-do-trees-use-water/

Raciti SM, Hutyra LR, Newell JD (2014) Mapping carbon storage in urban trees with multisource remote sensing data: relationships between biomass, land use, and demographics in Boston neighborhoods. Sci Total Environ 500:72–83. https://doi.org/10.1016/j.scitotenv.2014.08.070

Ramachandra TV, Aithal BH, Sreejith K (2015) GHG footprint of major cities in India. Renew Sust Energ Rev 44:473–495

Robine JM, Cheung SLK, Le Roy S, Van Oyen H, Griffiths C, Michel JP, Herrmann FR (2008) Death toll exceeded 70,000 in Europe during the summer of 2003. C R Biol 331:171–178

Rötzer T, Moser-Reischl A, Rahman MA, Hartmann C, Paeth H, Pauleit S, Pretzsch H (2021) Urban tree growth and ecosystem services under extreme drought. Agric For Meteorol 308–309: 108532

Saaroni H, Amorim JH, Hiemstra JA (2018) Urban green infrastructure as a tool for urban heat mitigation: survey of research methodologies and findings across different climatic regions. Urban Clim 24:94–110

Sæbø A, Popek R, Nawrot B, Hanslin HM, Gawronska H, Gawronski SW (2012) Plant species differences in particulate matter accumulation on leaf surfaces. Sci Total Environ 427–428:347–354

Santamouris M, Cartalis C, Synnefa A, Kolokotsa D (2015) On the impact of urban heat Island and global warming on the power demand and electricity consumption of buildings—a review. Energ Buildings 98:119–124

Sathre R, O'Connor J (2010) Meta-analysis of greenhouse gas displacement factors of wood product substitution. Environ Sci Policy 13:104–114. https://doi.org/10.1016/j.envsci.2009.12.005

Scarascia-Mugnozza G, Matteucci G (2014) The impact of temperature and drought on the carbon balance of forest ecosystems: the case of a beech forest in Central Italy. Agrochimica 58:34–39

Schraufnagel DE, Balmes JR, Cowl CT, De Matteis S, Jung SH, Mortimer K, Perez-Padilla R, Rice MB, Riojas-Rodriguez H, Sood A, Thurston GD, To T, Vanker A, Wuebbles DJ (2019) Air pollution and noncommunicable diseases: a review by the forum of international respiratory societies' environmental committee, part 1: the damaging effects of air pollution. Chest 155: 409–416

Schütt A, Becker JN, Gröngröft A, Schaaf-Titel S, Eschenbach A (2022) Soil water stress at young urban street-tree sites in response to meteorology and site parameters. Urban Forestry Urban Greening 75:127692

Schwaab J, Meier R, Mussetti G, Seneviratne S, Bürgi C, Davin EL (2021) The role of urban trees in reducing land surface temperatures in European cities. Nat Commun 12:6763. https://doi.org/10.1038/s41467-021-26768-w

Scott K, Simpson JR, McPherson EG (1999) Effects of tree cover on parking lot microclimate and vehicle emissions. J Arboric 25(3)

Sethi M, Lamb W, Minx J, Creutzig F (2020) Climate change mitigation in cities: a systematic scoping of case studies. Environ Res Lett 15:093008

Shindell D, Zhang Y, Scott M, Ru M, Stark K, Ebi KL (2020) The effects of heat exposure on human mortality throughout the United States. GeoHealth 3. https://doi.org/10.1029/2019GH000234

Tang YJ, Chen AP, Zhao SQ (2016) Carbon storage and sequestration of urban street trees in Beijing, China. Front Ecol Evol 4:53. https://doi.org/10.3389/fevo.2016.00053

Tiwari M, Sahu SK, Rathod T, Bhangare RC, Ajmal PY, Vinod Kumar A (2020) Measurement of size-fractionated atmospheric particulate matter and associated polycyclic aromatic hydrocarbons in Mumbai, India, and their dry deposition fluxes. Air Qual Atmos Health 13:939–949

Tsavachidis M, Le Petit Y (2022) Re-shaping urban mobility - key to Europe's green transition. J Urban Mobility 2:100014

United Nations DESA (2022) Goal 11: sustainable cities and human settlements. sdgs.un.org/topics/sustainable-cities-and-human-settlements

USDA-Forest Service (2022) I-Tree Tools, Forest and Community Trees. www.itreetools.org/about

Wang Y (2016) The effect of urban green infrastructure on local microclimate and human thermal comfort. PhD thesis, Wageningen University, NL

Wang C, Ren Z, Dong Y, Zhang P, Guo Y, Wang W, Bao G (2022) Efficient cooling of cities at global scale using urban green space to mitigate urban heat Island effects in different climatic regions. Urban Forestry Urban Greening 74:127635

WHO (2018) Climate Change and Health. World Health Organisation, Geneva. www.who.int/news-room/fact-sheets/detail/climate-change-and-health. Accessed 22 Feb 2020

WHO (2021) WHO global air quality guidelines. Particulate matter (PM2.5 and PM10), ozone, nitrogen dioxide, sulfur dioxide and carbon monoxide. World Health Organisation, Geneva

Wilkes P, Disney M, Boni Vicari M, Calders K, Burt A (2018) Estimating urban above ground biomass with multi-scale LiDAR. Carbon Balance Manag 13:10

Xing Y, Brimblecombe P (2020) Trees and parks as "the lungs of cities". Urban Forestry Urban Greening 48:126552

Young DT, Clark P, Hendry M, Barlow J (2015) Modelling radiative exchange in a vegetated urban street canyon model. ICUC9 - 9th international conference on urban climate jointly with 12th symposium on the urban environment. www.meteo.fr/icuc9/LongAbstracts/nomtm1-7-5891008_a.pdf

Yu Z, Yang G, Zuo S, Jørgensen G, Koga M, Vejre H (2020) Critical review on the cooling effect of urban blue-green space: a threshold-size perspective. Urban Forestry Urban Greening 49:126630

Zhang B, Brack CL (2021) Urban forest responses to climate change: a case study in Canberra. Urban For Urban Green 57:126910

Zhao SQ, Zhu C, Zhou DC, Huang D, Werner J (2013) Organic carbon storage in China's urban areas. PLoS One 8:e71975. https://doi.org/10.1371/journal.pone.0071975

Zupancic T, Westmacott C, Bulthuis M (2015) The impact of green space on heat and air pollution in urban communities: a mete-narrative systematic review. davidsuzuki.org/wp-content/uploads/2017/09/impact-green-space-heat-air-pollution-urban-communities.pdf. Accessed 5 Mar 2021

BioCities as Promotors of Health and Well-being

Mònica Ubalde-López, Mark Nieuwenhuijsen, Giuseppina Spano, Giovanni Sanesi, Carlo Calfapietra, Alice Meyer-Grandbastien, Liz O'Brien, Giovanna Ottaviani Aalmo, Fabio Salbitano, Jerylee Wilkes-Allemann, and Payam Dadvand

1 Introduction

The mainstream public health community often treats the natural environment with ambivalence. On one side, there are infectious agents, extreme weather, and catastrophic events such as floods, landslides, wildfires, storms, and earthquakes that directly or indirectly sicken, injure, or kill people (Hartig et al. 2014). On the other hand, human health is positively connected with the characteristics and quality of nature near to where people live. This ambivalence becomes crucial in cities where the living environment has peculiar characteristics both for humans and other living

M. Ubalde-López (✉) · M. Nieuwenhuijsen · P. Dadvand
Barcelona Institute for Global Health (ISGlobal), Barcelona, Spain
e-mail: monica.ubalde@isglobal.org

G. Spano · G. Sanesi
Università degli Studi di Bari, Bari, Italy

C. Calfapietra
National Research Council of Italy - Institute of Terrestrial Ecosystems Research, Monterotondo, Italy

A. Meyer-Grandbastien
Plante & Cité, Université de Rennes, CNRS, Rennes, France

L. O'Brien
Society and Environment Research Group, Forest Research, Surrey, UK

G. Ottaviani Aalmo
Norwegian Institute of Bioeconomy Research (NIBIO), Ås, Norway

F. Salbitano
University of Sassari, Sassari, Italy

J. Wilkes-Allemann
Bern University of Applied Sciences (BFH), Bern, Switzerland

G. E. Scarascia-Mugnozza et al. (eds.), *Transforming Biocities*, Future City 20,
https://doi.org/10.1007/978-3-031-29466-2_6

organisms. Indeed, there are many ways in which the urban environment can affect human health, positively or negatively. BioCities develop as dynamic socio-ecological systems hosted by nature. Therefore, addressing the issue of health according to an integrated and holistic approach, which reduces the negative effects of the natural environment and optimises its positive aspects, is a primary pillar in the construction of BioCities.

Two global approaches to health based on a more complex and sound understanding of human health and environmental health relationships have emerged in the last two decades: *One Health* and *EcoHealth* (Harrison et al. 2019). These two paradigms "*…posit that the epidemiological dynamics and stakeholders' actions that determine the health of animal and human populations need to be studied in their interconnected ecological, socioeconomic, and political contexts*" (Roger et al. 2016).

One Health is an approach that recognises that people's health is closely linked to the health of ecosystems and the characteristics of our shared environment. One Health is not a new concept. It has been developed and systematised in recent years in light of the fact that many physical, biotic, and social factors have changed the interactions between people, plants, animals, and our environment, with particular regard to the urban environment. It is a holistic approach aiming to understand the complex effects of these interactions in order to simultaneously improve human and environmental health.

Ecohealth is a field of research, education, and practice that is inspired by systemic approaches to promote the health of people and ecosystems in the multiscalar complex of social and ecological interactions. Health is seen as social, mental, spiritual, and physical well-being and not simply the absence of disease, as defined by the constitution of the World Health Organisation (WHO) in 1946 (IHC 2002).

The One Health approach is oriented towards biomedical issues, with an initial emphasis on zoonoses, or germ-caused diseases, and it is historically ascribed to the health sciences. On the other hand, the EcoHealth framework is defined as an ecosystemic approach to health, tending to focus on environmental and socioeconomic issues and initially developed by biologists and disease ecologists working in the field of biodiversity conservation, as a key aspect for the improvement of health (Roger et al. 2016). These two concepts are permeating the following discourses on health in BioCities. The two tend to overlap and converge as transdisciplinary approaches. According to the drivers and indicators developed to support the application of One Health framework (Box 1, Zhang et al. 2022), this chapter will focus on key topics relating to human health and urban environment. In terms of key issues and indicators, a special focus will be oriented to understand how nature in cities, namely green infrastructure components and ecosystem services, could improve interrelated positive effects.

Box 1 Drivers and indicators for One Health assessment

In order to assess the potential of One Health approach, Zhang et al. (2022) proposed a global One Health index based on drivers and indicators. They identified three categories of drivers (External, Intrinsic, and Core) and then associated key indicators and applied indicators. The index concerns the complex range of issues contributing to One Health improvement using 13 key indicators and 57 indicators.

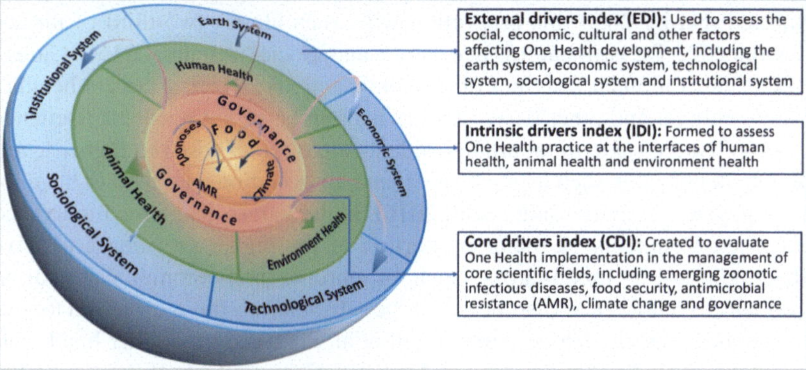

DRIVER	Key indicator	Indicator
A External	A1 Earth system	A1.1 Land
		A1.2 Forest
		A1.3 Water
		A1.4 Air
		A1.5 Natural disasters
	A2 Institutional system	A2.1 Justice
		A2.2 Governance
	A3 Economic system	A3.1 Finance
		A3.2 Work
		A3.3 Housing
	A4 Sociological system	A4.1 Demography
		A4.2 Education
		A4.3 Inequalities
	A5 Technological system	A5.1 Transport
		A5.2 Technology adoption
		A5.3 Consumption and production
B Intrinsic	B1 Human health	B1.1 Reproductive, maternal,
		B1.2 Infectious diseases
		B1.3 Non-communicable diseases and
		B1.4 Injuries and violence
		B1.5 Universal health coverage and
		B1.6 Health risk
	B2 Animal health and	B2.1 Animal epidemic diseases
		B2.2 Animal welfare
		B2.3 Animal nutritional status
		B2.4 Animal biodiversity
	B3 Environmental health	B3.1 Air quality and climate change
		B3.2 Land resources
		B3.3 Sanitation and water resources
		B3.4 Hazardous chemicals
		B3.5 Environmental biodiversity

DRIVER	Key indicator	Indicator
C Core	C1 Governance	C1.1 Participation
		C1.2 Rule of law
		C1.3 Transparency
		C1.4 Responsiveness
		C1.5 Consensus oriented
		C1.6 Equity and inclusiveness
		C1.7 Effectiveness and efficiency
		C1.8 Political support
	C2 Zoonotic diseases	C2.1 Source of infection
		C2.2 Route of transmission
		C2.3 Targeted population
		C2.4 Capacity building
		C2.5 Outcomes (case-studies)
	C3 Food security	C3.1 Food demand and supply
		C3.2 Food safety
		C3.3 Nutrition
		C3.4 Natural and social circumstances
		C3.5 Government support and
	C4 Antimicrobial	C4.1 AMR surveillance system
		C4.2 AMR laboratory network and
		C4.3 Antimicrobial control and
		C4.4 Improve awareness and
		C4.5 AMR rate for important
	C5 Climate change	C5.1 Government response
		C5.2 Climate change risks
		C5.3 Health outcome

Adapted from Zhang et al. 2022.

1.1 Public Health and the Urban Environment

Public health practice aims to maintain and improve the health of human populations based on the principles of social justice, attention to human rights and equity, evidence-informed policy and practice using the underlying determinants of health (Committee of Inquiry into the Future Development of the Public Health Function 1988). Natural environments can make an important contribution to maintaining and improving public health.

Currently, more than half of the global population is living in urban settings and it is projected that by 2050 this proportion will reach almost two-third of the world population (UN Department of Economic and Social Affairs 2015). Cities are sources of innovation and engines of economic activity, where access to healthcare, education, culture, and other basic services is often better for residents (Bettencourt et al. 2007). At the same time, urban residents are often exposed to higher levels of environmental hazards such as noise, heat, and air pollution, and tend to have lower levels of physical activity (Sallis et al. 2016), higher stress, and have limited access to natural environments. A major proportion of the higher prevalence of adverse health conditions in urban areas, such as chronic non-communicable diseases (NCDs) and psychological disorders, can be attributed to these urban-related environmental and lifestyle determinants (Cyril et al. 2013). On the other hand, natural environments including green spaces, have been shown to buffer adverse health effects of urban living by improving mental and physical health and well-being (Nieuwenhuijsen et al. 2017).

Considering the many benefits of green spaces, the health of residents in urban settings can be improved by increasing the amount and accessibility of natural environment and enhancing its quality (Nieuwenhuijsen et al. 2017; van den Bosch and Nieuwenhuijsen 2017). In this context, cities can be made healthier and more equitable for its residents by developing and enhancing its green infrastructure, such as having a park or urban forest close to where people live, planting trees in the streets, and introducing green roofs and urban gardens. Urban gardens, if implemented at a sufficiently large scale, have the additional benefits of local food production that contribute to more sustainable and self-efficient cities.

1.2 Conceptual Framework

For this chapter, we adapted the conceptual framework linking human health to the green space component of BioCities from the ones developed by Hartig et al. (2014) (Nieuwenhuijsen et al. 2017). The proposed framework helps to understand how the relationship between contact with the natural environment and health is mediated through several possible mechanisms: the enrichment of human microbiome and promotion of immune balance; the potential inhibition of cell signalling by the exposure to mixtured natural compounds; improving air quality, physical activity,

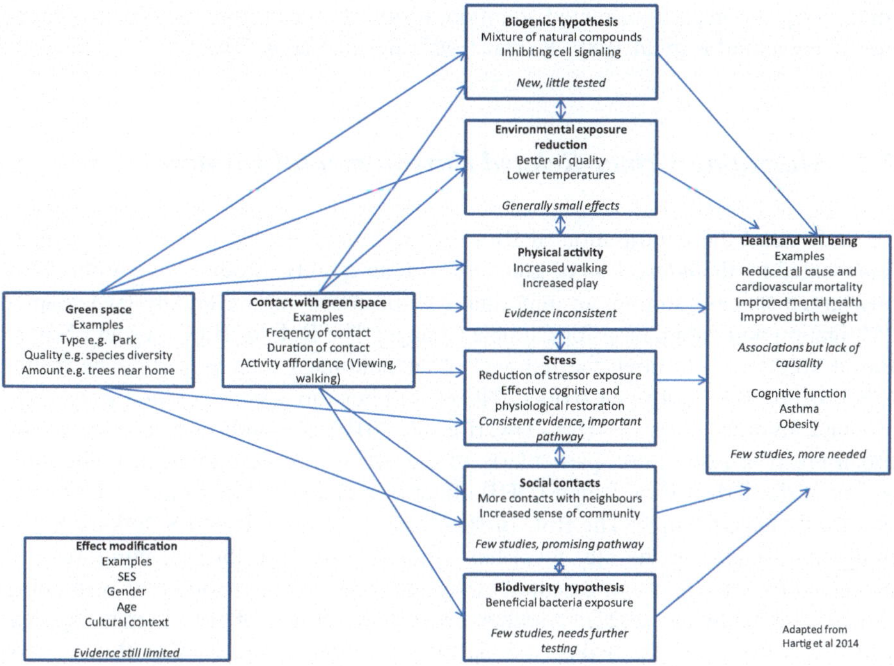

Fig. 1 Conceptual framework of green space, mechanisms, health effects, and current status of evidence (Source: Adapted from Hartig et al. 2014)

and social contacts; and reducing stress whilst restoring attention to nature. The mechanisms have several possible modifiers, such as quality of green spaces (including perceived safety, societal, and cultural context), gender, age, and socioeconomic status (Fig. 1).

2 Mechanisms Underlying Health and Well-being Benefits of Green Spaces

2.1 Stress Reduction/Attention Restoration

The capability of green spaces to restore attention function and reduce perceived stress has been consistently shown by prior studies (Nieuwenhuijsen et al. 2017). This could result in a wide range of health benefits (de Vries et al. 2013; Dadvand et al. 2016). *Stress Reduction Theory* postulates that properties of natural environment such as spatial openness, curving sightlines, and the presence of water and other natural features could induce recovery from stress and help to reduce states of negative thoughts and nervousness through psycho-physiological pathways (Ulrich 1984). *Attention Restoration Theory* proposes that pleasant stimuli provided by contact with nature could appeal to indirect attention restoration (i.e. effortless), minimising the need for directed attention, resulting in restoring the directed attention (Kaplan and Kaplan 1989; Kaplan 1995; Berman et al. 2008).

2.2 Mitigating Urban-Related Environmental Hazards

To understand the contribution of BioCities to human health, a key component is represented by the interaction of greenery with air quality. Studies have highlighted how green spaces can play an important role in improving air quality (see chapter "Mitigation and Adaptation for Climate Change: The Role of BioCities and Nature Based Solutions") through the removal of air pollutants such as particulate matter (PM), ozone (O_3), nitrogen oxides (NOx), sulphur dioxide (SO_2), and polycyclic aromatic hydrocarbons (PAHs) (Tiwari et al. 2019). An additional service of the urban green infrastructure, particularly of the urban forest component, is the mitigation of the Urban Heat Island (UHI) effect (see chapter "Mitigation and Adaptation for Climate Change: The Role of BioCities and Nature Based Solutions"). The mitigation of heat at the city level has a crucial effect on thermal comfort, well perceived by citizens as an essential ecosystem service and a source of well-being (Speak and Salbitano 2021). Another ecosystem service of urban green spaces, which has been less studied so far, is the reduction of noise (Van Renterghem 2019). This appears particularly important along roads and railways and includes the use of shrubs, trees, or perennial herbaceous species to buffer the noise. Given that traffic is the main source of both air pollution and noise in urban areas, their levels are strongly correlated and as expected the presence of leaves and woody vegetation can significantly reduce noise levels (Klingberg et al. 2017).

2.3 Enhancing Social Interaction and Cohesion

Green spaces represent an opportunity to offer an open environment in which people can interact with others. For this reason, green spaces play an important role in increasing social engagement, which in turn can positively affect social interaction and cohesion, reduce crime levels in cities and neighbourhoods (Kuo and Sullivan 2001), and increase social inclusion, especially in deprived communities (Sullivan et al. 2004; Weldon et al. 2007; Bell et al. 2008; Cohen et al. 2008). Furthermore, outdoor social-oriented activities can provide opportunities to decrease or avoid loneliness and to enhance perceived quality of life and life satisfaction (Comstock et al. 2010; Camps-Calvet et al. 2016; Dzhambov et al. 2018). This was highlighted in recent studies on the effects of banning accessing to green space during COVID-19 lockdowns (e.g. Maury-Mora et al. 2022; Sia et al. 2022).

Enhanced social interaction derived from green spaces can be considered as a potential pathway to the individual's health and well-being (Chuang et al. 2013; Jennings and Bamkole 2019). For example, a public green area can trigger a sense of belonging or attachment to a city or neighbourhood. This feeling of safety and comfort can encourage the individual to spend time in that green space to do physical activity, of which the health benefits are widely known (Lackey and Kaczynski 2009; Bocarro et al. 2015). Similarly, a city park can be a place for socialisation, which can reduce the level of perceived loneliness and stress of users, thus improving their general psychological health (Russell et al. 2013).

For children and adolescents, the available evidence is rather limited. Some studies showed that exposure to green space can improve a positive mood due to positive social interactions, emotional sharing, and cooperation (Amoly et al. 2014; Balseviciene et al. 2014; Richardson et al. 2017). For older people, urban green space represents an opportunity to face one of the most common problems for this age group, namely social isolation. It is well-known that a lack of social ties and networks can have detrimental health effects, such as depression and sedentary behaviour, especially after retirement (Cornwell and Waite 2009; Steptoe et al. 2013). At the same time, it appears that older people with stronger social ties are also the ones most likely to spend time in green spaces such as city parks (Enssle and Kabisch 2020), thus triggering a virtuous circle of further strengthening social interactions and psychological and physical health benefits.

2.4 *Increasing Physical Activity*

The available evidence has shown a substantial heterogeneity in the direction and strength of associations between green spaces and physical activity (de la Fuente et al. 2021). This inconsistency could be explained, at least in part, by the lack of taking account of the quality characteristics of green spaces in most of these studies, even though the use of green spaces for physical activity is affected by these characteristics (McCormack et al. 2010). Only a few studies have evaluated the mediation of health benefits of green spaces by physical activity, which suggests a modest mediation role of physical activity in these benefits (de Vries et al. 2013; Dadvand et al. 2016).

Sedentary behaviour is one of the main aspects of the urban lifestyle, where people are preferentially sedentary and increasingly use motorised transportation (Hallal et al. 2012; EEA 2022). For example, only 33% of Europeans meet the minimum recommended levels of physical activity (Gerike et al. 2016). At the same time, sedentary behaviour is a main risk factor for NCDs. As such, a crucial challenge of BioCities will be planning, designing, and developing easily accessible opportunities for moderate-intensity physical activity in the daily life of urban dwellers (Wengel and Troelsen 2020). Walking, cycling, playing, and various forms of informal and light exercise, carried out in outdoor public settings are considered as being decisive for improving human health in cities (WHO-ROE 2018). Walking and cycling for active commuting solely or in combination with public transport have a great potential to provide regular physical activity (Gerike et al. 2016). Additionally, active commuting solutions in green settings are increasingly assuming importance as key opportunities to enhance physical activity with the added benefit of simultaneously improving mental health. Xu et al. (2022) found that 91% of users of greenways, pathways in parks, and other greenspaces in Huangzou, China, indicate the active commuting opportunities in green settings as very positive for physical exercise. In this case, complete and continuous natural arrangements in green corridors were perceived as an important component for attracting users to outdoor activities whilst releasing psychological stress. Dallat et al. (2013) studied the potential health impact and the cost-effectiveness of the greenways in a major urban regeneration project that aimed to create improved opportunities for physical activity and active transport by constructing 19.4 km of new cycle and walkways and providing accessible and safe green space. The study showed that if 10% of those classified as 'inactive' became 'active', a total of 886 incident cases (1.2%) and 75 deaths (0.9%) from ischaemic heart disease, type 2 diabetes, stroke, colon, and breast cancer could be prevented by 2050. According to their cost-effectiveness analysis, the project could be cost-effective at improving physical activity levels plus delivering a wider set of co-benefits beyond health ones.

2.5 *Enriching Environmental Biodiversity*

Human health can be adversely affected by biodiversity loss when simplified ecosystems are less effective in providing ecosystem services, such as carbon sequestration, nutrient cycle performance, drought resistance and the regulation of the urban microclimate, and air pollution removal. In recent years, a consensus has emerged that ecosystem functions decline with biodiversity loss (Naeem et al. 2009). Changes in biodiversity have the potential to affect the risk of infectious disease exposure in plants and animals—including humans—because infectious diseases by definition involve interactions amongst species (Keesing et al. 2010). High biodiversity tends to reduce rates of pathogen transmission and mitigates disease risk for human beings, wildlife, livestock, and plants (Keesing et al. 2010). Land conversion from complex forests, grasslands, and wetlands to highly simplified cropping urban landscapes produces significant biodiversity losses. The health effects of simplification and homogenisation, particularly in cities where built environment barriers induce local shrinkage of habitats affect human health in various primary ways. A relevant example is the increased probability of interaction with disease hosts, vectors, and reservoirs (Romanelli et al. 2015). It derives a dramatic reduction in the widespread inhibitory effect of high biodiversity on pathogen transmission.

Green spaces in cities and their associated urban permeable soils are an important locus for environmental biodiversity, especially microbial diversity, within cities and beyond (Haaland and van den Bosch 2015). They are increasingly recognised as mitigating the negative impacts of urbanisation and contribute to halting the loss of global biodiversity at the genetic, species, and ecosystem levels (McDonnell and MacGregor-Fors 2016).

Studies have shown that biodiversity within green spaces can promote the benefits on health too (Jorgensen and Gobster 2010), especially in terms of psychological well-being and immune function (Fuller et al. 2007; Carrus et al. 2015). The health and well-being benefits of green spaces increase with plant and animal species diversity (Gerstenberg and Hofmann 2016; Marselle et al. 2014) as well as ecosystems diversity through landscape structural diversity (Voigt and Wurster 2015; Southon et al. 2017). Moreover, green spaces can enrich environmental microbiota, which can be translated to more biodiverse personal microbiota that, in turn, can enhance immunoregulation and hence reduce the risk of immunologic conditions such as asthma and atopia and enhance brain development (Rook 2013).

2.6 Climate Change and Health: Direct and Indirect Benefits

Climate change is affecting, to different degrees, all the urban environments worldwide and the predication is that its effects will increase in the future (Carter 2011; Guerreiro et al. 2018). The severity of its resulting impacts depends on the resilient and adaptive capacity of urban ecosystems as well as mitigating the climate change through reducing the emission of greenhouse gases. Extreme weather events and climate-related hazards such as heat waves, floods and droughts are projected to become more frequent and intense in many regions (Guerreiro et al. 2018). In this context, one of the main factors affecting the well-being in the cities of the world is the intensification of heat stress coupled with the urban heat island (UHI) effect. In the last two decades Europe has been affected by several episodes of extreme weather conditions (e.g. Nicholls and Alexander 2007). In 2003, in Greater London, it was estimated that the heat wave led to a 40% increase in mortality, compared with an excess of 16% in 1995 and 15% in 1976 (Johnson et al. 2005). These findings are consistent with other studies conducted in France (Poumadere et al. 2005), Portugal (Nogueira et al. 2005), and Italy (Michelozzi et al. 2005) to assess the effects of the same heat wave. In total, more than 70,000 excess deaths were attributed to this heatwave event from June to September of 2003 in 12 European countries (Baccini et al. 2008).

Elderly people are amongst the most vulnerable subpopulations at risk to heat-related mortality due to poorer physical health and the effects of cognitive impairment on the perception of heat-related health risk (Josseran et al. 2009). Moreover, urban residents are higher risk of heat-related health effects due to the urban heat island effect. The combination of the ongoing urbanisation and ageing population will therefore lead to an increasing number of vulnerable people to the health effects of extreme heat conditions. In addition to the elderly, all the people with chronic diseases and persons of lower socioeconomic status also have a high risk of heat-related mortality (Wolf et al. 2015). Health risks during heat extremes are also greater for people who are physically very active such as manual workers (Lucas et al. 2014).

Vegetation can mitigate the effects of both urban heat island and heat waves through shading and evapotranspiration, which can modify the energy balance and thermal comfort. However, mitigation is limited to green areas and the effects are scarce in the case of very high temperatures (e.g. Mariani et al. 2016; Aram et al. 2020; Tan et al. 2021) or whenever the tree canopy cover is limited or fragmented in the context of the actual urban morphology (Shinzato and Duarte 2012; Sodoudi et al. 2018; Speak and Salbitano 2022; Tamaskani et al. 2021). Greener cities that are also rich in green connections (e.g. rows of trees) between urban parks can guarantee a better adaptation against high temperatures. The City of Melbourne, for example has adopted an urban forest strategy in order not only to increase tree cover from 22% to 40% by 2040 (City of Melbourne 2011), but also to improve biodiversity, health, and soil conditions. Urban green areas also have an effect on human health and on perceived well-being during the summer period, even during heat waves. Several authors have highlighted these benefits and how they can be linked with the frequency of the visit (e.g. Lafortezza et al. 2009). Visiting urban green spaces in hot summers is also associated with a greater perceived well-being through less ego depletion (Panno et al. 2017). Based on these findings, urban green spaces, could be considered as "*climate refuges*" during heat waves, which could shelter the residents from the heat and, at the same time, improve their health and well-being.

3 Health and Well-being Benefits of Green Spaces

3.1 Mental Health, Well-being, and Quality of Life Benefits of Green Spaces Over the Life-Course

Green spaces have positive effects on mental health, well-being, and quality of life (Park et al. 2010; An et al. 2019). Several studies have shown that these benefits can vary over different life stages.

Children Studies have evidenced that being raised in greener neighbourhoods could have a beneficial impact on brain development (Dadvand et al. 2018a; Torquati et al. 2017; Bratman et al. 2012). It could also promote the integrity of the amygdala as an effect of stress reduction (Kühn et al. 2017). Green spaces, especially within playgrounds, also have a positive influence on the social competence of children by facilitating their communication and interactions (Seeland et al. 2009). Studies have also shown that green spaces can have a positive impact on self-satisfaction and social contacts amongst teenagers (Dadvand et al. 2019). The use of green spaces during childhood also strengthens the emotional development and connection to nature, which has been associated with enhanced psychological well-being (Shanahan et al. 2015; Wallner et al. 2018). Furthermore, the view on a green landscape from a school window has a positive impact on students' recovery from stress and mental fatigue (Li and Sullivan 2016).

Adults As green spaces allow oneself to detach from the thoughts and daily concerns that can come from work or responsibilities, green spaces help reduce the stress and anxiety in adults (Berto 2014). By reducing negative thoughts and promoting positive ones, a relationship has also been evidenced between the visit frequency of green spaces and a low prevalence of mood disorders such depression amongst adults (Cox et al. 2017). Moreover, exposure to green space could reduce rumination and subungual cortex activation in adults (Bratman et al. 2015). Furthermore, this exposure has been consistently related to improved perceived general health in adults (Gascon et al. 2015). Enhanced perceived social cohesion and, to less extent, enhanced physical activity are amongst the main mechanisms suggested to underlie this association (Dadvand et al. 2016; de Vries et al. 2013).

Elderly Several studies have shown that green spaces can help reduce the sense of loneliness and isolation that can be experienced by many elderly (Ward Thompson et al. 2016), which has been recognised as an aggravating factor of depressive disorders (Lay et al. 2018) and a predictor of mortality (Hawkley and Cacioppo 2003). Furthermore, available evidence is suggestive for a deceleration of cognitive ageing in elderly in association with exposure to neighbourhood green space (Ricciardi et al. 2022).

3.2 Physical Health Benefits of Green Spaces Over the Life-Course

Pregnancy Outcomes Foetal growth is the pregnancy outcome that has shown more consistent associations with exposure to green spaces. For instance, higher green space surrounding maternal residential address during pregnancy has been associated with the reduced risk of low birth weight and small for gestational age and increased birth weight in offspring (Akaraci et al. 2020). Conversely, the association of this exposure with the length of pregnancy is still inconsistent in the available evidence. Some studies have reported that higher green space surrounding maternal residential address is associated with a reduced risk of preterm birth (i.e. increased length of gestation) (Laurent et al. 2013; Hystad et al. 2014; Grazuleviciene et al. 2015; Nichani et al. 2017). In contrast, other studies have not supported this association (Dadvand et al. 2012a, b; Agay-Shay et al. 2014).

A few studies have also evaluated the relationship between exposure to green space and the risk of pregnancy complications (e.g. gestational diabetes and pregnancy-induced hypertensive disorders including preeclampsia). A recent systematic review and meta-analysis has found a statistically non-significant association between this exposure and pregnancy complications. However, the included studies generally supported a protective association.

Child Health and Development Contact with nature is thought to have a crucial role in brain development in children (Kahn and Kellert 2002; Kellert 2005). This is in accordance with the "biophilia hypothesis" that proposes evolutionary bonds of humans to nature (Wilson 1984; Kellert and Wilson 1993). Observational studies have revealed that higher availability of green space in the living environment and more time spent playing in these spaces in the long run could enhance cognitive development including attention and working memory (Ricciardi et al. 2022; Wells 2000; Dadvand et al. 2015a), induce beneficial anatomical changes in the developing brain (Dadvand et al. 2018b), and reduce symptoms of the attention deficit hyperactivity disorder (ADHD) (Ricciardi et al. 2022; Markevych et al. 2014; McCormick 2017). Moreover, green spaces, especially at school environment, has been associated with better academic performance (Ricciardi et al. 2022).

Further to the aforementioned benefits on mental and neurodevelopmental outcomes, contact with green spaces have also been associated with a number of physical health benefits such as reduction in blood pressure (Markevych et al. 2014) and blood sugar (Dadvand et al. 2018a) and a decelerated shortening of telomere length, an indicator of cellular ageing (Miri et al. 2020).

Communicable Diseases A communicable disease is any source of illness conveyed from one organism to another. People sometimes refer to communicable diseases as "infectious" or "contagious". Infectious disorders are caused by organisms, such as bacteria, viruses, fungi, or parasites. Many organisms live in and on our bodies. In large part, such organisms are not only harmless but even useful. Under certain conditions, some organisms can cause disease. Infectious diseases can be transmitted directly from person to person, or indirectly through vectors as insects or other animals, consuming contaminated food or water, or being exposed to organisms in the environment.

Whilst all communicable diseases are infectious, not all infections are communicable. Tetanus, for example can cause an infection, but a person with tetanus cannot spread it to other people.

Anyway, a communicable disease is a contagious one. The effect is external. If someone catches the illness, they can get sick and spread the pathogen—be it a cold, virus, or some other disease-causing agent—onto the next person. This can lead to small, isolated outbreaks or full-scale pandemics, as experienced with SARS-CoV-2. A dimension also to be considered in the discourses on the relationships between nature in cities and communicable diseases concerns the daily life microbiomes.

Questions relating this section to the BioCities approach include how urban environments and habitats, healthy or unhealthy, can interact, reducing or multiplying the relationships that generate communicable diseases? And what are the major evidence and research lines that support a better understanding of the ecological processes that foster a reduction in the incidence of communicable diseases in BioCities?

The ecological approach to the structure and functioning of BioCities goes far beyond the considerations strictly related to the recurring themes of urban ecology, such as climate change, urban heat island, regulation, and support ecosystem services. The domain of ecological conditions and dynamics of the ever-changing structural and functional relationships of the abiotic, biotic, and human component in the urban environment involves a series of fundamental aspects of the health of urban populations (WHO 2016). With the term "urban ecology", we must consider not only being linked to the presence of green spaces, but also whether they are more or less natural. This impacts the overall metabolism of cities.

One way to address issues related to infectious and communicable diseases and how the dynamic pattern of BioCities can influence their performance and effectiveness for human populations is to observe the health of urban (eco)systems and the meaning that this can have in improving, directly or indirectly, the incidence and severity of communicable diseases.

The Dilution Effect Reservoir species tend to persist when diversity is reduced, increasing in relative abundance relative to species that are more sensitive to disturbance. Conversely, communities with high ecological diversity will have a diluted reservoir effect and a reduced disease risk.

In disease systems where species vary in their susceptibility to infection by a pathogen, higher diversity often results in lower disease risk (Keesing and Ostfeld 2021a, b). This is termed "the dilution effect" and acts on processes at different levels of the disease cycle (Khalil et al. 2016). The dilution effect framework (Keesing and Ostfeld 2021a, b) in zoonotic systems was developed for the tick-borne Lyme disease (Schmidt and Ostfeld 2001). The "dilution effect" of high diversity has been studied in plant and wildlife diseases but is also known to be widespread in human pathogens (Khalil et al. 2016). Its operation was framed by Ostfeld (2017) according to the following reflections:

1. Most human infectious diseases are zoonotic in origin. The zoonotic pathogens are capable of infecting multiple species of host.
2. Although zoonotic pathogens can originate in many different species of vertebrates, certain mammalian orders are over-represented as pathogen sources. Rodents, in particular, a synanthropic species well adapted to urban low-biodiverse environment where populations of predators are controlled, carry more zoonotic pathogens than any other mammalian order.
3. Fast life histories, as found in many species of rodents, are typically associated with commonness and resilience to disturbance, suggesting that competent reservoir species will often predominate in low-diversity communities.

Associations between urbanisation and the prevalence of pathogens in populations of free-ranging wildlife have been described for a wide taxonomic range of host species and pathogens. Evidence suggests that through altered habitat structure and changes to resource availability, urbanisation results in significant changes to the structure of wildlife communities, which are subsequently characterised by low biodiversity with proportional increases in abundance of certain generalist species. From a landscape-scale perspective, this results in a declining trend in species richness from rural areas to urban centres (biotic homogenisation) with synanthropic species occurring at higher densities in urban and suburban environments than less-disturbed areas.

Habitat fragmentation drives species into relative isolation, as seen with tick carrier species in peri- and urban areas. Taking a landscape approach to managing green space in BioCities can help to minimise habitat fragmentation in favour of ecological connectivity to minimise the isolation of species.

3.2.1 The Microbiomes Approach: Reflections and Research Development

Due to the potentially fatal effect of human–pathogenic microbes, the public health mainstream supports limited contact with harmful microbes, through infrastructural and socio-cultural practices or the use of pharmaceutical drugs targeting infectious microorganisms. However, the human microbiome may also mediate positive effects of biodiversity on human health (Marselle et al. 2021), as negative correlations between microbial or environmental diversity and the incidence of non-communicable diseases, and in particular those that are autoimmune (Marselle et al. 2021).

Microbiome research has evolved rapidly over the past few decades but still lacks a clear definition of the term "microbiome". Berg et al. (2020) defined *microbiome* as a "characteristic microbial community occupying a reasonably well-defined habitat which has distinct physio-chemical properties. The microbiome not only refers to the microorganisms involved but also encompass their arena of activity, which results in the formation of specific ecological niches. The microbiome, which forms a dynamic and interactive micro-habitat prone to change in time and scale, is integrated in macroecosystems including eukaryotic hosts, and here crucial for their functioning and health".

Consequently, the *microbiota* consist "of the assembly of microorganisms belonging to different kingdoms while their theatre of activity includes microbial structures, metabolites, mobile genetic elements (such as transposons, phages, and viruses), and relic DNA embedded in the environmental conditions of the habitat" (Berg et al. 2020). According to this definition, a critical aspect in cities is the fitness that microbiomes and microbiota could perform in relation to the limited ecosystem-level exchange due to isolation factors.

Mills et al. (2017) propose the *Microbiome Rewilding Hypothesis*, where the restoration of biodiverse habitat in urban green spaces can rewild the environmental microbiome to a state that benefits human health by primary prevention as an ecosystem service. Microorganisms, both resident and colonising, are immune system inducers and pacifiers, capable of both positive and negative immunomodulation that results in the adjustment of immune responses to normal levels in healthy mammals. The relationship between the soil microbiome, plants, and humans is synthesised by Mills et al. (2017). According to the scheme presented in Fig. 2.

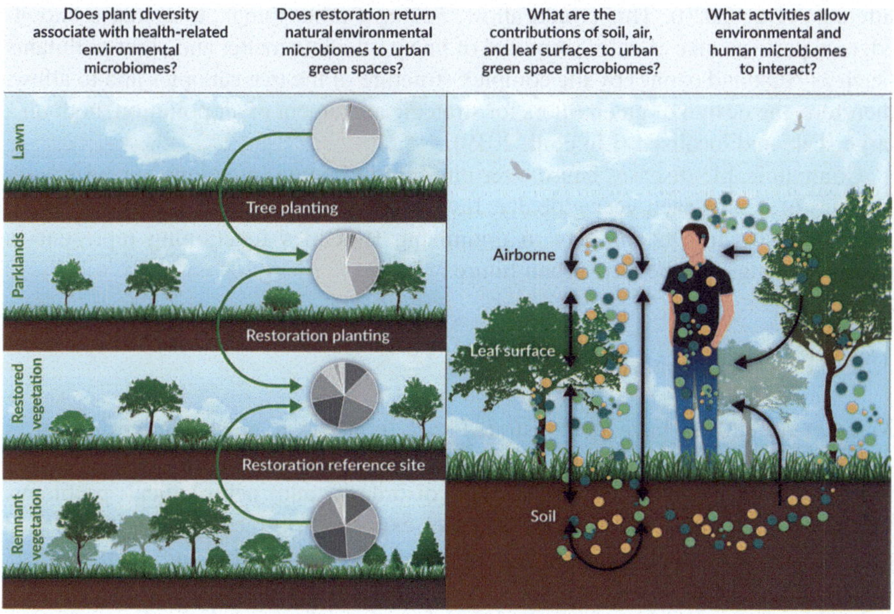

Fig. 2 Urban habitat restoration provides a human health benefit through microbiome rewilding: the Microbiome Rewilding Hypothesis (Source: Mills et al. 2017)

Under the Microbiome Rewilding Hypothesis, the relationship between plant communities and the environmental microbiome is core to restoring adequate human exposure to beneficial microbial communities. Although more evidence is needed to establish this hypothesis, it is a critical and innovative research issue that might be addressed in co-creating future BioCities.

3.2.2 Green Barriers for Agents and Vectors of Communicable Diseases

Improving the green side of BioCities could strengthen the indirect effect that vegetation- and nature-based solutions might play in mitigating the effects of communicable diseases. Following on from the COVID-19 pandemic of 2020, studies are starting to shed light on the potential biological mechanisms that may explain the relationship between air pollution and viral infection outcomes (Ruiz-Gil et al. 2020). For example, it has been hypothesised that chronic exposure to PM2.5 causes alveolar angiotensin-converting enzyme 2 (ACE-2) receptor overexpression and impairs host defences (Wu et al. 2020). This could cause a more severe form of COVID-19 in ACE-2-depleted lungs, increasing the likelihood of poor outcomes, including death. The mitigation effect of vegetation on this issue sounds rather promising. Traces of COVID-19 have been found on atmospheric particulate matter, but also in sewage water. In this context, the leaves of the trees on which the fine particles are deposited as natural filters could be sentinels of possible foci of infection (Day 2020). This would allow, amongst other things, to increase knowledge on performance effective removal of fine particulate matter and other pollutants (such as NOx and ozone) by the complex structure of the tree canopies and to allow, therefore, the design of green filters for strategic abatement of air pollution, both on a large scale and localised (Liu et al. 2019).

Communicable diseases cause over one billion human infections per year, with millions of deaths each year globally. Investigating on the role that components of green infrastructure could play in optimising BioCities functioning represents a major challenge for the next urban future.

3.2.3 Physical Health in Adults

Maas et al. (2009) have demonstrated that living in a greener environment is associated with reduced risk of a wide range of physical morbidities, such as musculoskeletal conditions, respiratory disorders, and neurological problems. Other recent studies have associated higher exposure to green space with reduced risk of cardiovascular conditions such as hypertension, ischaemic heart disease, myocardial infarction, and stroke (Seo et al. 2019; Bauwelinck et al. 2020). Similarly, this exposure has been reported to reduce the risk of metabolic disorders such as diabetes and metabolic syndrome (de la Fuente et al. 2021; de Keijzer et al. 2019a). A limited number of studies have also evaluated the potential association of exposure to green space with cancers and have shown inconsistent results. Whilst some have reported a reduced risk of cancers (i.e. prostate cancer), others have shown no association or increased risk of cancers (i.e. skin cancers) (Zare Sakhvidi et al. 2022).

3.2.4 Healthy Ageing

There is evidence highlighting the exposure to green space to promote healthy ageing, well-being, and physical health of elderly (de Keijzer et al. 2020). A number of longitudinal studies, conducted in the ageing Whitehall II cohort, have shown higher residential surrounding green space is associated with a decelerated cognitive ageing (de Keijzer et al. 2018) and physical functioning decline (de Keijzer et al. 2019b). Other studies have also reported that long-term green space exposure is associated with better physical functioning, cognitive function, and well-being and lower risk of morbidities (de Keijzer et al. 2020). Additionally, such exposure has been related to a better direct attentional capacity and lower concentration problems (de Keijzer et al. 2016). A limited body of evidence has also associated this exposure with slower cognitive decline in elderly (Ricciardi et al. 2022).

3.2.5 Mortality

Higher availability of green spaces in the living environment has been found to have an impact on the reduction of all-cause premature mortality as well as cardiovascular mortality in two systematic reviews and meta-analyses (Gascon et al. 2016; Rojas-Rueda et al. 2019). The association between this exposure and mortality has been reported to be mediated by increased physical activity, lower exposure to air pollution, enhanced social cohesion, and improved mental health (James et al. 2016).

3.2.6 Health Risks of Green Spaces

In addition to the health benefits, green spaces might also potentially induce a number of adverse health outcomes such as asthma and allergic conditions, infectious diseases, and accidental injuries. This also includes the emissions from management interventions (CO_2, PM, and NOx from combustion engines used; noise and mechanical vibrations by machineries, etc.) (Roman et al. 2021).

Asthma and Allergic Conditions The available evidence on the impact of greenspace on the development and/or episodes of asthma and allergic conditions in children is inconsistent. Such an inconsistency could be partially explained by the differences in the type of greenspace and the bioclimatic properties of the study region. For example, a study showed that whilst natural green spaces (e.g. forests) did not show any relationship with asthma or allergic attack, urban parks were related to a higher risk of these respiratory problems (Dadvand et al. 2014). Another study conducted in seven birth cohorts in Australia, Canada, Germany, Netherlands, and Sweden showed a notable heterogeneity in terms of the direction and strength of associations (Fuertes et al. 2016). More recently, a systematic review has suggested a potentially protective association of the green spaces with asthma in children (Hartley et al. 2020).

Exposure to Pesticides Individuals living nearby agricultural fields and/or use these areas could be exposed to the pesticides, which eventually could lead to several adverse conditions in nervous, reproductive, endocrine, and immune systems as well as cancers (Blair et al. 2015).

Vector-Based and Zoonotic Disease The risk of vector-borne diseases transferred by sandflies (e.g. leishmaniasis), mosquitoes (e.g. dengue fever or malaria), or ticks (e.g. Lyme disease and tick-borne encephalitis) could be increased by their reservoirs being hosted by poorly managed green spaces. Moreover, exposure to animal faeces in green spaces could lead to zoonotic infections such as toxocariasis or toxoplasmosis (WHO Regional Office for Europe 2016).

Accidental Injuries Users of green spaces, especially children and green space workers, could experience drowning, falls, or injuries related to slippery leaves, falling branches, or chain saw use. However, accidental injuries in natural environments account for a very tiny proportion of accidental injuries at the population level. Moreover, through proper design and preventive measures these events are preventable to a large extend (WHO Regional Office for Europe 2016).

4 Green Space as an Integral Part of Healthy Living in BioCities

4.1 Transportation

Green spaces and transportation system have the potential to promoting healthy urban living. Roads and parking areas are a major part of public space in our currently car-dominated cities, which otherwise could be used for developing green spaces. Reducing space for cars and the number of cars could have the additional advantage as people switch to public and active transportation and thereby reduce the major urban-related environmental hazards. The major urban-related environmental hazards (i.e. air pollution and noise) in cities would be reduced by limiting the number of cars and space devoted to cars (Nieuwenhuijsen 2020). Reducing number of cars and their allocated space could have the additional advantage by increasing physical activity in switching to public and active transportation (Nieuwenhuijsen and Khreis 2016). Walking or cycling for commuting should be promoted with routes containing natural green space features. This could have mental health benefits as exposure to green space can reduce stress and improve mental health as described earlier. One of the few available studies found, especially for active commuters, better mental health associated with daily commuting through "green" (Zijlema et al. 2018). These findings suggest that cities should invest in commuting routes for cycling and walking within green areas.

4.2 Greening School Environments

Public spaces such as school environments can provide opportunities to reconnect with nature by implementing nature-based solutions to promote equitable health, and reach out to different socioeconomic levels across neighbourhoods. Children spend a substantial amount of their day in classrooms and schoolyards. As such, school environments are amongst the most crucial settings to ensure their health, well-being, and effective learning (Silvers et al. 1994; Dorizas et al. 2013). Surprisingly, many school environments are lacking in green spaces. Taking advantage of schools as a point of action for planting more trees, building green walls, or increasing the presence of water features, offers a good strategy to improve environmental exposures and related health and well-being outcomes in an equitative way. These adaptations will have a positive impact on both social cohesion and the physical and mental health and well-being of children and other citizens for whom these spaces are made accessible. Moreover, higher exposure to green space has been associated with better school performance and academic achievement (Browning and Rigolon 2019). Green views from the classroom can positively influence school performance (Matsuoka 2010). Greening school grounds can also provide opportunities for physical activity and more diverse play (Dyment and Bell 2008).

4.3 Greenery Along Roads, Waterways, Railways; Street Trees, Green Walls, and Green Roofs

Greenery along roadways, waterways and railways, and on roofs and the walls of buildings come in many sizes and guises including a variety of different species of trees, shrubs, and smaller plants. These important components of green infrastructure perform critical functions in the urban BioCity landscape and are an important part of everyday lives. In BioCities, these types of green infrastructure will be as important as other infrastructure such as transport, communication, and water treatment. In a BioCity, there will be a recognition of their role in reducing heat stress by cooling air in the summer and providing shade (Anderson and Gough 2021), improving biodiversity which can result in better mental health outcomes (Beute et al. 2021), providing aesthetically attractive environments conducive to physical activity, stress reduction and improving mental health, as well as improving air quality (Manso et al. 2021; Tomson et al. 2021). Further research is needed, however, in many of these areas. In a systematic review of the characteristics of green spaces that have an impact on mental health and well-being, the review found a clear relationship between trees and better mental health (Beute et al. 2021). Street greenery was found in a study in China to have a positive association with physical activity in older adults (He et al. 2020). The reduction in speed that street trees provide, and consequently harmful emissions, also contribute to promote health benefits and traffic safety (Manman et al. 2022).

Green walls and roofs are becoming increasingly common, with green roofs providing spaces for residents and workers of BioCities and increasing the aesthetic benefits of the everyday environment. The role of green roofs in reducing the impacts of heatwaves on people's health has been illustrated by Marvuglia et al. (2020). There is increasing research on green walls and roofs as a feature in new buildings but also in retrofitting existing ones. They can contribute to energy savings, reduce sound transmission into buildings, increase property values, and improve air quality. Their specific contribution to health and well-being, however, has not yet been quantified (Manso et al. 2021).

4.4 Greening Housing and Business Developments: Greening Healthcare Settings, Prisons, and Care Homes

Green spaces, green networks, and pathways provide a strategic network of green infrastructure which is important in a range of everyday settings, including housing, business developments, schools, health and social care settings, and around prisons and other institutions. Hospitals and healthcare settings provide important opportunities for greening to improve patients', staffs', and visitors' health and well-being. A classic study by Ulrich (1984) showed postoperative patients assigned to rooms with a nature view had shorter hospital stays. Demonstration projects in Scotland are highlighting the potential for green spaces to be incorporated into the design of new healthcare settings, but also retrofitted into existing healthcare facilities (Partnership 2014).

Green spaces in prison environments have been reported to be able to contribute to lower self-harm and violence (Moran et al. 2020), enhance self-reported restoration (Moran 2019), and induce similar health effects as those found in healthcare facilities (Moran and Turner 2019). Greening residential housing can lead to gentrification and marginalisation of some groups who do not always benefit equally from greening. However, greening vacant land around residences could contribute to a reduction in crime density (Hadavi et al. 2021). Similarly, green street views in neighbourhoods were found to reduce fear of crime (Jing et al. 2021). Mechanisms for these observations are suggested to include social interaction, community perception, and stress reduction (Shepley et al. 2019).

5 Interventions, Enabling, and Indicators

5.1 Available Therapies, Protocols, and Programmes: Forest Therapy and Healing Gardens

Increasing attention is being paid to the use of outdoor natural environments as therapy settings for human health and well-being. The use of these therapeutic activities dates back to the 1980s thanks to the Japanese medical tradition of the so-called "Shinrin-yoku" or "forest bathing", which recognises the therapeutic effects of staying and/or walking in the forest (Ohtsuka et al. 1998). More recently the health benefits of such activities have been supported by scientific evidence.

Forest-based interventions and therapies have been reported on several physical health outcomes. A number of systematic reviews (Oh et al. 2017; Rajoo et al. 2020; Stier-Jarmer et al. 2021) have supported the usefulness of this type of interventions for improving physiological responses. Forest therapy, for example has been shown to be able to regulate both systolic and diastolic blood pressure levels (Lee and Lee 2014), positively impact the immune response (Jia et al. 2016; Lyu et al. 2019) and inflammatory response (Mao et al. 2012), and heart and pulmonary functions (Lee and Lee 2014). Furthermore, the stress level was found to be lower both when self-perceived and by measuring cortisol levels in people engaged in these activities (Sonntag-Öström et al. 2015).

There has been a particular interest in this research field to the potential benefits of forest therapy on depression (Lee et al. 2017). Evidence seems to support the use of forest therapy as a strategy for the prevention and treatment of depression thanks to a considerable decrease in depressive symptoms compared to non-forest-based interventions (e.g. urban areas and hospitals). However, these protective effects were observed only in real forest therapy interventions or targeted for mental health rather than during simple walks in the forest or green exercise (Rosa et al. 2021). In contrast, mixed results are reported on other psychological outcomes such as anxiety, mood, and quality of life probably due to the low quality of studies and reviews available on these topics (Stier-Jarmer et al. 2021). However, significant improvements were found on perceived restoration, relaxation, creativity, and sociality (Lee et al. 2017; Bielinis et al. 2018; Yu and Hsieh 2020; Spano et al. 2020).

A popular type of environment designed for green therapy is the "healing or therapeutic garden". They are valued in therapeutic and/or rehabilitative contexts such as hospitals and nursing homes. In order to be defined as therapeutic gardens, they must be designed according to a series of very specific characteristics and planned for each type of user, such as older people with age-related diseases, terminally ill patients, and children with special needs (Scartazza et al. 2020; Gueib et al. 2020).

5.2 *Green Spaces as Treatment for Disabled/Marginalised People*

According to the World Health Organisation (2021), about 15% of the world's population lives with some form of disability and this percentage is expected to rise because of the ageing world population and the incidence of chronic diseases. Disabilities and their interplay with different personal traits such as gender, age, ethnicity, religion, or belief lead to additional discrimination and marginalisation (UNDESA 2016). Half of the population with disabilities live in urban areas, and by 2050 it is estimated that 70% of the world's population will live in cities. Therefore, the number of people with disabilities living in urban areas will also increase, intensifying the need for accessible green spaces to serve their needs (Seeland and Nicolè 2006). The health related benefits induced by green spaces have been reported by both cross-sectional (Lee and Maheswaran 2011) but also experimental studies (Müller-Riemenschneider et al. 2020) on different segments of the population. These studies are aimed, in general, at filling knowledge gaps from a non-disabled point of view. Literature is lacking on the disabled and marginalised persons' view and their experience of green spaces (Pini et al. 2016). As an additional treatment, ecotherapy has shown positive effects on the health status of patients using secondary and tertiary mental healthcare services (Wilson et al. 2008). Ecotherapy as a nature-based method of physical and psychological healing represents a new form of psychotherapy, often carried out in natural settings (Buzzell and Chalquist 2009) and addressing the positive human–nature relationship in therapeutic programmes and prescriptions (Chalquist 2009). Eco-therapy has also been reported to complement recovery strategies from substance addictions (Berry et al. 2021) as well as to help people cope with stress, anxiety, and mood disorders (Burls 2007). Pain induced by different forms of disabilities can also be reduced by green space exposure (Stanhope et al. 2020).

5.3 The Way Ahead: Enabling Environment, Institutional Tools, Actions, and Knowledge

The recently promoted conceptual and operational framework of urban health includes the relational and dynamic characteristics of the physical and social environment which, individually or in synergy with each other, influence the well-being and quality of life of citizens within an urban context (Galea and Vlahov 2005). The physical and built environment of cities (open spaces, structures, and urban infrastructures) can generate impacts on health, especially if there are critical issues regarding water quality, thermal and acoustic comfort, wastewater, atmospheric pollution, and the dynamic quasi-equilibrium of biological and microbiological diversity. Urban health approach has several implications for the development of BioCities, first of all claiming for health in all policies (Ramirez-Rubio et al. 2019).

The research and application development of urban health towards BioCities could follow one of two obvious pathways. In a first, practitioners of urban health would develop a catalogue of the myriad of ways in which human health is affected by anthropogenic changes to the environment and list potential actions to be implemented. Although such an exercise might be valuable, it would not be novel, and the information by itself might not lead to solutions to the critical problems affecting cities. In the second pathway, urban health researchers and practitioners would use the many adverse health consequences of anthropogenic environmental constraints as grounds to advocate for better environmental protections. Although there is no doubt that advocacy is necessary to prevent ever more rapid degradation of environmental health, this pathway would de-emphasise the scientific mission of the *One Health* unified and integrated approach that aims at a sustainable balance of the health of people, animals, and ecosystems, and emphasise the political one.

Urban health advocates could also document the many cases in which human health and environmental health are simultaneously improved by the same policy or management actions. A thorough exploration of these win–win situations, with careful analysis of the mechanisms that underlie co-benefits to environmental and human health, could uncover key principles and inform new applications, whilst providing concrete options for policy and management.

Box 2 Towards a guidance on green spaces and health in BioCities: Criteria for decision makers, professionals, and scientists

As synthesis of the former issues, we propose a list of guiding criteria to be checked in formulating green policies in BioCities and to be developed through urban governance, planning, design, and management.

Policy and legal framework ▪ Ensure that recommendations provided by WHO and other international bodies on the availability, accessibility, quality and security of healthy urban landscapes and public green spaces are incorporated and reflected in national and city policies and regulatory/legal documents. ▪ Develop strategic documents and actions plans fostering green/blue infrastructures of biocities as natural systems of public health (sensu Calaza Martinez, 2016) ▪ Promote and adopt strategies and financial and administrative mechanisms to maximize the potential for green spaces to deliver positive environmental and health benefits. ▪ Propose and support regional (as the European Medicines Agency EMA) and national agencies to provide working protocols to be currently adopted by health services, physicians, and practitioners. ▪ Promote collaboration and information-sharing among the sectors involved in environmental planning and design, urban greening, urban planning and healthcare. ▪ Ensure the inclusion of health and well-being objectives in multisectoral policies on urban environment. ▪ Ensure that savings in healthcare costs generated by ecosystem services provided by landscape-based solutions are taken into account in relevant policies and duly incorporated in the financial accounts of governments.	**Planning, design and management** ➢ Adopt sound standards for the design and management of urban environment with the aim of encouraging physical activities and improving mental health. ➢ Optimize the availability, accessibility, proximity, permeability and security of urban green spaces and nature based solutions to promote the use of such resources by all citizens. ➢ Maximize the thermal comfort, pollution filtration and noise reduction functions of green/blue infrastructure components when planning and designing public as private spaces and streets. ➢ Consider including in the design and management of outdoors setting the production of fresh and nutritious food as well as natural and traditional remedies for use by local communities. ➢ Include nature-based solutions in the planning and design of hospitals and schools for their proven therapeutic and psychological benefits for patients and children. ➢ Develop greenways/blueways to increase alternative mobility (e.g. bicycles) as a way of promoting physical and mental health and reducing pollution. ➢ Minimize the potential undesirable impacts of living organisms and habitat on human health and well-being in designing and managing biocities. ➢ Co-production (design, implementation, and management) of demonstration projects with the community
Key monitoring criteria ❑ Availability, accessibility and proximity of green spaces ❑ Perceived attractiveness of outdoors environment and healthy indoor settings to citizens ❑ Pollution levels in the city ❑ Suitability of biocities and their green/blue components to host sporting and other leisure activities ❑ Number of awareness raising and communication programmes activated in the civil society ❑ Number of "green" prescriptions ❑ Perceived thermal/physical/psychological comfort	**Key competencies and skills** ❖ Design and management of urban forests and other green spaces for restorativeness, illness prevention, therapy and rehabilitation ❖ Design and management of urban forests to maximize their capacity to remove traffic pollutants and optimize thermal comfort for pedestrians ❖ Planning and design of green spaces to encourage sporting and other leisure activities ❖ Planning and design of green spaces to encourage cycling and walking as healthy transport modes ❖ Management of potentially allergenic urban tree species

5.4 Success Stories and Good Practices

A review of success stories has been recently carried out by the European Commission in relation to EU-funded projects and the potential of green infrastructure and nature-based solutions (NBS) to improve human health and well-being (Calfapietra 2020). A strong focus emerged in relation to this research with the reduction of heat in cities, such as London, where vegetation was measured to mitigate temperatures by 3 °C (Lindberg and Grimmond 2011), and Barcelona, where the reduction of UHI allowed a consistent saving of the emission of greenhouse gases (Baró et al. 2014).

In other case studies, green infrastructure emerged to promote health by reducing air pollution and inducing positive effects on mental stress. For instance, Dadvand et al. (2015a) observed an improvement in working memory and a reduction in inattentiveness in children living or attending green schools, which was in part related to the reduction in exposure to air pollution. Unfortunately, the links between epidemiological studies and levels of green spaces are often characterised by high variability. It is therefore crucial to have more robust interdisciplinary case studies with a high variety of environmental, social, and NBS conditions, whilst at the same time collect data involving a large number of individuals.

6 Outcomes and Concluding Remarks

In BioCities, urban planning needs to be understood as a powerful public health tool. The promotors of health and well-being in urban settings should not come solely from the health system, but also from urban designers and planners. In this context, BioCities, through enhancing the abundance, quality, and equitable access to natural environment, could eliminate, limit, or mitigate the adverse health effects of urban living such as higher exposure to heat, air pollution, noise, or sedentary and stressful lifestyle. Such goals can be achieved, at least in part, by transforming public spaces, currently devoted to grey infrastructure, into green infrastructure, adapting buildings with green components (e.g. green walls or roofs), and expanding the quality of greenness and biodiversity around the city.

BioCity green infrastructure does not only provide for the social and environmental welfare of a city, but also the health and well-being of its urban residents. To maximise the positive impacts of green infrastructure on human health and well-being, we need to accurately quantify the contribution of new green planning strategies and have indicators that help private and public investors to understand the environmental, social, and economic benefits for society. Moreover, urban planning in BioCities is a powerful public health tool.

Finally, the following issues should be considered to bridge the knowledge gap between practice and research:

- Develop contextualised Indicators for monitoring and evaluating the effects of nature and green space exposure on human health and well-being.
- Foster innovative research on interrelated effects of healthy environment and human health concerning communicable and non-communicable diseases.
- Test and communicate the effectiveness of urban forests and other green spaces in therapies and recovery programmes related to mental health and non-communicable diseases.
- Integrate knowledge of urban fabric and green infrastructure components for filtering noise and atmospheric pollution caused by road traffic and industries, as well as for cooling the urban environment.
- Assess soft transportation interacting with green settings (e.g. cycling, walking in greenways, and green pathways).

References

Agay-Shay K, Peled A, Crespo AV, Peretz C, Amitai Y, Linn S, Friger M, Nieuwenhuijsen MJ (2014) Green spaces and adverse pregnancy outcomes. Occup Environ Med 71(8):562–569

Akaraci S, Feng X, Suesse T, Jalaludin B, Astell-Burt T (2020) A systematic review and meta-analysis of associations between green and blue spaces and birth outcomes. Int J Environ Res Public Health 17.2949

Amoly E, Dadvand P, Forns J, López-Vicente M, Basagaña X, Julvez J, Alvarez-Predrerol M, Nieuwenhuijsen MJ, Sunyer J (2014) Green and blue spaces and behavioral development in Barcelona schoolchildren: the BREATHE project. Environ Health Perspect 122(12):1351–1358

An B, Wang D, Liu XJ, Guan HM, Wei HX, Ren ZB (2019) The effect of environmental factors in urban forests on blood pressure and heart rate in university students. J For Res 24(1):27–34. https://doi.org/10.1080/13416979.2018.1540144

Anderson V, Gough WA (2021) Nature-based cooling potential: a multi-type green infrastructure evaluation in Toronto, Ontario, Canada. Int J Biometeorol 66(3)

Aram F, Solgi E, Baghaee S, García EH, Mosavi A, Band SS (2020) How parks provide thermal comfort perception in the metropolitan cores; a case study in Madrid Mediterranean climatic zone. Clim Risk Manag 30:100245

Baccini M, Biggeri A, Accetta G, Kosatsky T, Katsouyanni K, Analitis A, Anderson HR, Bisanti L, D'Ippoliti D, Danova J, Forsberg B, Medina S, Paldy A, Rabczenko D, Schindler C, Michelozzi P (2008) Heat effects on mortality in 15 European cities. Epidemiology 19:711–719

Balseviciene B, Sinkariova L, Grazuleviciene R, Andrusaityte S, Uzdanaviciute I, Dedele A, Nieuwenhuijsen MJ (2014) Impact of residential greenness on preschool children's emotional and behavioral problems. Int J Environ Res Public Health 11(7):6757–6770

Baró F, Chaparro L, Gomez-Baggerthun E, Langemeyer J, Nowak DJ, Terradas J (2014) Contribution of ecosystem services to air quality and climate change mitigation policies: the case of urban forests in Barcelona, Spain. Ambio 43:466–479

Bauwelinck M, Zijlema WL, Bartoll X, Vandenheede H, Cirach M, Lefebvre W, Vanpoucke C, Basagana X, Nieuwenhuijsen MJ, Borrell C, Deboosere P, Dadvand P (2020) Residential urban green space and hypertension: a comparative study in two European cities. Environ Res 191: 110032

Bell S, Hamilton V, Montarzino A, Rothnie H, Travlou P, Alves S (2008) Green space and quality of life: a critical literature review. GreenSpace Scotland, Stirling

Berendsen RL, Pieterse CMJ, Bakker PAHM (2012) The rhizosphere microbiome and plant health. Trends Plant Sci 17(8):478–486

Berg G, Rybakova D, Fischer D, Cernava T, Champomier Verges MC, Charles T, Chen X, Cocolin L, Eversole K, Corral GH, Kazou M, Kinkel L, Lange L, Lima N, Loy A, Macklin JA, Maguin E, Mauchline T, McClure R, Mitter B, Ryan M, Sarand I, Smidt H, Schelkle B, Schloter M (2020) Microbiome definition re-visited: old concepts and new challenges. Microbiome 8:103. https://doi.org/10.1186/s40168-020-00875-0

Berman MG, Jonides J, Kaplan S (2008) The cognitive benefits of interacting with nature. Psychol Sci 19(12):1207–1212

Berry MS, Rung JM, Crawford MC, Yurasek AM, Ferreiro AV, Almog S (2021) Using greenspace and nature exposure as an adjunctive treatment for opioid and substance use disorders: preliminary evidence and potential mechanisms. Behav Process 186:104344

Berto R (2014) The role of nature in coping with psycho-physiological stress: a literature review on restorativeness. Behav Sci 4(4):394–409. https://doi.org/10.3390/bs4040394

Bettencourt LMA, Lobo J, Helbing D, Kühnert C, West GB (2007) Growth, innovation, scaling, and the pace of life in cities. Prec Natl Acad Sci 104(17):7301–7306

Beute F, Andreucci MB, Lammel A, Davies Z, Glanville J, Keune H, Marselle MR, O'Brien L, Olszewska-Guizzo A, Remmen R, Russo A, de Vries S (2021) Types and characteristics of urban and peri-urban green spaces having an impact on human mental health and wellbeing. Technical report prepared by an EKLIPSE Expert Working Group, Wallingford

Bielinis E, Takayama N, Boiko S, Omelan A, Bielinis L (2018) The effect of winter forest bathing on psychological relaxation of young polish adults. Urban Forestry Urban Green 29:276–283

Blair A, Ritz B, Wesseling C, Beane Freeman L (2015) Pesticides and human health. Occup Environ Med 72(2):81–82

Bocarro JN, Floyd MF, Smith WR, Edwards MB, Schultz CL, Baran P, Moore RA, Cosco N, Suau LJ (2015) Social and environmental factors related to boys' and girls' park-based physical activity. Prev Chronic Dis 12:E97

Bratman GN, Hamilton JP, Daily GC (2012) The impacts of nature experience on human cognitive function and mental health. Ann N Y Acad Sci 1249(1):118–136. https://doi.org/10.1111/j.1749-6632.2011.06400

Bratman GN, Hamilton P, Hahn KS, Daily GC, Gross JJ (2015) Nature experience reduces rumination and subgenual prefrontal cortex activation. Prec Natl Acad Sci 112(28):8567–8572. https://doi.org/10.1073/pnas.1510459112

Browning MH, Rigolon A (2019) School green space and its impact on academic performance: a systematic literature review. Int J Environ Res Public Health 16(3):429

Burls A (2007) People and green spaces: promoting public health and mental well-being through ecotherapy. J Public Ment Health

Buzzell L, Chalquist C (2009) Ecotherapy: healing with nature in mind. Sierra Club Books, San Francisco

Calfapietra C (2020) Nature-based solutions for microclimate regulation and air quality. Analysis of EU-funded Projects European Commission Directorate-General for Research and Innovation 32 pp

Camps-Calvet M, Langemeyer J, Calvet-Mir L, Gómez-Baggethun E (2016) Ecosystem services provided by urban gardens in Barcelona, Spain: insights for policy and planning. Environ Sci Policy 62:14–23

Carrus G, Scopelliti M, Lafortezza R, Colangelo G, Ferrini F, Salbitano F, Agrimi M, Portoghesi L, Semenzato P, Sanesi G (2015) Go greener, feel better? The positive effects of biodiversity on the well-being of individuals visiting urban and peri-urban green areas. Landsc Urban Plan 134:221–228

Carter JG (2011) Climate change adaptation in European cities. Curr Opin Environ Sustain 3:193–198

Chalquist C (2009) A look at the ecotherapy research evidence. Ecopsychology 1(2):64–74

Chuang YC, Chuang KY, Yang TH (2013) Social cohesion matters in health. Int J Equity Health 12(1):1–12

City of Melbourne (2011) Urban forest strategy: making a great City Greener 2012–2032. https://www.melbourne.vic.gov.au/SiteCollectionDocuments/urban-forest-strategy.pdf

Cohen DA, Inagami S, Finch B (2008) The built environment and collective efficacy. Health Place 14:198–208

Committee of Inquiry into the Future Development of the Public Health Function (1988) Public Health in England: report of the Committee of Inquiry into the future development of the public health function (Cm. 289). The Stationery Office, London

Comstock N, Dickinson LM, Marshall JA, Soobader MJ, Turbin MS, Buchenau M, Litt JS (2010) Neighborhood attachment and its correlates: exploring neighborhood conditions, collective efficacy, and gardening. J Environ Psychol 30(4):435–442

Cornwell EY, Waite LJ (2009) Social disconnectedness, perceived isolation, and health among older adults. J Health Soc Behav 50(1):31–48

Cox DT, Shanahan DF, Hudson HL, Plummer KE, Siriwardena GM, Fuller RA, Anderson K, Hancock S, Gaston KJ (2017) Doses of neighborhood nature: the benefits for mental health of living with nature. Bioscience 67(2):147–155. https://doi.org/10.1093/biosci/biw173

Cyril S, Oldroyd JC, Renzaho A (2013) Urbanisation, urbanicity, and health: a systematic review of the reliability and validity of urbanicity scales. BMC Public Health 13:513

Dadvand P, de Nazelle A, Triguero-Mas M, Schembari A, Cirach M, Amoly E, Figueres F, Basagaña X, Ostro B, Nieuwenhuijsen MJ (2012a) Surrounding greenness and exposure to air pollution during pregnancy: an analysis of personal monitoring data. Environ Health Perspect 120(9):1286–1290

Dadvand P, de Nazelle A, Figueras F, Basagaña X, Su J, Amoly E, Jerrett J, Vrijheid M, Sunyer J, Nieuwenhuijsen MJ (2012b) Green space, health inequality and pregnancy. Environ Int 40:110–115

Dadvand P, Villanueva CM, Font-Ribera L, Martinez D, Basagaña X, Belmonte J, Vrijheid M, Grazulevinciene R, Kogevenas M, Nieuwenhuijsen MJ (2014) Risks and benefits of green spaces for children: a cross-sectional study of associations with sedentary behavior, obesity, asthma, and allergy. Environ Health Perspect 122(12):1329–1325

Dadvand P, Nieuwenhuijsen MJ, Esnaola M, Forns J, Basagaña X, Alvarez-Pedrerol M, Rivas I, López-Vicente M, De Castro PM, Su J, Jerrett J, Querol X, Sunyer J (2015a) Green spaces and cognitive development in primary schoolchildren. Proc Natl Acad Sci 112(26):7937–7942

Dadvand P, Rivas I, Basagaña X, Alvarez-Pedrerol M, Su J, De Castro PM, Amato F, Jerret M, Querol X, Sunyer J, Nieuwenhuijsen MJ (2015b) The association between greenness and traffic-related air pollution at schools. Sci Total Environ 523:59–63

Dadvand P, Bartoll X, Basagaña X, Dalmau-Bueno A, Martinez D, Ambros A, Cirach M, Tiguero-Mas M, Gascon M, Borrell C, Nieuwenhuijsen MJ (2016) Green spaces and general health: roles of mental health status, social support, and physical activity. Environ Int 91:161–167

Dadvand P, Poursafa P, Heshmat R, Motlagh ME, Qorbani M, Basagaña X, Kelishadi R (2018a) Use of green spaces and blood glucose in children; a population-based CASPIAN-V study. Environ Pollut 243:1134–1140

Dadvand P, Pujol J, Macià D, Martínez-Vilavella G, Blanco-Hinojo L, Mortamais M, Alvarez-Pedrerol M, Fenoll R, Esnaola M, Dalmau-Bueno A, López-Vicente M, Basagaña X, Jerrett J, Nieuwenhuijsen MJ, Sunyer J (2018b) The association between lifelong green space exposure and 3-dimensional brain magnetic resonance imaging in Barcelona schoolchildren. Environ Health Perspect 126(2):027012. https://doi.org/10.1289/EHP1876

Dadvand P, Hariri S, Abbasi B, Heshmat R, Qorbani M, Motlagh ME, Basagaña X, Kelishadi R (2019) Use of green spaces, self-satisfaction and social contacts in adolescents: a population-based CASPIAN-V study. Environ Res 168:171–177

Dallat MAT, Soerjomataram I, Hunter RF, Tully MA, Cairns KJ, Kee F (2013) Urban greenways have the potential to increase physical activity levels cost-effectively. Eur J Pub Health 24(2): 190–195. https://doi.org/10.1093/eurpub/ckt035

Day BH (2020) The value of greenspace under pandemic lockdown. Environ Resour Econ 76: 1161–1185

de Keijzer C, Gascon M, Nieuwenhuijsen MJ, Dadvand P (2016) Long-term green space exposure and cognition across the life course: a systematic review. Curr Environ Health Rep 3(4): 468–477

de Keijzer C, Tonne C, Basagaña X, Valentín A, Singh-Manoux A, Alonso J, Antó JM, Nieuwenhuijsen MJ, Sunyer J, Dadvand P (2018) Residential surrounding greenness and cognitive decline: a 10-year follow-up of the Whitehall II Cohort. Environ Health Perspect 126(7):077003. https://doi.org/10.1289/EHP2875. PMCID: PMC6108840

de Keijzer C, Basagaña X, Tonne C, Valentín A, Alonso J, Antó JM, Nieuwenhuijsen MJ, Kivimaki M, Singh-Manoux A, Sunyer J, Dadvand P (2019a) Long-term exposure to green space and metabolic syndrome: a Whitehall II study. Environ Pollut 255:113231

de Keijzer C, Tonne C, Sabia S, Basagaña X, Valentín A, Singh-Manoux A, Antó JM, Alonso J, Nieuwenhuijsen MJ, Sunyer J, Dadvand P (2019b) Green and blue spaces and physical functioning in older adults: longitudinal analyses of the Whitehall II study. Environ Int 122: 346–356

de Keijzer C, Bauwelinck M, Dadvand P (2020) Long-term exposure to residential green space and healthy aging: a systematic review. Curr Environ Health Rep 7:65–88

De la Fuente F, Saldías MA, Cubillos C, Mery G, Carvajal D, Bowen M, Bertoglia MP (2021) Green space exposure association with type 2 diabetes mellitus, physical activity, and obesity: a systematic review. Int J Environ Res Public Health 18(1):97

de Vries S, van Dillen SME, Groenewegen PP, Spreeuwenberg P (2013) Streetscape greenery and health: stress, social cohesion and physical activity as mediators. Soc Sci Med 94:26–33

Dorizas PV, Kapsanaki-Gotsi E, Assimakopoulos MN, Santamouris M (2013) Correlation of particulate matter with airborne fungi in schools in Greece. Int J Vent 12(1):1–16

Dyment JE, Bell AC (2008) Grounds for movement: green school grounds as sites for promoting physical activity. Health Educ Res 23(6):952–962

Dzhambov AM, Markevych I, Hartig T, Tilov B, Arabadzhiev Z, Stoyanov D, Gatseva P, Dimitrova DD (2018) Multiple pathways link urban green-and bluespace to mental health in young adults. Environ Res 166:223–233

Enssle F, Kabisch N (2020) Urban green spaces for the social interaction, health and well-being of older people—an integrated view of urban ecosystem services and socio-environmental justice. Environ Sci Policy 109:36–44

European Environment Agency (2022) Transport and environment report 2021: decarbonising road transport—the role of vehicles, fuels and transport demand. EEA Report No 02/2022. Publications Office of the European Union, Luxembourg. https://doi.org/10.2800/68902

Fuertes E, Markevych I, Bowatte G, Gruzieva O, Gehring U, Becker A, Berdel D, von Berg A, Bergström A, Brauer M, Brunekreef B, Brüske I, Carlsten C, Chan-Yeung M, Dharmage SC, Hoffmann B, Klümper C, Koppelman GH, Kozyrskyj A, Korek M, Kull I, Lodge C, Lowe A, MacIntyre E, Pershagen G, Standl M, Sugiri D, Wijga A, Heinrich J (2016) Residential greenness is differentially associated with childhood allergic rhinitis and aeroallergen sensitisation in seven birth cohorts. Allergy 71(10):1461–1471. https://doi.org/10.1111/all.12915:n/a-n/a

Fuller RA, Irvine KN, Devine-Wright PH, Warren PH, Gaston KJ (2007) Psychological benefits of green space increase with biodiversity. Biol Lett 3:390–394. https://doi.org/10.1098/rsbl.2007.0149

Galea S, Vlahov D (2005) Urban health: evidence, challenges, and directions. Annu Rev Public Health 26:341–365. https://doi.org/10.1146/annurev.publhealth.26.021304.144708

Gascon M, Triguero-Mas M, Martínez D, Dadvand P, Forns J, Plasància A, Nieuwenhuijsen MJ (2015) Mental health benefits of long-term exposure to residential green and blue spaces: a systematic review. Int J Environ Res Public Health 12(4):4354–4379

Gascon M, Triguero-Mas M, Martínez D, Dadvand P, Rojas-Rueda D, Plasència A, Nieuwenhuijsen MJ (2016) Residential green spaces and mortality: a systematic review. Environ Int 86:60–67

Gerike R, de Nazelle A, Nieuwenhuijsen MJ (2016) Physical activity through sustainable transport approaches (PASTA): a study protocol for a multicentre project. BMJ Open 6:e009924. https://doi.org/10.1136/bmjopen-2015-009924

Gerstenberg T, Hofmann M (2016) Perception and preference of trees: a psychological contribution to tree species selection in urban areas. Urban Forestry Urban Green 15:103–111. https://doi.org/10.1016/j.ufug.2015.12.004

Grazuleviciene R, Danileviciute A, Dedele A, Vencloviene J, Andrusaityte S, Uzdanaviciute I, Nieuwenhuijsen MJ (2015) Surrounding greenness, proximity to city parks and pregnancy outcomes in Kaunas cohort study. Int J Hyg Environ Health 218(3):358–365

Gueib C, Pop A, Bannay A, Nassau E, Fescharek R, Gil R, Luc A, Rivasseau Jonveaux T (2020) Impact of a healing garden on self-consciousness in patients with advanced Alzheimer's disease: an exploratory study. J Alzheimers Dis 75(4):1283–1300

Guerreiro SB, Dawson RJ, Kilsby C, Lewis E, Ford A (2018) Future heatwaves, droughts and floods in 571 European cities. Environ Res Lett 13:034009

Haaland C, van den Bosch CK (2015) Challenges and strategies for urban green space planning in cities undergoing densification: a review. Urban Forestry Urban Green 14:760–771. https://doi.org/10.1016/j.ufug.2015.07.009

Hadavi S, Rigolon A, Gobster PH, Stewart WP (2021) Resident-led vacant lot greening and crime: do ownership and visual condition-care matter? Landsc Urban Plan 211:104096

Hallal PC, Andersen LB, Bull FC, Guthold R, Haskell W, Ekelund U (2012) Global physical activity levels: surveillance progress, pitfalls, and prospects. Lancet 380(9838):247–257

Harrison S, Kivuti-Bitok K, Macmillan A, Priest P (2019) EcoHealth and One Health: a theory-focused review in response to calls for convergence. Environ Int 132:105058

Hartig T, Mitchell R, de Vries S, Frumkin H (2014) Nature and health. Annu Rev Public Health 35: 207–228

Hartley K, Ryan P, Brokamp C, Gillespie GL (2020) Effect of greenness on asthma in children: a systematic review. Public Health Nurs 37:453–460

Hawkley LC, Cacioppo JT (2003) Loneliness and pathways to disease. Brain Behav Immun 17 (Suppl 1):S98–105. https://doi.org/10.1016/S0889-1591(02)00073-9

He H, Lin X, Yang Y, Lu Y (2020) Association of street greenery and physical activity in older adults: a novel study using pedestrian-centered photographs. Urban Forestry Urban Green 55: 126789

Hystad P, Davies HW, Frank L, Van Loon J, Gehring U, Tamburic L, Brauer M (2014) Residential greenness and birth outcomes: evaluating the influence of spatially correlated built-environment factors. Environ Health Perspect 122(10):1095–1102

International Health Conference (2002) Constitution of the World Health Organisation. 1946. Bulletin of the World Health Organisation, 80(12):983–984. World Health Organisation. apps.who.int/iris/handle/10665/268688

James P, Hart JE, Banay RF, Laden F (2016) Exposure to greenness and mortality in a nationwide prospective cohort study of women. Environ Health Perspect 124(9):1344–1352

Jennings V, Bamkole O (2019) The relationship between social cohesion and urban green space: an avenue for health promotion. Int J Environ Res Public Health 16(3):452

Jia BB, Yang ZX, Mao GX, Lyu YD, Wen XL, Xu WH, Lyu XL, Cao WB, Wang GF (2016) Health effect of forest bathing trip on elderly patients with chronic obstructive pulmonary disease. Biomed Environ Sci 29(3):212–218

Jing F, Liu L, Zhou S, Song J, Wang L, Zhou H, Wang Y, Ma R (2021) Assessing the impact of street-view greenery on fear of neighborhood crime in Guangzhou, China. Int J Environ Res Public Health 18(1):1–17

Johnson H, Kovats RS, McGregor G, Stedman J, Gibbs M, Walton H (2005) The impact of the 2003 heat wave on daily mortality in England and Wales and the use of rapid weekly mortality estimates. Euro Surveill 10(7):168–171

Jorgensen A, Gobster PH (2010) Shades of green: measuring the ecology of urban green space in the context of human health and well-being. Nat Cult 5:338–363. https://doi.org/10.3167/nc.2010.050307

Josseran L, Caillère N, Brun-Ney D, Rottner J, Filleul L, Brucker G, Astagneau P (2009) Syndromic surveillance and heat wave morbidity: a pilot study based on emergency departments in France. BMC Med Inform Decis Mak 9(1):1–9

Kahn PH, Kellert SR (2002) Children and nature: psychological, sociocultural, and evolutionary investigations. Massachusetts Institute of Technology Press, Cambridge

Kaplan S (1995) The restorative benefits of nature: toward an integrative framework. J Environ Psychol 15(3):169–182

Kaplan R, Kaplan S (1989) The experience of nature: a psychological perspective. Cambridge University Press, New York

Keesing F, Ostfeld RS (2021a) Impacts of biodiversity and biodiversity loss on zoonotic diseases. Proc Natl Acad Sci 118(17):e2023540118. https://doi.org/10.1073/pnas.2023540118

Keesing F, Ostfeld RS (2021b) Dilution efects in disease ecology. Ecol Lett 24(11):2490–2505. https://doi.org/10.1111/ele.13875

Keesing F, Belden LK, Daszak P et al (2010) Impacts of biodiversity on the emergence and transmission of infectious diseases. Nature 468:647–652

Kellert SR (2005) Building for life: designing and understanding the human-nature connection. Island Press, Washington, DC

Kellert SR, Wilson EO (1993) The biophilia hypothesis. Island Press, Washington, DC

Khalil H, Ecke F, Evander M, Magnusson M, Hörnfeldt B (2016) Declining ecosystem health and the dilution effect. Sci Rep 6:31314. https://doi.org/10.1038/srep31314

Klingberg J, Broberg M, Strandberg B, Thorsson P, Pleijel H (2017) Influence of urban vegetation on air pollution and noise exposure - a case study in Gothenburg, Sweden. Sci Total Environ 599–600:1728–1739. https://doi.org/10.1016/j.scitotenv.2017.05.051. Epub 2017 May 20

Kühn S, Düzel S, Eibich P, Kerkel C, Wustemann H, Kolbe J, Martensson J, Goebel J, Gallinat J, Wagner GG, Lindenburger U (2017) In search of features that constitute an "enriched environment" in humans: associations between geographical properties and brain structure. Sci Rep 7: 11920. https://doi.org/10.1038/s41598-017-12046-7

Kuo FE, Sullivan WC (2001) Environment and crime in the inner city: does vegetation reduce crime? Environ Behav 33(3):343–367. https://doi.org/10.1177/0013916501333002

Lackey KJ, Kaczynski AT (2009) Correspondence of perceived vs. objective proximity to parks and their relationship to park-based physical activity. Int J Behav Nutr Phys Act 6(1):1–9

Lafortezza R, Carrus G, Sanesi G, Davies C (2009) Benefits and well-being perceived by people visiting green spaces in periods of heat stress. Urban Forestry Urban Green 8(2):97–108

Laurent O, Wu J, Li L, Milesi C (2013) Green spaces and pregnancy outcomes in Southern California. Health Place 24:190–195

Lay JC, Scott SB, Hoppmann CA (2018) Social relationship quality buffers negative affective correlates of everyday solitude in an adult lifespan and an older adult sample. Psychol Aging 33(5):728–738. https://doi.org/10.1037/pag0000278

Lee AC, Maheswaran R (2011) The health benefits of urban green spaces: a review of the evidence. J Public Health 33(2):212–222

Lee JY, Lee DC (2014) Cardiac and pulmonary benefits of forest walking versus city walking in elderly women: a randomised, controlled, open-label trial. Eur J Integrative Med 6(1):5–11

Lee I, Choi H, Bang KS, Kim S, Song M, Lee B (2017) Effects of forest therapy on depressive symptoms among adults: a systematic review. Int J Environ Res Public Health 14(3):321

Li D, Sullivan WC (2016) Impact of views to school landscapes on recovery from stress and mental fatigue. Landsc Urban Plan 148:149–158. https://doi.org/10.1016/j.landurbplan.2015.12.015

Lindberg CF, Grimmond SB (2011) The influence of vegetation and building morphology on shadow patterns and mean radiant temperatures in urban areas: model development and evaluation. Theor Appl Climatol 105:311–323

Liu L, Zhong Y, Ao S, Wu H (2019) Exploring the relevance of green space and epidemic diseases based on panel data in China from 2007 to 2016. Int J Environ Res Public Health 16(14):2551. https://doi.org/10.3390/ijerph16142551

Lucas RA, Epstein Y, Kjellstrom T (2014) Excessive occupational heat exposure: a significant ergonomic challenge and health risk for current and future workers. Extreme Physiol Med 3(1): 1–8

Lyu B, Zeng C, Xie S, Li D, Lin W, Li N, Jiang M, Liu S, Chen Q (2019) Benefits of a three-day bamboo forest therapy session on the psychophysiology and immune system responses of male college students. Int J Environ Res Public Health 16(24):4991

Maas J, Van Dillen SME, Verheij RA, Groenewegen PP (2009) Social contacts as a possible mechanism behind the relation between green space and health. Health Place 15(2):586–595

Manman Zhu NN, Sze SN (2022) Effect of urban street trees on pedestrian safety: a micro-level pedestrian casualty model using multivariate Bayesian spatial approach. Accid Anal Prev 176: 106818. https://doi.org/10.1016/j.aap.2022.106818. ISSN 0001-4575

Manso M, Teotonio I, Silva CM, Cruz CO (2021) Green roof and green wall benefits and costs: a review of the quantitative evidence. Renew Sust Energ Rev 135:110111

Mao GX, Lan XG, Cao YB, Chen ZM, He ZH, Lv YD, Wang YZ, Hu XL, Wang GF, Yan J (2012) Effects of short-term forest bathing on human health in a broad-leaved evergreen forest in Zhejiang Province, China. Biomed Environ Sci 25(3):317–324

Mariani L, Parisi SG, Cola G, Lafortezza R, Colangelo G, Sanesi G (2016) Climatological analysis of the mitigating effect of vegetation on the urban heat Island of Milan, Italy. Sci Total Environ 569:762–773

Markevych I, Thiering E, Fuertes E, Sugiri D, Berdel D, Koletzko S, von Berg A, Bauer CP, Heinrich J (2014) A cross-sectional analysis of the effects of residential greenness on blood pressure in 10-year old children: results from the GINIplus and LISAplus studies. BMC Public Health 14:477

Marselle MR, Irvine KN, Warber SL (2014) Examining group walks in nature and multiple aspects of well-being: a large-scale study. Ecopsychology 6(3):134–147. https://doi.org/10.1089/eco.2014.0027

Marselle MR, Hartig T, Cox DTC, de Bell S, Knapp S, Lindley S, Triguero-Mas M, Böhning-Gaese K, Braubach M, Cook PA, de Vries S, Heintz-Buschart A, Hofmann M, Irvine KN, Kabisch N, Kolek F, Kraemer R, Markevych I, Martens D, Müller R, Nieuwenhuijsen M, Potts JM, Stadler J, Walton S, Warber SL, Bonn A (2021) Pathways linking biodiversity to human health: a conceptual framework. Environ Int 150:106420. https://doi.org/10.1016/j.envint.2021.106420

Marvuglia A, Koppelaar R, Rugani B (2020) The effect of green roofs on the reduction of mortality due to heatwaves: results from the application of a spatial microsimulation model to four European cities. Ecol Model 438:109351

Matsuoka RH (2010) Student performance and high school landscapes: examining the links. Landsc Urban Plan 97(4):273–282

Maury-Mora M, Gómez-Villarino MT, Varela-Martínez C (2022) Urban green spaces and stress during COVID-19 lockdown: a case study for the city of Madrid. Urban Forestry Urban Green 69:127492

McCormack GR, Rock M, Toohey AM, Hignell D (2010) Characteristics of urban parks associated with park use and physical activity: a review of qualitative research. Health Place 16(4):712–726

McCormick R (2017) Does access to green space impact the mental well-being of children: a systematic review. J Pediatr Nurs 37:3–7. https://doi.org/10.1016/j.pedn.2017.08.027

McDonnell MJ, MacGregor-Fors I (2016) The ecological future of cities. Science 352(6288): 936–938. https://doi.org/10.1126/science.aaf3630

Michelozzi P, De Donato F, Bisanti L, Russo A, Cadum E, DeMaria M, D'Ovidio M, Costa G, Perucci CA (2005) The impact of the summer 2003 heat waves on mortality in four Italian cities. Eur Secur 10(7):11–12

Mills JG, Weinstein P, Gellie NJC, Weyrich LS, Lowe AJ, Breed MF (2017) Urban habitat restoration provides a human health benefit through microbiome rewilding: the microbiome rewilding hypothesis. Restor Ecol 25(6):866–872. https://doi.org/10.1111/rec.12610

Miri M, de Prado-Bert P, Alahabadi A, Najafi ML, Rad A, Moslem A, Aval HE, Ehrampoush MH, Bustamante M, Sakhvidi MJZ, Nawrot T, Sunyer J, Dadvand P (2020) Association of green space exposure with telomere length in preschool children. Environ Pollut 266:115228

Moran D (2019) Back to nature? Attention restoration theory and the restorative effects of nature contact in prison. Health Place 57:35–43

Moran D, Turner J (2019) Turning over a new leaf: the health-enabling capacities of nature contact in prison. Soc Sci Med 231:62–69

Moran D, Jones PI, Jordaan JA, Porter AE (2020) Does nature contact in prison improve well-being? Mapping land cover to identify the effect of green space on self-harm and violence in prisons in England and Wales. Ann Am Assoc Geogr 111(6):1779–1795

Müller-Riemenschneider F, Petrunoff N, Yao J, Ng A, Sia A, Ramiah A, Tai BC, Uijtdewilligen L (2020) Effectiveness of prescribing physical activity in parks to improve health and wellbeing-the park prescription randomized controlled trial. Int J Behav Nutr Phys Act 17(1):1–14

Naeem S, Bunker D, Hector A, Loreau M, Perrings C (2009) Biodiversity, ecosystem functioning, and human wellbeing: an ecological and economic perspective. Oxford University Press

Nichani V, Dirks K, Burns B, Bird A, Morton S, Grant C (2017) Green space and pregnancy outcomes: evidence from growing up in New Zealand. Health Place 46:21–28

Nicholls N, Alexander L (2007) Has the climate become more variable or extreme? Progress 1992-2006. Prog Phys Geogr 31(1):77–87

Nieuwenhuijsen MJ (2020) Urban and transport planning pathways to carbon neutral, liveable and healthy cities; a review of the current evidence. Environ Int 140:105661

Nieuwenhuijsen MJ, Khreis H (2016) Car free cities: pathway to healthy urban living. Environ Int 94:251–262

Nieuwenhuijsen MJ, Khreis H, Triguero-Mas M, Gascon M, Dadvand P (2017) Fifty shades of green: pathway to healthy urban living. Epidemiology 28(1):63–71

Nogueira PJ, Falcão JM, Contreiras MT, Paixão E, Brandão J, Batista I (2005) Mortality in Portugal associated with the heat wave of August 2003: early estimation of effect, using a rapid method. Eur Secur 10(7):5–6

Oh B, Lee KJ, Zaslawski C, Yeung A, Rosenthal D, Larkey L, Back M (2017) Health and well-being benefits of spending time in forests: systematic review. Environ Health Prev Med 22(1): 1–11

Ohtsuka Y, Yabunaka N, Takayama S (1998) Shinrin-yoku (forest-air bathing and walking) effectively decreases blood glucose levels in diabetic patients. Int J Biometeorol 41(3):125–127

Ostfeld RS (2017) Biodiversity loss and the ecology of infectious disease. Lancet Planetary Health 1(1):E2–E3. https://doi.org/10.1016/S2542-5196(17)30010-4

Panno A, Carrus G, Lafortezza R, Mariani L, Sanesi G (2017) Nature-based solutions to promote human resilience and wellbeing in cities during increasingly hot summers. Environ Res 159: 249–256

Park BJ, Tsunetsugu Y, Kasetani T, Kagawa T, Miyazaki Y (2010) The physiological effects of Shinrin-yoku (taking in the forest atmosphere or forest bathing): evidence from field experiments in 24 forests across Japan. Environ Health Preven Med 15:18. https://doi.org/10.1007/s12199-009-0086-9

Partnership GE (2014) Innovative NHS green space in Scotland. Scotland Forestry Commission, Edinburgh

Pini B, Morris D, Mayes R (2016) Rural youth: mobilities, marginalities, and negotiations. Space Place Environ 3:463

Poumadere M, Mays C, Le Mer S, Blong R (2005) The 2003 heat wave in France: dangerous climate change here and now. Risk Anal 25(6):1483–1494

Rajoo KS, Karam DS, Abdullah MZ (2020) The physiological and psychosocial effects of forest therapy: a systematic review. Urban Forestry Urban Green 54:126744

Ramirez-Rubio O, Daher C, Fanjul G, Gascon M, Mueller N, Pajín L, Plasencia A, Rojas-Rueda D, Thondoo M, Nieuwenhuijsen MJ (2019) Urban health: an example of a "health in all policies" approach in the context of SDGs implementation. Global Health 15:87. https://doi.org/10.1186/s12992-019-0529-z

Ricciardi E, Spano G, Lopez A, Tinella L, Clemente C, Elia G, Dadvand P, Sanesi G, Bosco A, Caffò AO (2022) Long-term exposure to greenspace and cognitive function during the lifespan: a systematic review. Int J Environ Res Public Health 19(18):11700. https://doi.org/10.3390/ijerph191811700

Richardson EA, Pearce J, Shortt NK, Mitchell R (2017) The role of public and private natural space in children's social, emotional and behavioural development in Scotland: a longitudinal study. Environ Res 158:729–736

Roger F, Caron A, Morand S, Pedrono M, de Garine-Wichatitsky M, Chevalier V, Tran A, Gaidet N, Figuié M, de Visscher MN, Binot A (2016) One Health and EcoHealth: the same wine in different bottles? Infect Ecol Epidemiol 17(6):30978. https://doi.org/10.3402/iee.v6.30978

Rojas-Rueda D, Nieuwenhuijsen MJ, Gascon M, Perez-Leon D, Mudu P (2019) Green spaces and mortality: a systematic review and meta-analysis of cohort studies. Lancet Planet Health 3: e469–e477

Roman LA, Conway TM, Eisenman TS, Koeser AK, Ordóñez Barona C, Locke DH, Jenerette GD, Östberg J, Vogt J (2021) Beyond 'trees are good': disservices, management costs, and tradeoffs in urban forestry. Ambio 615–630. https://doi.org/10.1007/s13280-020-01396-8

Romanelli C, Cooper D, Campbell-Lendrum D, Maiero M, Karesh WB, Hunter D, Golden CD (2015) Connecting global priorities: biodiversity and human health. A state of the knowledge review. World Health Organisation and Secretariat of the Convention on Biological Diversity
Rook GAW (2013) Regulation of the immune system by biodiversity from the natural environment: an ecosystem service essential to health. Proc Natl Acad Sci 110(46):18360–18367
Rosa CD, Larson LR, Collado S, Profice CC (2021) Forest therapy can prevent and treat depression: evidence from meta-analyses. Urban For Urban Green 57:126943
Ruiz-Gil T, Acuña JJ, Fujiyoshi S, Tanaka D, Noda J, Maruyama F, Jorquera MA (2020) Airborne bacterial communities of outdoor environments and their associated influencing factors. Environ Int 145:106156. https://doi.org/10.1016/j.envint.2020.106156
Russell R, Guerry AD, Balvanera P, Gould RK, Basurto X, Chan KM, Klain S, Levine J, Tam J (2013) Humans and nature: how knowing and experiencing nature affect well-being. Annu Rev Environ Resour 38:473–502
Sallis JF, Cerin E, Conway TL, Adams MA, Frank LD, Pratt M, Salvo D, Schipperijn J, Smith G, Cain KL, Davey R, Kerr J, Lai PC, Mitáš J, Reis R, Sarmiento OL, Schofield G, Troelsen J, Van Dyck D, De Bourdeaudhuij I, Owen N (2016) Physical activity in relation to urban environments in 14 cities worldwide: a cross-sectional study. Lancet 387(10034):2207–2217. https://doi.org/10.1016/S0140-6736(15)01284-2
Scartazza A, Mancini ML, Proietti S, Moscatello S, Mattioni C, Costantini F, Di Baccio D, Villani F, Massacci A (2020) Caring local biodiversity in a healing garden: therapeutic benefits in young subjects with autism. Urban Forestry Urban Green 47:126511
Schmidt KA, Ostfeld RS (2001) Biodiversity and the dilution effect in disease ecology. Ecology 82: 609–619
Seeland K, Nicolè S (2006) Public green space and disabled users. Urban For Urban Green 5(1): 29–34
Seeland K, Dübendorfer S, Hansmann R (2009) Making friends in Zurich's urban forests and parks: the role of public green space for social inclusion of youths from different cultures. Forest Policy Econ 11(1):10–17. https://doi.org/10.1016/j.forpol.2008.07.005
Seo S, Choi S, Kim K, Kim SM, Park SM (2019) Association between urban green space and the risk of cardiovascular disease: a longitudinal study in seven Korean metropolitan areas. Environ Int 125:51–57
Shanahan DF, Fuller RA, Bush R, Lin BB, Gaston KJ (2015) The health benefits of urban nature: how much do we need? BioScience 65(5) 476–485. https://doi.org/10.1093/biosci/biv032
Shepley M, Sachs N, Sadatsafavi H, Fournier C, Peditto K (2019) The impact of green space on violent crime in urban environments: an evidence synthesis. Int J Environ Res Public Health 16(24)
Shinzato P, Duarte D (2012) Microclimatic effect of vegetation for different leaf area index-LAI. In: Proceedings of the 28th international PLEA conference: opportunities, limits & needs towards an environmentally responsible architecture, Lima, Peru, 7–9 November 2012; vol 6
Sia A, Tan PY, Meng Wong JC, Araib S, Ang WF, Boon KHE (2022) The impact of gardening on mental resilience in times of stress: a case study during the COVID-19 pandemic in Singapore. Urban Forestry Urban Green 68:127448
Silvers A, Florence BT, Rourke DL, Lorimor RJ (1994) How children spend their time: a sample survey for use in exposure and risk assessments. Risk Anal 14(6):931–944
Sodoudi S, Zhang H, Chi X, Müller F, Li H (2018) The influence of spatial configuration of green areas on microclimate and thermal comfort. Urban Forestry Urban Green 34:85–96
Sonntag-Öström E, Stenlund T, Nordin M, Lundell Y, Ahlgren C, Fjellman-Wiklund A, Jarvholm LS, Dolling A (2015) Nature's effect on my mind – patients' qualitative experiences of a forest-based rehabilitation programme. Urban Forestry Urban Green 14(3):607–614
Southon GE, Jorgensen A, Dunnett N, Hoyle H, Evans KL (2017) Biodiverse perennial meadows have aesthetic value and increase residents' perceptions of site quality in urban green-space. Landsc Urban Plan 158:105–118. https://doi.org/10.1016/j.landurbplan.2016.08.003

Spano G, D'Este M, Giannico V, Carrus G, Elia M, Lafortezza R, Panno A, Sanesi G (2020) Are community gardening and horticultural interventions beneficial for psychosocial well-being? A meta-analysis. Int J Environ Res Public Health 17(10):3584

Speak AF, Salbitano F (2021) Thermal comfort and perceptions of the ecosystem services and disservices of urban trees in Florence. Forests 12:1387

Speak AF, Salbitano F (2022) Summer thermal comfort of pedestrians in diverse urban settings: a mobile study. Build Environ 208:108600. https://doi.org/10.1016/j.buildenv.2021.108600

Stanhope J, Breed MF, Weinstein P (2020) Exposure to greenspaces could reduce the high global burden of pain. Environ Res 187:109641

Steptoe A, Shankar A, Demakakos P, Wardle J (2013) Social isolation, loneliness, and all-cause mortality in older men and women. Proc Natl Acad Sci 110(15):5797–5801

Stier-Jarmer M, Throner V, Kirschneck M, Immich G, Frisch D, Schuh A (2021) The psychological and physical effects of forests on human health: a systematic review of systematic reviews and meta-analyses. Int J Environ Res Public Health 18(4):1770

Sullivan WC, Kuo FE, DePooter SF (2004) The fruit of urban nature: vital neighbourhood spaces. Environ Behav 36(5):678–700

Tamaskani Esfehankalateh A, Ngarambe J, Yun GY (2021) Influence of tree canopy coverage and leaf area density on urban heat Island mitigation. Sustainability 13:7496. https://doi.org/10.3390/su13137496

Tan JK, Belcher RN, Tan HT, Menz S, Schroepfer T (2021) The urban heat Island mitigation potential of vegetation depends on local surface type and shade. Urban Forestry Urban Green 62:127128

Tiwari A, Kumar P, Baldauf R, Zhang KM, Pilla F, Di Sabatino S, Brattich E, Pulvirenti B (2019) Considerations for evaluating green infrastructure impacts in microscale and macroscale air pollution dispersion models. Sci Total Environ 672:410–426

Tomson M, Kumar P, Barwise Y, Perez P, Forehead H, French K, Moraska L, Watts JF (2021) Green infrastructure for air quality improvement in street canyons. Environ Int 146:106288

Torquati J, Schutte A, Kiat J (2017) Attentional demands of executive function tasks in indoor and outdoor settings: behavioral and neuroelectrical evidence. Children Youth Environ 27(2): 70–92. https://doi.org/10.7721/chilyoutenvi.27.2.0070

Ulrich R (1984) View through a window may influence recovery. Science 224(4647):224–225

UNDESA, U (2016) The world cities in 2016

UN Department of Economic and Social Affairs (2015) World urbanisation prospects; the 2014 revision. United Nations, New York

van den Bosch M, Nieuwenhuijsen M (2017) No time to lose - green the cities now. Environ Int 99: 343–350

Van Renterghem T (2019) Towards explaining the positive effect of vegetation on the perception of environmental noise. Urban Forestry Urban Green 40:133–144

Voigt A, Wurster D (2015) Does diversity matter? The experience of urban nature's diversity: case study and cultural concept. Ecosyst Serv 12:200–208. https://doi.org/10.1016/j.ecoser.2014.12.005

Wallner P, Kundi M, Arnberger A, Eder R, Allex B, Weitensfelder L, Hutter HP (2018) Reloading pupils' batteries: impact of green spaces on cognition and wellbeing. Int J Environ Res Public Health 15(6):1205. https://doi.org/10.3390/ijerph15061205

Ward Thompson C, Aspinall P, Roe J, Robertson L, Miller D (2016) Mitigating stress and supporting health in deprived urban communities: the importance of green space and the social environment. Int J Environ Res Public Health 13(4):440. https://doi.org/10.3390/ijerph13040440

Weldon S, Bailey C, O'Brien L (2007) New pathways to health and wellbeing: summary of research to understand and overcome barriers to accessing woodland. Forestry Commission, Scotland

Wells NM (2000) At home with nature effects of greenness on children's cognitive functioning. Environ Behav 32(6):775–795

Wengel TTT, Troelsen J (2020) How the urban environment impacts physical activity: a scoping review of the associations between urban planning and physical activity. The Danish Health Authority, Copenhagen, 86 pp
WHO (2016) Urban green spaces and health - a review of evidence. Available: www.euro.who.int/en/health-topics/environment-and-health/urban-health/publications/2016/urban-green-spaces-and-health-a-review-of-evidence-2016
WHO (2018) Towards more physical activity in cities: transforming public spaces to promote physical activity – a key contributor to achieving the sustainable development goals in Europe. World Health Organisation. Regional office for Europe www.euro.who.int/data/assets/pdf_file/0018/353043/2017_WHO_Report_FINAL_WEB.pdf
Wilson EO (1984) Biophilia. Harvard University Press, Cambridge
Wilson N, Ross M, Lafferty K, Jones R (2008) A review of ecotherapy as an adjunct form of treatment for those who use mental health services. J Public Ment Health
Wolf T, Lyne K, Martinez GS, Kendrovski V (2015) The health effects of climate change in the WHO European region. Climate 3(4):901–936
Wu X, Nethery RC, Sabath MB, Braun D, Dominici F (2020) Air pollution and COVID-19 mortality in the United States: strengths and limitations of an ecological regression analysis. Sci Adv 6(45):eabd4049. https://doi.org/10.1126/sciadv.abd4049
Xu B, Shi Q, Zhang Y (2022) Evaluation of the health promotion capabilities of greenway trails: a case study in Hangzhou, China. Land 11(4):547. https://doi.org/10.3390/land11040547
Yu CPS, Hsieh H (2020) Beyond restorative benefits: evaluating the effect of forest therapy on creativity. Urban Forestry Urban Green 51:126670
Zare Sakhvidi MJ, Yang J, Mehrparvar AH, Dzhambov AM, Ebrahimi A, Dadvand P, Jacquemin B (2022) Exposure to greenspace and cancer incidence, prevalence, and mortality: a systematic review and meta-analyses. Sci Total Environ 10:838 (Pt 2):156180. https://doi.org/10.1016/j.scitotenv.2022.156180
Zhang XX, Liu JS, Han LF, Xia S, Li SZ, Li OY, Kassegne K, Li M, Yin K, Hu QQ, Xiu LS, Zhu YZ, Huang LY, Wang XC, Zhang Y, Zhao HQ, Yin JX, Jiang TG, Li Q, Fei SW, Gu SY, Chen FM, Zhou N, Cheng ZL, Zhou XN (2022) Towards a global One Health index: a potential assessment tool for One Health performance. Infect Dis Poverty 11:57. https://doi.org/10.1186/s40249-022-00979-9
Zijlema WL, Avila-Palencia I, Triguero-Mas M, Gidlow C, Maas J, Kruize H, Andrusaityte S, Grazuleviciene R, Nieuwenhuijsen MJ (2018) Active commuting through natural environments is associated with better mental health: results from the phenotype project. Environ Int 121:721–727

Forests, Forest Products, and Services to Activate a Circular Bioeconomy for City Transformation

Giovanna Ottaviani Aalmo, Divina Gracia P. Rodriguez, Lone Ross Gobakken, and Fabio Salbitano

1 Introduction

In the context of BioCities, the circular bioeconomy has transformative potential in rethinking urban areas, especially using urban, peri-urban, and rural forestry as a nature-based solution. These can be seen as an interconnected forest network providing essential and high-value services, such as health benefits and climate resilience, and sustainable products, principally through rural and peri-urban forests.

The circular bioeconomy has been gaining considerable attention across the globe, particularly in many public and private sector strategies (Ghisellini et al. 2016; Baumgartner and Rauter 2017). Circular bioeconomy may be defined as 'an economy where the basic building blocks for materials, chemicals, and energy are derived from renewable biological resources' (McCormick and Kautto 2013), and the products produced are recycled, reused, and/or repurposed to ensure they remain within the system as long as possible. As the circular bioeconomy allows society to use the unexploited value of biological products and processes, these will also create new growth and welfare benefits for citizens and nations. These benefits are represented by productivity gains (in the agriculture and health sectors), enhancement effects (in the health and nutrition sectors), and substitution effects (in the environmental, industrial, and energy sectors); with additional benefits derived from more eco-efficient and sustainable use of natural resources to provide goods and services to an ever-growing global population (Barañano et al. 2021). Circular bioeconomy can address the grand societal challenges by meeting the societal

G. O. Aalmo (✉) · D. G. P. Rodriguez · L. Ross Gobakken
Norwegian Institute of Bioeconomy Research (NIBIO), Ås, Norway
e-mail: giovanna.ottaviani.aalmo@nibio.no

F. Salbitano
University of Sassari, Sassari, Italy

G. E. Scarascia-Mugnozza et al. (eds.), *Transforming Biocities*, Future City 20,
https://doi.org/10.1007/978-3-031-29466-2_7

needs for energy, chemicals, food, and raw materials whilst integrating science with business and society.

Since the role of forests and the forest/wood sector is often viewed as providers of timber, wood-based products, pulp, paper, and bioenergy (de Arano et al. 2018), the potential contribution of the non-wood forest products (NWFPs) has been neglected in the past (Inazio Martinez de Arano 2021), particularly in the industrialised economies. NWFPs include cork, resins, gums, wild mushrooms, aromatic and medicinal plants, and wild nuts and berries.[1] NWFPs can contribute to human nutrition, renewable materials, and cultural and experiential services, as well as create job and income opportunities in both urban and rural areas (Weiss et al. 2020). Many NWFP enterprises remain in the informal sector (i.e. those businesses not managed through formal arrangements) and in-depth understanding of underlying factors remains limited (Meinhold and Darr 2019).

In the context of the forest and wood sector, the circular bioeconomy involves the principle of 'cascading in value' (Jarre et al. 2020), which prioritises the highest possible use (value) of wood over the whole life cycle to optimise the social, economic, and environmental benefits, whilst minimising concomitant trade-offs (Toppinen et al. 2020). In this regard, forest and forest/wood-based products and services can play a key role in activating a circular bioeconomy in urban communities, by providing a renewable source of raw materials needed to manufacture, maintain, improve, and sustain the goods and services required for the proper functioning of a BioCity, hence reducing the dependence on non-renewable materials (Antikainen et al. 2017) and reaching other UN Sustainable Development Goals (European Commission 2018, 2019). For example, forest biomass is increasingly being used in the production of textiles, bioplastics, chemicals and intelligent packaging, pharmaceuticals, and construction (Hetemäki and Hurmekoski 2016; Ladu et al. 2020). Forest-based sector companies and businesses 'need to restructure their business and create novel business models along with the demands set by the surrounding environment' (Näyhä 2020). However, small and medium enterprises (SMEs) weakly recognised the concept of circular bioeconomy in business as its profitability was perceived as dependent on government subsidies (D'Amato et al. 2019).

Whilst wood is abundant in cities around the world, its potential contribution to a circular bioeconomy has received little attention not only by the general public, but also by the scientific community (Kampelmann 2020). Biomass from trees felled in cities, for example in North America and European, is mostly used as mulch and firewood instead of further exploiting its versatile nature. Forest biomass can be turned into biofuel, cosmetics and perfumes, food additives, and nutritional supplements, as in the case of lignin that can be converted into chemicals adding smoky

[1]FAO Zola, A. (1999). 'EC-FAO PARTNERSHIP PROGRAMME (1998–2000) Funded in part by the Tropical Forestry Budget line B7–6201 PROJECT GCP/INT/679/EC '.Define NWFP as '… products [that] consist of goods of biological origin other than wood, derived from forests, other wooded land and trees outside forests'.

flavours to foods (Mathew and Zakaria 2015), as well as more traditional uses like construction timber, wood products, and pulp and paper.

To address the challenges related to the consequences of climate change that have emerged over the past decade and their negative effects, especially in urban environments, the concept of circular economy has been gaining significance as an efficient mitigation strategy to tackle carbon emissions. As a bioeconomy, through its principles, is a renewable component of the circular economy, cities can become BioCities by adopting this concept and hence solve many of their socioeconomic development challenges.

How can this be achieved? Firstly, strengthening the linkages between the circular bioeconomy, forestry, and the wood sector in both urban and peri-urban contexts helps to define the framework of BioCities. Secondly, acknowledging the fundamental interdependence between urban and rural communities in relation to forest management. And thirdly, innovative forest-based solutions using forest and non-wood forest products (i.e. food, fodder, fibres, fragrances for perfumes; ornamental pods and seeds; resins; and oils) and services that contribute, secure, and increase the bio-circularity of the economy, thus accounting for the benefits and trade-offs which are often intertwined.

2 Trends in the Reuse of Materials in Architectural and Urban Development

The concept of the 'cascade chain' was introduced for the first time by Sirkin and ten Houten (1994). 'Wood cascading' can be defined as a 'strategy of using raw materials or products made in time-sequential steps for as long, as often and as efficiently as possible and only using them energetically at the end of the product life cycle' (Kosmol et al. 2012). In principle, this should result in environmental benefits because less virgin material is required to provide the same function(s), and the carbon in the cascaded wood is stored for a longer time as timber in buildings (see chapter "Innovative Design, Materials, and Construction Models for BioCities"). In addition, more value is derived per unit of material because it is used in more product cycles, although each cycle produces less value for a given amount of material, compared with the previous cycle (value hierarchy).

The role of cascading of wood waste in the bioeconomy is highlighted in the Circular Economy Strategy (Camilleri 2021) and the European Union Forest Strategy (ECC 2013). The European Waste Framework Directive (European Parliament 2008) describes a waste hierarchy where reuse and recovery are considered more favourable options compared with energy recovery. It applies reuse and recycling targets of 50% (by weight) of household waste and 70% of non-hazardous construction and demolition waste (including waste wood) by 2020. By 2030, it is likely that the demand for wood in the EU will exceed supply, mainly due to increasing requirements of biomass for energy (Thonemann and Schumann 2018). This means that strategies must be developed that ensure the most efficient use of a resource for which there will be increasing competition.

As mentioned above, one strategy is the reuse of the material by cascading it through the value chain. The idea behind the circular economy is to retain the value of materials, products, and services in the European economy for as long as possible. Current levels of cascading of wood waste differ greatly from country to country. Countries with a small forest resource base and a high population density exhibit higher levels of cascading (e.g. Netherlands, Italy, and the UK), whereas countries with a large forest resource base and lower population densities (e.g. Nordic countries) have low levels of cascading. One reason for this difference is the tendency for wood-based panel board industries to be located close to consumer markets (Vis et al. 2016). In countries with a high level of timber provision, the requirements of the wood-based panel products industry can be easily met by utilising processing residues, though the use of post-consumer waste for this purpose could mean to gather and move these residues away from the markets. Furthermore, in countries with dispersed populations, the long transport distances involved in transporting post-consumer wood waste to a few particleboard plants can potentially lead to negative environmental consequences. The main destination for wood waste in new products is currently limited to the particleboard industry, which utilises clean wood from various sources, including pallets and furniture as well as wood recovered from construction and demolition.

3 Ecosystem Services Provided by Urban and Peri-Urban Forests

Forests are not only about providing wood, as they also offer valuable forest ecosystem services and other benefits for the well-being of the people (MEA 2005). Urban and peri-urban forests perform a set of ecological functions that can be assessed through their composition, structure, or ecological processes (e.g. productivity and energy flows). Together, these functions originate what is called ecosystem services (ES), which generate direct benefits for the users, and in the case of urban forests, this will be the city residents. These benefits and services influence human well-being and can be evaluated through different approaches (i.e. ecological, sociological, and economical) (Haase et al. 2014). The most immediate benefits provided by urban and peri-urban forests are those related to the provisioning and regulatory services: food, fuel, and wood provision, as well as carbon storage, temperature and noise regulation and mitigation, and the water cycle (Davies et al. 2017); not to mention erosion control, climate regulation, and precipitation. Other services related to the cultural sphere include spiritual, aesthetic, religious, and recreational uses (Fish and Church 2014). The Coronavirus pandemic raised awareness of the unbalanced ratio between the excessive spaces devoted to cars in urban areas to the availability of green spaces. When lockdowns were enforced throughout Europe from March 2020, citizens invaded green spaces whilst car movements were reduced to a minimum, highlighting the new possibilities for the BioCity. Enforced confinement undeniably accentuated the need for accessible green spaces in the city, especially when leaving the city was impossible. The pandemic raised questions about current urban planning models, which are not focused on the sustainability of local biodiversity to ensure a healthy relationship between humans, species, and green networks. Forests have become hubs for many of the European regions linking rural and urban areas, giving people the possibility to access otherwise limited ecosystem services and to develop existing forms of businesses, or create new ones.

Case Study 1 The FORESTAMI Project (Politecnico di Milano)

Forestami is a project promoted by the Metropolitan City of Milan, the Municipality of Milan, the Lombardy Region, the North Milan Park, the South Milan Agricultural Park, ERSAF, and the Milan Community Foundation. The project aims at planting 3 million trees by 2030 to provide clean air, improve the life of the greater Milan area, and counteract the effects of climate change, whilst involving public and private entities. Developed from research at the Politecnico di Milano and supported by the Falck Foundation and FS Sistemi Urbani, Forestami builds a strategic vision of the role of green areas in the metropolitan area of Milan.

Forestami gives life to a process of census, enhancement, and implementation of all green, permeable, and tree-lined systems, to promote policies and projects that support urban forestry activities and build a Metropolitan Park in the Milan area. The project will stimulate and leverage the physical and mental well-being of the people who live there. Multiplying the number of plants along streets, squares, and courtyards, on the roofs and facades of houses, is the most effective, economic, and engaging way to intervene on global urban warming, energy consumption, and air pollution.

4 Urban Food Production (Agroforestry)

Forests contribute to food security and nutrition in several ways (Fanzo et al. 2017): (1) direct provision of food from forests contributes to dietary quality and diversity; (2) forest foods (e.g. fruits and vegetables) are rich in micronutrients; (3) provision of energy to process agricultural and forest foods for consumption; (4) forest and forest products provides income and employment that allow for the purchase of food and other necessities; and (5) provision of products and services that people consumed or used to serve their needs (e.g. forest plants, fungi, and animals).

Agroforestry is defined as a dynamic and ecologically based land-use system, integrating woody perennials, leafy crops, and/or livestock on the same plot of land. These agroforestry systems can be managed differently, at the same time, or managed on a rotating basis over several years depending on the implementation context and the expectations of the end-users (Leakey 1996). A well-managed agroforestry system is able to improve the biological, physical, and ecological interaction of its elements, whilst increasing the environmental, social, and economic benefits for end-users at all levels (Lundgren and Raintree 1983). As a practice, agroforestry was utilised for centuries before the demand for products slowly intensified, evolving into specialised monocultures (Lassoie et al. 2009; Leakey 2010).

Landscape designers and urban planners are starting to integrate food production into city planning by using a multifunctional landscape approach that recognises the value of ecological and cultural functions, beyond the simple metric of food production (Lovell 2010). Urban forestry and urban agriculture have remained relatively separated in their practice and science (Clark and Nicholas, 2013). 'Urban food forestry' (UFF), proposed by Clark and Nicholas (2013), is the 'intentional and strategic use of woody perennial food producing species in urban edible landscapes' (p. 1652). This is the use of food trees (i.e. fruit and nut trees) for their multifunctionality in urban landscapes and encompasses any forms or use of food trees in urban landscapes. Urban food forestry encompasses a range of different food tree systems and practices from street trees to orchards to multistory polyculture systems that have food trees in urban landscapes. It is primarily focused on food-producing trees planted across a landscape, highlighting the contribution of food trees in increasing landscape multifunctionality through the provision of food and other benefits that urban trees generally provide (e.g. air quality improvement, temperature, and stormwater runoff control). This is different from food forestry, which involves complex vegetation structure and composition (Park et al. 2018). Urban food forestry aims to improve local food security and is supposed to 'combine elements of urban agriculture, urban forestry, and agroforestry' (Clark and Nicholas 2013).

The contribution of forest and forest-based products in relation to the bioeconomy to provide food for the nutritional needs of the increasing human population in urban landscapes, however, has not been fully realised. In the BioCity, food harvesting (foraging) and food production from the urban forest and green spaces can be promoted as a set of bottom-up initiatives with strong support from co-governance of urban spaces. Thus, food production will be an important co-benefit from urban forests and green spaces, but the main benefits of food production in urban areas may well be related to the many social connections created between different groups of the cities (Leakey 2010).

Case Study 2 The City of Havana

Foto: http://www.fao.org/ag/agp/greenercities/en/ggclac/havana.html

Havana has 2.1 million inhabitants and is an outstanding example of urban agriculture (forestry) on a large scale. After the end of the Soviet era, Havana fell into the worst economic crisis in its history. Since 1994, the Cuban government drafted a strategy that has been transforming Havana into one of the most successful examples of urban agriculture worldwide.

More than half of the consumed food is grown organically on-site (Hoornweg and Munro-Faure 2008; Baumgartner and Rauter 2017). Yields were low at first, owing to the lack of farming experience and inputs. But with strong government support, urban agroforestry has rapidly transformed from a spontaneous response to food insecurity to a national priority.

Urban farmers in Havana use predominately low-tech methodologies to achieve yields of up to 20 kg per m^2, a greater value than commonly achieved in mixed-stand small-scale agriculture (Fermont and Benson 2011). The most common technologies in use are drip irrigation, organoponics, regular addition of compost, and other common horticultural practices (e.g. use of well-adapted varieties, crop mixing, crop rotation, and integrated pest management). Despite the lack of advanced technology, the City's urban and peri-urban agriculture sector development now includes five agricultural enterprises, managing some 700 crop farms, 170 cattle farms, and 27 tree production units, two provincial companies specialising in pig and livestock production, 29 agricultural cooperatives, and 91 credit and service cooperatives that grow flowers, vegetables, and raise small animals (Somarriba et al. 2012; Borelli et al. 2017).

5 Urban–Rural Community Linkages

The interpretation of *rural* and *urban* is subjective, dependant on economic, social, environmental, and political variables (Lynch 2005). Nevertheless, as a mainstream simplistic conceptualisation, rural has been perceived as equating to farming areas, with urban as crowded population settlements (Braun 2007). This view represents the basis for treating these two types of communities differently. Further, it underestimates the contribution that each space has to provide for development, employment, poverty reduction, entrepreneurship, and environmental–social impacts in both areas.

There is generally a marked difference between rural and urban areas both in population density and in the availability and access to social services. Urban areas are places where many people live in a limited amount of space with better and improved access to social services, whilst rural areas are places where people live in a dispersed space with diversified access, in space and time, to social services. This variation decreases when moving closer to urban areas. The population density translates into a higher density of land consumption and soil sealing for infrastructure, residential, commercial, and industrial construction. Urban and rural areas are strongly interdependent (Steinberg 2014), however, due to the functional links between and across sectors, services, and because of the constant flow of people (Gebre and Gebremedhin 2019).

The necessity for broadening the bio-based economic models with a regional focus has been dramatically increasing in the past years (chapter "From BioCities to BioRegions and Back: Transforming Urban-Rural Relationships"). This trend has opened up the potential for regional, urban, and rural development to make all regions more equitable, inclusive, and resilient, based on bio-based value chains and products. Particular attention has been given to unused and residual streams from the agricultural and forestry, strengthening the linkages between urban and rural communities and making value chains more efficient and competitive by adding higher economic value. To increase the sustainability of bio-based industries, new value chains crossing the boundaries of different economic sectors are needed. For example, biorefineries transform rural-originated products into energy for urban areas within facilities located in urban areas, and hence make urban areas the main market and service centres for rural-related energy businesses (Hjalager 2017).

The local context is also fundamental in making urban–rural linkages work and should inform and guide the translation of global agendas such as the 2030 Agenda for Sustainable Development (including the UN Sustainable Development Goals [SDGs]), the New Urban Agenda (NUA), and others (Desa 2016; Hosagrahar et al. 2016; Caprotti et al. 2017). National and regional commitments should have policy coherence and integrated actions across the local territory. Such translations can mainstream urban–rural linkages and integrated territorial development, and it should help local authorities and regional actors to take the lead in overcoming social, economic, and environmental inequalities, whilst also leveraging the comparative advantages of the flows of people, goods, and services across the urban–rural continuum.

Strengthening governance mechanisms by incorporating urban–rural linkages into multisectoral, multi-level, and multi-stakeholder governance is key to the delivery of the UN SDGs and to address human needs. These require policies, strategies, and action plans that are: (1) horizontally integrated across spatial scales in metropolitan regions, adjacent cities, and towns, including rural hinterlands; (2) sectorally integrated with the public and private sectors, civil society organisations, research and professional institutions, formal and informal civic associations; and (3) vertically integrated across different levels of engagement and official decision-making.

Mapping the linkages between urban and rural communities, and their diverging perceptions of forest management and forest issues, is key to increasing circularity opportunities in cities. Filling the gap of understanding between forestry and urban communities about the values and uses of the forest areas is paramount to engage society, especially the urban population, to support policies needed in a circular bioeconomy strategy and for acceptance in policy implementation. The European Commission's plan for developing a long-term vision for rural areas (EU rural vision) (MCELDOWNEY 2021) was officially proposed in July 2019 as part of the European Commission's next key priorities, and has a central concern about urban–rural linkages. It aims to mobilise policymakers, rural stakeholders and citizens, and urban actors more widely in a dialogue on the future of Europe. The ultimate aim will be to provide a holistic vision up to 2040 that will allow the development and implementation of innovative, inclusive, and sustainable solutions tailored to rural–urban linkages in light of the climate crisis, the ongoing digital transformation, and the recovery from the COVID-19 pandemic.

6 The Crucial Role of EU-Funded Research to Solve the Rural–Urban Dilemma

Many EU-funded projects under the Horizon 2020 programme have already put in place the skills and expertise to address rural–urban issues and to enhance their potential, in order to seize opportunities and contribute to the future of Europe. For example, the EU-funded project ROBUST (https://rural-urban.eu) has studied new ways to enhance synergies in the governance of urban and rural communities by activating living labs on specific solutions (i.e. *The Lisbon Living Lab—A Territorial Economy of Proximity*). This project reflects the importance of approaching the regional territory in a unitary way, where the rural and urban dimensions are complementary without any hierarchical relationship. The strengthening and enhancement of the relationship between these dimensions, through new business models and the promotion of sustainable food systems that capitalise on ecosystem services, are the main objectives of this Living Lab.

7 Enabling the Circular Economy: Bottlenecks and Trade-Offs

A circular economy (and bioeconomy) aims to transform the actual linear fossil-based economy into a more efficient and waste recirculating one, and in the case of the bioeconomy, on renewable biological resources reducing production costs and promoting innovation and competitiveness (Guenster et al. 2011). In order to *biologise* the urban economy, new ways to remain competitive must be found and adapted to existing biogeographical, economic, and social specifics to maximise economic, social, and environmental benefits (de Arano et al. 2018).

The amount of wood harvested in cities is substantial. In the Randstad example, a polycentric urban complex in which the four largest cities of the Netherlands can be found, 2000 m^3 of exploitable roundwood are produced per year (Stadshout 2021). The potential contribution of wood to the circular bioeconomy in these cities, however, has been overlooked (Kampelmann 2020). Current practices are mostly limited to low-value uses of wood (e.g. mulch or firewood). What is needed are initiatives to develop the local value chain to make the best use of wood, including the coordination of different types of actors and the later phases of the wood supply chain.

The use values of both forest products and NWFPs are often underestimated in terms of their total contribution to sustainable development. To be able to make better-informed decisions and strengthen policy formulation for achieving high level of circularity in the forest sector, the values derived from forest products and NWFPs, in terms of ecosystem services, need to be quantified, which is often difficult. The economic value can, in fact, be related to the recreational services provided by the forest rather than to the timber price, and combining different values can also be challenging. Moreover, the expected trade-offs between economic gains and ecological losses from the production and consumption of forest and forest-based products, which are typically uncertain and vary at temporal and spatial scales, are also difficult to measure and the methods to do so also have limitations. For example, the production of industrial grade wood from the boreal forests in Canada has led to the degradation of ecological functions and services in boreal zones (Brandt et al. 2013).

The large-scale production and market penetration of forest and forest-based products remain major challenges to be addressed (Clark et al. 2012). For example, forest-based biofuel is a promising solution to increase the share of renewable and sustainable energy in the transportation sector. Currently, biofuel is mainly produced from food crops and palm oil (Afiff et al. 2013). Advanced forest-based biofuels are regarded as a sustainable alternative (European Commission 2021). Whilst the use of raw wood material for biofuel production will lead to less competition with food production in terms of land use, it can increase fuel competition since it is currently used in both the heat and power sectors, not to mention its use in the traditional forest industries (Bryngemark 2019).

Policies are also a pivotal element in transitioning from linear to circular economies. Policy interventions for the circular economy can be categorised into five different groups: (1) regulatory frameworks setting requirements or bans; (2) market-based policies dealing with existing incentives; (3) information policies related to raising awareness at large; (4) public procurement and infrastructure broadly used to act on materials lifespan and their disposal; and (5) innovation support schemes. This last type of policy has been promoted and adopted in many cities and regions, although the impact of the activities supported is often not assessed. Furthermore, there are a wide range of forest-related policies dependent on national strategies, but these are fragmented across sectors and lack a shared European vision (Ollikainen 2014). The EU's Bioeconomy Action Plan (European Commission 2012), for example fails to recognise the role and nature of the forest sector as a high-tech biomass utilising sector and omits its current challenge to renew the product matrix from forest biomass as a response to the decreasing demand for paper. Bio-based materials from the forest sector can easily adapt to circular product designs, developing products that are used more than once (e.g. wood residuals can be used to produce bioenergy and materials) (Ladu et al. 2020). There is a need for a supportive policy framework that is aligned with the synergies in circular bioeconomy, especially for assessing urban transitions. Cities hold a central role in the bioeconomy, as urban dwellers increasingly use biogenic materials.

To ultimately achieve sustainability, sets of bio-circularity indicators that fit all products and industries should be optimised and adopted. These indicators are important for assessing the effects of a circular (bio)-economy in terms of the so-called five capitals (natural, human, social, financial, and physical) on profitability, job creation, and environmental impacts, just to name three. These indicators should therefore include environmental and sustainability elements, as well as new socioeconomic indicators (Kardung et al. 2021).

8 Forest and Forest Products as Carbon Sinks and Substitutes for Fossil Fuels in Cities

Most important suppliers of carbon-neutral renewable materials and products for the bioeconomy are the national forest and wood-based sectors (Ranacher et al. 2020). Wood is a renewable and carbon-storing resource. Wood products can displace fossil fuel emissions by substituting for other functionally equivalent materials with a higher carbon footprint. Wood material most often requires less processing energy compared with alternative materials such as concrete, steel, aluminium, or plastic. Transferring and applying the basic principles and mechanisms from nature through the advancement of polymer chemistry and nanotechnology has given contemporary businesses a range of hybrid wood materials suitable for multiple purposes.

9 Forests as Providers of Inclusive Growth and Services

Forests have the potential to generate income and wealth across social strata, in rural–urban communities, and reaching both the wealthier and poorer. Identifying and valuing ecosystem services offers an opportunity to improve the environmental and economic sustainability at the smallholder level, increasing their income-generating capacity (Milder et al. 2010; Vignola et al. 2015). Understanding and quantifying ecosystem services is needed to increase urban–rural interdependency and boost possibilities for increasing circularity (Elliot et al. 2019). Since the first decades of the past century, society, and forestry have gone through paradigm changes, where human rights, equity, poverty, and environmental concerns have become more and more relevant. To this end, the forestry sector has also evolved to be more inclusive and participatory, and this new way is reflected in the evolution of its governance, including a more systemic approach (Weiss et al. 2021).

10 Conclusions

Wood and NWFP, in addition to representing carbon sinks, support Biocities principles by fostering substitution of fossil fuels. The forestry sector is, in fact, the main supplier of carbon renewable materials for the bioeconomy, and wood products can displace fossil fuel emissions by substituting other functionally equivalent materials with a higher carbon footprint.

The COVID-19 pandemic policy frameworks and the consequent behavioural changes exhibited by civil society during lockdowns have been drawing attention to bioeconomic thinking, offering the possibility to refocus and strengthen sustainable management of renewable resources. The bioeconomy discourse first started by focusing mainly on biotechnology and to replace fossil fuels with biofuels, but more recently it has gained a new role in addressing the challenges in various fields including economic development, urban services, and food security.

BioCities will need to fully incorporate the bioeconomy by emphasising the potential of urban and peri-urban forests in upgrading and converting their biological raw materials, but also in its bioecological values where the recirculation of existing materials plays a more relevant role. Furthermore, the rural–urban forestry nexus will be important as BioCities and their rural hinterlands are fundamentally connected in ways that both the citizenry and policymakers can understand. The evolution of the bioeconomy discourse in the past decades has been shifting towards addressing broader challenges in the forestry sector, but will need further reframing to provide sustainable and novel policy solutions. Amongst these is a rejection of simple 'urban vs. rural' or 'product vs. service' thinking and an embracing of a more holistic approach.

References

Afiff S, Wilkenson J, Carriquiry M, Jumbe C, Searchinger T (2013) Biofuels and food securit: a report by the high level panel of experts on food security and nutrition of the committee on world food security. Royal Netherlands Academy of Arts and Sciences

Antikainen R, Dalhammar C, Hildén M, Judl J, Jääskeläinen T, Kautto P, Koskela S, Kuisma M, Lazarevic D, Mäenpää I (2017) Renewal of forest based manufacturing towards a sustainable circular bioeconomy. Reports of the Finnish Institute

Barañano L, Garbisu N, Alkorta I, Araujo A, Garbisu C (2021) Contextualisation of the bioeconomy concept through its links with related concepts and the challenges facing humanity. Sustainability 13(14):7746

Baumgartner RJ, Rauter R (2017) Strategic perspectives of corporate sustainability management to develop a sustainable organisation. J Clean Prod 140:81–92

Borelli S, Conigliaro M, Quaglia S, Salbitano F (2017) Urban and peri-urban agroforestry as multifunctional land use. In: Dagar JC, Tweari VP (eds) Agroforestry. Springer, pp 705–724

Brandt JP, Flannigan M, Maynard D, Thompson I, Volney W (2013) An introduction to Canada's boreal zone: ecosystem processes, health, sustainability, and environmental issues. Environ Rev 21(4):207–226

Bryngemark E (2019) Second generation biofuels and the competition for forest raw materials: a partial equilibrium analysis of Sweden. Forest Policy Econ 109:102022

Camilleri MA (2021) A circular economy strategy for sustainable value chains: a European perspective. In: Vertigans S, Idowu SO (eds) Global challenges to CSR and sustainable development, Springer, pp 141–161

Caprotti F, Cowley R, Datta A, Broto VC, Gao E, Georgeson L, Herrick C, Odendaal N, Joss S (2017) The new urban agenda: key opportunities and challenges for policy and practice. Urban Res Pract 10(3):367–378

Clark KH, Nicholas KA (2013) Introducing urban food forestry: a multifunctional approach to increase food security and provide ecosystem services. Landsc Ecol 28(9):1649–1669

Clark JH, Luque R, Matharu AS (2012) Green chemistry, biofuels, and biorefinery. Annu Rev Chem Biomol Eng 3(1):183–207

D'Amato D, Korhonen J, Toppinen A (2019) Circular, green, and bio economy: how do companies in land-use intensive sectors align with sustainability concepts? Ecol Econ 158:116–133

Davies H, Doick K, Handley P, O'Brien L, Wilson J (2017) Delivery of ecosystem services by urban forests. Research Report-Forestry Commission, UK(026)

de Arano IM, Muys B, Topi C, Pettenella D, Feliciano D, Rigolot E, Lefevre F, Prokofieva I, Labidi J, Camus JM (2018) A forest-based circular bioeconomy for southern Europe: visions, opportunities and challenges. European Forest Institute

Desa U (2016) Transforming our world: the 2030 agenda for sustainable development. United Nations Department of Economic and Social Affairs

ECC (2013) A new EU forest strategy: for forests and the forest-based sector. European Commission

Elliot T, Babí Almenar J, Niza S, Proença V, Rugani B (2019) Pathways to modelling ecosystem services within an urban metabolism framework. Sustainability 11(10):2766

European Commission (2012) Innovating for sustainable growth: a bioeconomy for Europe. European Commission

European Commission (2018) A sustainable bioeconomy for Europe: strengthening the connection between economy, society and the environment. https://knowledge4policy.ec.europa.eu/publication/sustainable-bioeconomy-europe-strengthening-connection-between-economy-society_en

European Commission (2019) A European green deal: Striving to be the first climate-neutral continent. https://ec.europa.eu/info/strategy/priorities-2019-2024/european-green-deal_en

European Commission (2021) Biofuels. https://energy.ec.europa.eu/topics/renewable-energy/bioenergy/biofuels_en

European Parliament (2008) Directive 2008/98/EC of the European Parliament and of the council on waste and repealing certain directives. Off J Eur Union 312:3–30

Fanzo J, Arabi M, Burlingame B, Haddad L, Kimenju S, Miller G, Nie F, Recine E, Serra-Majem L, Sinha D (2017) Nutrition and food systems. A report by the High Level Panel of Experts on Food Security and Nutrition of the Committee on World Food Security

Fermont A, Benson T (2011) Estimating yield of food crops grown by smallholder farmers. Int Food Policy Res Instit 1:68

Fish R, Church A (2014) Cultural ecosystem services: stretching out the concept. J Instit Environ Scientist:31–44

Gebre T, Gebremedhin B (2019) The mutual benefits of promoting rural-urban interdependence through linked ecosystem services. Glob Ecol Conserv 20:e00707

Ghisellini P, Cialani, Ulgiati S (2016) A review on circular economy: the expected transition to a balanced interplay of environmental and economic systems. J Clean Prod 114:11–32

Guenster N, Bauer R, Derwall J, Koedijk K (2011) The economic value of corporate eco-efficiency. Eur Financ Manag 17(4):679–704

Haase D, Larondelle N, Andersson E, Artmann M, Borgström S, Breuste J, Gomez-Baggethun E, Gren Å, Hamstead Z, Hansen R (2014) A quantitative review of urban ecosystem service assessments: concepts, models, and implementation. Ambio 43(4):413–433

Hetemäki L, Hurmekoski E (2016) Forest products markets under change: review and research implications. Curr Forestry Rep 2(3):177–188

Hjalager AM (2017) Rural–urban business partnerships - towards a new trans-territorial logic. Local Econ 32(1):34–54

Hoornweg D, Munro-Faure P (2008) Urban agriculture for sustainable poverty alleviation and food security. Position paper, FAO. Africa

Hosagrahar J, Soule J, Girard LF, Potts A (2016) Cultural heritage, the UN sustainable development goals, and the new urban agenda. BDC - Bollettino Del Centro Calza Bini 16(1):37–54

Jarre M, Petit-Boix A, Priefer C, Meyer R, Leipold S (2020) Transforming the bio-based sector towards a circular economy-what can we learn from wood cascading? Forest Policy Econ 110: 101872

Kampelmann S (2020) Wood works: how local value chains based on urban forests contribute to place-based circular economy. Urban Geogr 41(6):911–914

Kardung M, Cingiz K, Costenoble O, Delahaye R, Heijman W, Lovrić M, van Leeuwen M, M'barek R, van Meijl H, Piotrowski S (2021) Development of the circular bioeconomy: drivers and indicators. Sustainability 13(1):413

Kosmol J, Kanthak J, Herrmann F, Golde M, Alsleben C, Penn-Bressel G, Schmitz S, Gromke U (2012) Glossary of natural resource [*Glossar zum Ressourcenschutz*]. Umweltbundesamt

Ladu L, Imbert E, Quitzow R, Morone P (2020) The role of the policy mix in the transition toward a circular forest bioeconomy. Forest Policy Econ 110:101937

Lassoie JP, Buck LE, Current D (2009) The development of agroforestry as an integrated land use management strategy. North American Agroforestry, pp 1–24

Leakey R (1996) Definition of agroforestry revisited. Agrofor Today 8:5–5

Leakey RR (2010) Agroforestry: a delivery mechanism for multi-functional agriculture. In: Kellimore RL (ed) Handbook on agroforestry: management practices and environmental impact, pp 461–471

Lovell ST (2010) Multifunctional urban agriculture for sustainable land use planning in the United States. Sustainability 2(8):2499–2522

Lundgren B, Raintree T (1983) Sustained agroforestry. Agricultural research for development: potenials and challenges in Asia. ICRAF reprint #3, The Hague

Lynch K (2005) Rural-urban interaction in the developing world. Routledge, 224 p

Martinez de Arano I, Maltoni S, Picardo A, Mutke S (2021) Non-wood forest products for people, nature and the green economy. European Forestry Institute

Mathew S, Zakaria ZA (2015) Pyroligneous acid—the smoky acidic liquid from plant biomass. Appl Microbiol Biotechnol 99(2):611–622

McCormick K, Kautto N (2013) The bioeconomy in Europe: an overview. Sustainability 5(6): 2589–2608
McEldowney J (2021) Long-term vision for rural areas: European Commission communication. European Parliamentary Research Service
Meinhold K, Darr D (2019) The processing of non-timber forest products through small and medium enterprises—a review of enabling and constraining factors. Forests 10(11):1026
Milder JC, Scherr SJ, Bracer C (2010) Trends and future potential of payment for ecosystem services to alleviate rural poverty in developing countries. Ecol Soc 15(2)
Millenium Ecosystem Assessment (2005) Ecosystems and human well-being. World Resources Institute, Island Press
Näyhä A (2020) Finnish forest-based companies in transition to the circular bioeconomy-drivers, organisational resources and innovations. Forest Policy Econ 110:101936
Ollikainen M (2014) Forestry in bioeconomy–smart green growth for the humankind. Scand J For Res 29(4):360–366
Park S, Yun S, Kim H, Kwon R, Ganser J (2018) Forestry monitoring system using lora and drone. In: Proceedings of the 8th international conference on web intelligence, mining and semantics
Ranacher L, Wallin I, Valsta L, Kleinschmit D (2020) Social dimensions of a forest-based bioeconomy: a summary and synthesis. Ambio 49(12):1851–1859
Sirkin T, ten Houten M (1994) The cascade chain: a theory and tool for achieving resource sustainability with applications for product design. Resour Conserv Recycl 10(3):213–276
Somarriba E, Beer J, Alegre-Orihuela J, Andrade HJ, Cerda R, DeClerck F, Detlefsen G, Escalante M, Giraldo LA, Ibrahim M (2012) Mainstreaming agroforestry in Latin America. In: Ramachandran Nair PK, Garrity D (eds) Agroforestry: the future of global land use. Springer, pp 429–453
Stadshout (2021) Wood of the city, by the city, for the city. http://stadshout.nu/
Steinberg F (2014) Rural–Urban Linkages: an urban perspective. Working paper series N° 128. Working group: development with territorial cohesion. Territorial cohesion for development program. Rimisp, Santiago, Chile
Thonemann N, Schumann M (2018) Environmental impacts of wood-based products under consideration of cascade utilisation: a systematic literature review. J Clean Prod 172:4181–4188
Toppinen A, D'amato D, Stern T (2020) Forest-based circular bioeconomy: matching sustainability challenges and novel business opportunities? Forest Policy Econ 110:102041
Vignola R, Harvey CA, Bautista-Solis P, Avelino J, Rapidel B, Donatti C, Martinez R (2015) Ecosystem-based adaptation for smallholder farmers: definitions, opportunities and constraints. Agric Ecosyst Environ 211:126–132
Vis M, Mantau U, Allen B (eds) (2016) Study on the optimised cascading use of wood. #394/PP/ENT/RCH/14/7689. Final Report, Brussels. 337 p
Weiss G, Ludvig A, Živojinović I (2020) Four decades of innovation research in forestry and the forest-based industries: a systematic literature review. Forest Policy Econ 120:102288
Weiss G, Hansen E, Ludvig A, Nybakk E, Toppinen A (2021) Innovation governance in the forest sector: reviewing concepts, trends and gaps. Forest Policy Econ 130:102506
Zola A (1999) Non-wood forest product statistics: Angola. EC-FAO Partnership Programme

Innovative Design, Materials, and Construction Models for BioCities

Daniel Ibañez, Michael Salka, Vicente Guallart, Stefano Boeri, Livia Shamir, Maria Lucrezia De Marco, Sofia Paoli, Maria Chiara Pastore, Massimo Fragiacomo, Lone Ross Gobakken, and Sylvain Boulet

1 Introduction and Statement

> Our buildings generate 40% of our emissions. They need to become less wasteful, less expensive and more sustainable. And we know that the construction sector can even be turned from a carbon source into a carbon sink, if organic building materials like wood and smart technologies like AI are applied.
>
> Ursula von der Leyen, President of the European Commission, State of the Union Address, September 2020 (von der Leyen 2020).

Realising the transition to BioCities, which function as holistically integrated systems similar to natural ecosystems, will require not just the creation of green and blue infrastructure (GI) (Davies et al. 2015), associated land-use patterns, and urban forests (UF) (Konijnendijk et al. 2006), but also utilising appropriate design of built structures and landscapes that support nature-based solutions (NBS) and ecosystem services (ESS).

D. Ibañez (✉) · M. Salka · V. Guallart
Institute for Advanced Architecture of Catalonia (IAAC), Barcelona, Spain

S. Boeri · L. Shamir · M. L. De Marco · S. Paoli
Stefano Boeri Architetti (SBA), Milan, Italy

M. C. Pastore
Politecnico of Milan (PoliMi), Milan, Italy

M. Fragiacomo
University of L'Aquila (UNIVAQ), L'Aquila, Italy

L. R. Gobakken
Norwegian Institute of Bioeconomy Research (NIBIO), Ås, Norway

S. Boulet
Institut Technologique Forêt Cellulose Bois-construction Ameublement (FCBA), Champs-sur-Marne, France

G. E. Scarascia-Mugnozza et al. (eds.), *Transforming Biocities*, Future City 20,
https://doi.org/10.1007/978-3-031-29466-2_8

The basic structural materials which came to define the modernist era of architecture (i.e. the twentieth century), and continue to dominate the construction industry worldwide (namely, concrete, iron, and steel), account for about half of industrial direct CO_2 emissions and comprise 20% of total global emissions (IEA 2020). Consequently, rethinking the material composition of BioCities is a primary task. The solution is, however, not merely a simple replacement of materials with a large carbon footprint for those with reduced impacts on the environment. Such a shallow substitution-based mentality threatens to perpetuate the core problem of how buildings and cities are regarded; which is to say as definitively composed, constructed, static objects, rather than temporary coalescences of different materials and complex, interrelated production processes (Ibañez et al. 2019).

BioCities must represent a new understanding of not only what buildings and cities are made of, but also their forms and functions, as well as the methods of their conception, production, operation, and decommissioning. Additionally, barriers must be overcome concerning the perceived risks and benefits of innovative design and construction models, along with the perceived strength, durability, and desirability of bio-based building materials, to better align public awareness with recent science and promote widespread acceptance. As Stefano Boeri, architect and urban planner, stated in June of 2021: '*Just as the cities of the last century were born from steel and concrete, it is time to imagine that the cities of today can be born from forests, wood, and the extraordinary economy that it can fuel, triggering a universe of inspirations and economic and cultural activities of great potential. A regeneration that would give a great boost to the many forest districts and communities around the world*'. This vision is supported by the recent global timber production trends of industrial round wood and wood-based panels, which have been growing at 1.4 and 11% annually, respectively, over the last decade (FAO 2021).

2 Key Issues

There are overarching challenges, as well as potential solutions, both theoretical and practical examples, associated with innovative design, materials, and construction models for BioCities. A central question is how embodied carbon, life cycle analysis (LCA), and the prioritisation of resilience, alongside health and well-being, can lead to the development of a built environment appropriate to BioCities. Furthermore, exploration of the distinguishing features of the future buildings and built urban spaces of BioCities is needed, in addition to exploration of the processes or tools (including physical, social, economic, and governmental infrastructures) required for successful planning, design, construction, management, and disassembly. The use of prefabricated components made of engineered timber products ought to be evaluated as a possible way to improve embodied carbon and LCA indicators, reduce construction times (and consequently cost), and eventually to achieve sustainable multistorey buildings secured against fire and natural hazards such as earthquakes, strong winds, and floods, but at the same time are also aesthetically pleasing, durable, and supportive of health and well-being.

A variety of adaptive approaches, techniques, systems, and technologies can be used to explore alternative ways to meet human habitational production, maintenance, and management objectives. Special emphasis is placed in this chapter on identifying key barriers (e.g. performative or logistical stipulations, perceptions, disciplinary silos, educational shortcomings, regulations, and economics), which can be understood as cultural/organisational constructions that negate long-term architectural sustainability, and potential solutions addressing multiple aspects of both quantitative/technical and qualitative/experiential requirements. Continuous learning and communication are needed too, based on transdisciplinary approaches, within academic fields, organisational structures, and professional practices. Keeping this in mind, what are the foremost prospective alternatives based on the current state of knowledge?

3 State of the Art

In light of the pressures imposed by global climate change, the advent of energy efficient building systems and components has shifted the focus of design optimisation from the energy expended during the operational phase of buildings to the embodied energy expended during the construction and material production phases (Dixit et al. 2010). Moreover, embodied carbon has gained precedence over embodied energy, in recognition of the fact that not all energy is equally clean (Hammond and Jones 2008). These reprioritisations may at first seem misguided, considering the 10% of global greenhouse gas (GHG) emissions embodied in building materials and construction processes is less than half of the 27% from building operations (Architecture 2030 2022). However, four critical factors must be considered (IEA and UNEP 2018):

1. Operational energy can be reduced over time with energy efficiency updates and the introduction of renewable energy, whereas embodied carbon can never be diminished after a building is completed (Architecture 2030 2018).
2. Due to improved energy efficiency standards, new construction until 2050 will see the proportion of embodied carbon to operational energy jump to almost 1:1 (Ibid).
3. GHGs released until 2050 will determine whether or not the goals of the 2015 Paris Climate Accord are met (UNFCCC 2015).
4. Though some advocate for constructing new buildings only as a last resort in favour of remodelling and renovation, since avoiding the use of new materials eliminates their impacts altogether (Preservation Green Lab et al. 2011), the construction industry will inevitably be compelled to build many new buildings by 2050 to meet rapidly rising demands.

Overall, there is a demand to refine the performance of buildings in terms of both embodied and operating energy and carbon footprint (Padilla-Rivera et al. 2018). It has become apparent, however, that historic interpretations of embodied energy and carbon have been problematically vague and variable, and associated databases are likewise inconsistent and incomparable (Dixit et al. 2010). Reliable templates, standards, and protocols are therefore necessary to translate this demand into quantifiable actions. The leading state-of-the-art solution is life cycle assessment or analysis (LCA). As defined by the International Organisation for Standardisation (ISO), '*LCA studies the environmental aspects and potential impacts throughout a product's life cycle (i.e., cradle-to-grave) from raw materials acquisition through production, use and disposal. The general categories of environmental impacts needing consideration include resource use, human health, and ecological consequences*' (Saling and ISO Technical Committee 207/SC 5 2006). Over the past decade, LCA has progressed from being used to a very limited extent in the building sector to incorporation in several prevalent green building rating systems (GBRS), such as *LEED*, *BREEAM*, *Green Star*, and the *Living Building Challenge* (LBC) (Malmqvist et al. 2011; Sartori et al. 2021). Still, researchers recommend governments mandate for improved data quality, and support the development of a transparent and simplified methodology in order to address the current lack of implementation of the considerable body of academic work on LCA within industry practice (De Wolf et al. 2017).

4 Automated Life Cycle Analysis

LCA includes the carbon footprint, but also extends to measure many more impact categories to fully understand the effects on an ecosystem. Global warming potential (GWP), for instance, represents all greenhouse gas emissions, not just carbon, in accordance with the standards of the European Environment Agency (EEA 2021b). The adoption of LCA by building professionals came about as a response to increasing awareness of the environmental impacts of buildings followed by frustration with vague or falsified eco-labelling (i.e. greenwashing) (Dahl 2010). LCA can be paired with life cycle costing (LCC), a compatible methodology for evaluating material costs and savings over a building's entire life cycle (US GSA 2019). Together, these tools can help substantially in designing buildings more sustainably from environmental and financial perspectives.

There are limits, however, to LCA/LCC from the point of view of a building designer, which inhibit implementation and undermine the prospective resulting gains; primarily, ease of use, the time required, cost, and the associated availability of properly trained personnel (Bayer et al. 2010). Until recently, LCA/LCC were expensive and time consuming, requiring weeks to months to complete. Favourably, these hurdles have been greatly reduced thanks to advances in automating LCA/LCC with digital tools.

4.1 *Building Information Modelling, Material Passports, and Cascading Waste Streams*

This new generation of automated LCA/LCC tools rely on accurately detailed virtual building information modelling (BIM), and comprehensive environmental product declaration (EPD) databases (Ingwersen et al. 2019). The development of these prerequisites has empowered automated LCA/LCC to gain prominence, albeit not fast enough. Stronger incentives and legislative initiatives are needed, taking precedence from European programmes such as *Level(s)* (EC 2018), which promotes standardisation of LCA/LCC indicators, or the EU *Construction Products Regulation* (CPR), which mandates constructions be designed, built, demolished, and recycled ensuring the sustainable use of natural resources (UNECE and FAO 2014).

Building stocks and infrastructures constitute the largest stores of materials in industrial economies (Kovacic et al. 2019); in order to minimise the use of primary resources and the dependence on imports, it is necessary to recycle these urban stocks. The construction industry's adoption of distributed ledger technologies (DLT), including blockchain, a decentralised database managed by multiple participants or nodes who propose and validate transactions, promises great potential for enhancing resource management in the built environment (Li et al. 2019).

Coupling BIM material passports and DLT will make it possible to create an accessible yet trustworthy material inventory updated in real time, and establish a common understanding about the material assets of buildings and of collective building stocks. This record will be essential to systematise a network of material deposits in a territory in order to improve circularity. For example, when building timber can finally no longer be used as a construction material, it can proceed through composting and anaerobic digestion processes, producing biogas useful for heating, cooking, or manufacturing operations, and organic fertiliser useful for returning nutrients to soils. In this way of cascading circular economics, everything that was collected from the tree is eventually returned to the earth and contributes to the regeneration of new timber.

The large number of waste products generated by construction processes and the architecture, engineering, and construction (AEV) industries is a sweeping problem with dramatic negative impacts on the environment at large (of all global waste in 2016, 67.6% came from prospective construction material stocks; largely demolished buildings and infrastructures), and new resources are increasingly difficult or costly to find or extract (PACE 2021). Using resources in cascades is therefore increasingly supported by legislative bodies. In Europe, the reuse and recycling of materials are often given priority over incineration for energy production, following the principle of waste hierarchy depicted by the 'Lansink Ladder' (EC 2019).

For bio-based materials, such as timber, the immediate challenges are in adapting practices and technologies to tap into the digitally driven data revolution currently changing how business and operations are conducted across all industrial sectors (i.e. Industry 4.0) (BMBF 2021). The objective of using technology to track the flows of every piece of timber, through the harvested wood products (HWP) value chain from tree to product, and through service life to ultimate disposal, has not yet been fully realised (Singh et al. 2021; Fatima et al. 2018). Fortunately, rapid developments in DLT, remote sensors, 3D scanning, drones, artificial intelligence (AI), machine learning, databasing, 5G or satellite-enabled networks, and the linking of these technologies contribute to making this process newly feasible (BAMB 2019).

Reclaimed wood is a heterogeneous material group, so to manufacture new products out of reclaimed wood, such as wood-based panels and engineered wood products, sorting, and processing are needed based on the quality and quantity of the subcategories of reclaimed wood (Hoennige 2018; Hegnes et al. 2019). Rising to these challenges will, like the rollout of DLT-enabled BIM material passports, entail the creation of new jobs as the industry expands its scope to manage circular flows. A recent study found that one-third of the wood recovered from buildings is suitable for high-value recycling, proving that the potential amount of waste wood for cascading is considerably higher than currently utilised (Höglmeier et al. 2017).

5 Wood and Engineered Timber

Wood and engineered timber should be considered a principal material for the built environment of BioCities due to: (1) reduction of GHG emissions through the substitution of concrete and steel, (2) mitigation of fossil fuel dependency through the substitution of concrete and steel, (3) unparalleled potential for combating global climate change through carbon sequestration, and (4) biodiversity and ESS enhancement when supported by sustainable forest management, although a wise balance between timber supply and ecological benefits should be continuously preserved.

As outlined in chapter "Mitigation and Adaptation for Climate Change: The Role of BioCities and Nature Based Solutions" about the key role of BioCities for climate change mitigation, replacing other energy and carbon-intensive construction materials like steel, concrete, and brick with wood could reduce global CO_2 emissions by 14%–31% (Oliver et al. 2014), on top of the present contribution to GHG mitigation by the forests of the world which already offset 29% of anthropogenic CO_2 emissions worldwide (Friedlingstein et al. 2019). Moreover, wood replacement for other construction materials would also reduce global fossil fuel consumption by 12%–19% (Oliver et al. 2014). A share of global carbon emissions reduction gained by using wood for construction refers to CO_2 stored in wood products, generally for many decades, as carbon is naturally metabolised by the source trees during their growth. It is estimated by Churkina et al. (2020) that timber utilised for newly constructed buildings globally could store 0.01–0.68 gigatonnes of CO_2 (GtC) per annum over the coming years, depending on different development scenarios and average floor area per capita ratios.

Engineered timber used for wood constructions features a vast range of timber-based technologies, primarily cross-laminated timber (CLT), glue-laminated timber (glulam), laminated-veneer lumber (LVL), parallel strand-lumber (PSL), engineered timber joists (I-joist), and particle/chip/fibreboards. The unifying property of all these products is manufactured by binding or fixing wood strands, veneers, boards, particles, or fibres, together with adhesives or alternative fixation methods to form a composite material. Research has also been conducted into 3D printing wood fibre biocomposites (Le Duigou et al. 2016). Each engineered timber product has distinctly suitable uses and respective architectural, legislative, social, or environmental constraints.

5.1 *Prefabrication and Design for Disassembly at the Urban Scale*

Whilst buildings of many different scales will have to be designed and constructed to meet the needs of future cities, mid- and high-rise buildings will have to be safeguarded against earthquakes in vulnerable regions, and, more universally, against the increasing threats posed by climate changes like stronger storms and hurricanes (EC 2021). Because they are large and costly enough to benefit measurably from economies of scale in the production of their components, mid- and high-rise buildings are well suited to prefabrication. Shortening the erection time through prefabrication can be seen as a key measure to reduce the total cost of construction, and in some cases can even counterbalance the use of more expensive structural materials.

Box 1

An example is provided by the *Stadthaus* building located in Hackney, 24 Murray Grove, London, UK and completed in 2009 (Yates et al. 2008). This nine-storey building, designed by Waugh Thistleton Architects, was initially conceived as a cast-in-situ reinforced concrete construction. At a later stage, a proposal was made to employ prefabricated CLT panels to replace the reinforced concrete structure. The panels had to be manufactured in Austria and transported to London, as at that time this engineered timber product was not yet produced within the UK. Still, as formwork placement, removal and the ‘dead’ times for concrete curing were no longer necessary, a significant reduction in the erection schedule was achieved. The economic calculation showed that the increase in material cost due to the use of timber panels in place of reinforced concrete was more than counterbalanced by the cost savings due to the reduction in erection time, leading the client to elect the prefabricated timber panels.

An important requirement of contemporary and future architecture is demountability. The possibility to construct a building at a certain location and, at a later stage, disassemble it and rebuild it at another location (or to repurpose its components in an altogether different construction) allows a greater degree of flexibility compared to current practices (Privett 2020). Engineered timber products, like those leveraged by the *Stadthaus* building (see Box 1), as opposed to conventional materials and the now dominate light-frame techniques for building with wood, are exceptionally well suited for prefabrication due to the easy machinability of the wood substrates, the relatively low toxicity of substances involved, and, crucially, the lightness of the base materials and resulting components. This quality of lightness enables numerous positive feedback loops by optimising the capacity for transport and rapid assembly (or disassembly) of large modules, as well as mitigating the demand for resources and energy-intensive investments in heavy foundations, structural cores, and so forth (Lowe 2020). Updates will also be required of regulations (e.g. timber structure storey limits) and financial practices (e.g. land valuations, building loans, and insurance policies) to more accurately reflect the true risks and benefits of prefabricated engineered timber systems.

5.2 Technical Performance

Engineered timber products have been developed to improve upon the mechanical properties of basic sawn timber, more specifically to reduce anisotropy, defined as the marked difference in mechanical properties parallel (excellent) and perpendicular (poor) to the grain; and to reduce the influence of defects such as knots and grain deviations on the strength of structural members (Blaß and Sandhaas 2017). Amongst these products, CLT has played a major role in timber engineering in the last decade due to its indisputable advantages, principally its strength-to-weight ratio, which is comparable to concrete despite being five times lighter. Since CLT was incorporated into the International Building Code (IBC) in 2015, it has been widely used as an alternative, sustainable construction material worldwide.

CLT is typically manufactured in a prefabrication plant as two-dimensional panels up to three metres wide and 15 m long. Each panel is made of a variable odd number of layers of timber boards (the number and thickness of layers depend on the required performance), with the adjacent layers glued perpendicularly under pressure (Fig. 1). Interestingly, for CLT manufacturing it is possible to use low-grade timber, as the influence of defects such as knots and grain deviation is reduced due to the lamination process, which makes CLT a suitable use for locally grown timber and enables the potential development of short supply chains (Sciomenta et al. 2021).

CLT panels can be used effectively in multistorey buildings. The panels can be prefabricated off-site, cut to size using CNC machines, then transported to the building site and craned to position using temporary props until connected together. A significant advantage of using only 'dry' elements (e.g. timber panels, metal plates, metal fasteners, bolts, and dowels) is the rapidity of erection, which also

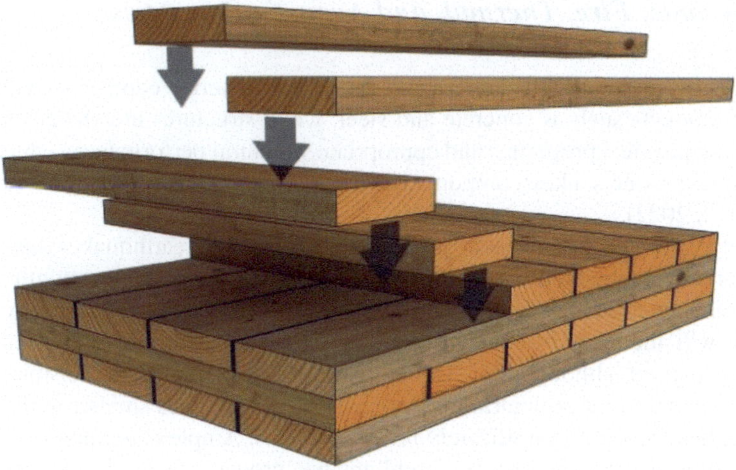

Fig. 1 Conventional layout of a cross-laminated timber panel. © Deitrich Buck et al. (Buck et al. 2016)

Fig. 2 Erection of a CLT building with platform construction system. © Lendlease (Malone 2016)

leads to reduced costs. Another advantage is the possibility to attain a fully demountable construction, not possible with a cast-in situ reinforced concrete construction.

For CLT buildings, a platform construction system is generally used (Vassallo et al. 2018), with the floor panels supported on the underlying wall panels, and the above storey's wall panels supported atop the floor panels (Fig. 2).

5.3 Seismic, Fire, Thermal, and Acoustic Properties

Wood is not just a highly sustainable structural material compared with other building products such as concrete and steel, wood structures are characterised by outstanding physical properties and appropriate structural performances which make mass timber products a key component of transformative change in the urban fabric (Bazli et al. 2022).

Wood as a building material is quite capable of resisting earthquakes due in large part to its previously cited lightness. Timber boasts a compressive strength comparable to concrete, but has a strength-to-density ratio five times as large. A timber structure will therefore be about five times lighter than an analogous reinforced concrete structure, although their volume is nearly the same. Since seismic actions are proportional to the structural mass, they will be five times smaller in the timber structure, hence causing significantly fewer problems. Ample experimental 'shaking table' testing of entire multistorey buildings has proven seismic behaviour can be further improved by designing the metal connections between timber elements to dissipate seismic energy through plasticisation of select connections (Follesa et al. 2018; Ceccotti et al. 2013).

Meanwhile, the combustibility of wood is indisputable and, to date, no product or system has been developed to make it entirely incombustible, though it must be clearly stated that combustibility does not directly equate to a lack of fire resistance. An advantage of CLT and other mass timber systems (defined as buildings in which the primary loadbearing structure is made of either solid or engineered wood), however, compared to the present standard of light timber frame constructions (made of regularly spaced, small, dimensional lumber) is the higher fire resistance rating, even when the structure is not encapsulated by supplemental materials. The massive cross-sections used in CLT construction ensure that the residual cross-section left unburnt at the end of a fire event will be capable of resisting the design loads without collapse (Fragiacomo et al. 2013; Buchanan and Abu 2017). Where needed, even higher fire resistance ratings can easily be attained by using thermally insulative protective claddings.

Accordingly, the solution to the critical issue of fire resistance is the use of mass timber members together with a proper performance-based design. Such designs can be standardised by developing joint initiatives (such as *COST Action FP 1402* in Europe, and the *Global Network on Fire Safe Use of Wood in the World*) amongst different countries aimed at removing the barriers in terms of prescriptive regulations that prevent the use of timber members in multistorey buildings, to reconcile policies with the results of extensive research (FSUW 2008; ETH Zurich et al. 2021).

Wood has a very low thermal conductivity compared to concrete and to other structural materials, which facilitates compliance with stringent standards for heat conduction losses (Craig et al. 2021). The higher self-weight of CLT compared to light timber frame construction is also an advantage for heating and cooling: the consequently higher thermal inertia leads to reduced operational energy. Still, hybrid systems, including layers of non-structural concrete (Jensen et al. 2020), or natural fibre insulation including wood fibre, cellulose, hemp, flax, cotton, and wool (Sutton et al. 2011), could further improve the thermal behaviour of wood constructions. Fruitful research has also examined the prospect of designing engineered timber products as passive heat exchangers, made of porous panels through which outside air flows into the building whilst being warmed by crossing the mass timber, or by a thermally active surface warmed with circulated water. This could lead to valuable reductions in insulation requirements and mechanical air-conditioning system loads (Craig et al. 2021).

On the other hand, achieving an effective acoustic separation between apartment units may be an issue due to the lower density of timber compared to other building materials (Praeger 2019). To overcome this challenge, a number of effective details have been developed and used to ensure satisfactory performance, for example suspended ceilings, floating layers, or natural fibre insulation infills or panels which optimise acoustic performance.

5.4 Wood Façades

Wood exposed to weather and not covered with a film-forming finish (i.e. stains and paints) gradually becomes grey, whatever the species or pre-treatment (e.g. preservation treatment, thermally-, or chemically modified wood). The development of the grey tint results from the combined action of UV radiation and water. Known as greying, this phenomenon appears after several months or years. As it develops more quickly on facades most exposed to harsh weather, the overall aesthetic of the building is unequally affected. Over time the building will present a heterogeneous aesthetic, often not appreciated, which harms the public image of wood materials and impedes development of the sector (Fig. 3). Avoiding the change in colour from new wood to grey wood is possible by giving the material a grey tint before installation. Manufacturers of finishing products have developed bio-inspired solutions to obtain pre-greyed wood whilst maintaining a very natural appearance.

Wood cladding is increasingly being used as façade material in larger and taller buildings for both public and commercial activities in Northern Europe. Wood is, by nature, designed to deteriorate, but under ideal conditions can have an almost indefinite service life. Wooden buildings that are properly designed to shed exterior water, and to avoid trapping moisture from interior sources, can exhibit a service life of more than 100 years (Williams et al. 2000).

Five main principles should be employed when seeking the right wood material and treatment for a wooden façade to achieve the expected longevity: (1) protection by design; (2) exploitation of the natural durability of the wood species; (3) wood modification; (4) wood impregnation; and (5) surface treatment. Unpainted wood façades are often chosen due to the lower maintenance required, and both untreated and treated wood can be suitable options in this context (Zimmer et al. 2020). In compliment, research has also evidenced the ability to entrap organic essential oil biocides in lignin nanoparticles extracted from sawdust, which can then be used as

Fig. 3 Heterogeneous ageing of wooden cladding due to differing solar orientation and roof overhang. ©NIBIO (Zimmer et al. 2020)

surface treatments protecting wood from visible ageing and parasite infestations (Zikeli et al. 2020, 2022). As the sawdust substrate employed would otherwise be a high-volume waste product, this accomplishment further serves as an ideal example of circular bioeconomy.

5.5 *Regulation, Perception, and Certification*

A politically navigable difficulty for bio-based building products is that some national technical regulations limit their use in buildings higher than a certain number of storeys (Build-in-Wood 2019). For example, in England the maximum number of storeys allowed for a light-frame timber building is six, whereas in Finland the limit is four; whilst in other countries, like Italy, there is no such limitation. Similarly, the use of combustible materials is often prohibited along corridors and egress paths. These restrictions stem from understandable, if overly simplistic, worries about the combustibility of bio-based materials (addressed above), and some fire-spread accidents occurring mostly during construction of light-frame buildings.

Regulation can also be used as a crucial tool in balancing environmental and social development strategies with the economic reality of variable urban land values. Mandates for optimised embodied carbon or energy, operational energy, and other LCA/LCC factors are gaining traction in numerous places and may significantly alter the form and quality of architectural projects. Policymakers can also influence the heavily regulated banking and insurance industries to support expanded bio-based construction by making loans, rates, and premiums more accurately reflective of the latest science regarding risks, rather than artificially inflated due to a past lack of information (WoodWorks 2021). Recent 'wood first policies' in building and construction at the national level (e.g. in Japan and various EU countries), as well as the local level (e.g. various Canadian provinces, Australian states, or London Borough Councils) provide strong examples (Ramage et al. 2017).

Apart from restrictive formal regulations, previous sections in this chapter have alluded to the fact that bio-based materials continue to suffer from negative perceptions (Ramage et al. 2017), though this trend is changing thanks to new data, exemplars, and priorities. These misperceptions are not contained to the public unengaged with the processes of architectural development, but rather affect architects and contractors as well, perhaps with greater consequence. Although a nation with its own legacy of building with wood, a study in Sweden revealed a low probability of architects or contractors selecting bio-based materials for residential buildings, due mainly to insufficient incentives, lack of knowledge and experience, bad examples, issues regarding performance, and construction-related culture and habit (Markström et al. 2016). Encouragingly, the same study indicated a shift in attitudes in a more positive direction, notably emboldened by green building certificates, along with other environmental standards and regulations, as well as with measures to increase the incentives to select bio-based materials. Evidence that these bio-based materials maintain a certain quality over time was also identified as an important aspect to enhance their use for construction, together with educational support from municipalities.

5.6 Health and Well-Being

The topic of health is addressed more comprehensively in chapter "BioCities as Promotors of Health and Wellbeing": *BioCities to Promotors of Health and Well-Being*. Here, it is necessary to highlight particularities of health as they relate to the materials, design, and construction of the built environment. As indicated, bio-based materials (including wood) are commonly perceived (Markström et al. 2016) by the building sector as inferior in terms of strength, durability, and desirability, despite burgeoning evidence to the contrary (Ramage et al. 2017). In parallel, the concept of health has historically been treated in the negative sense, as a problem to be solved, because it is linked to health risks such as inflammation of the skin, mucous membranes, or pulmonary system, nausea, cancers, etc. (e.g. sick building syndrome) (Joshi 2008). That said, it must be noted that these problems have almost

Fig. 4 Akershus University Hospital in Nordbyhagen, Norway. © Jørgen True & Torben Eskerod (CF Møller 2015)

Fig. 5 ESEAN—Aftercare and rehabilitation centre for children and adolescents in Nantes, France. © Philippe Ruault (ESEAN 2010)

always stemmed from additives to wood products (i.e. finishes and glues) rather than wood itself, so can be relieved with healthier product designs and installation methods (Adamová et al. 2020).

Despite outmoded conventions, wood is increasingly associated with architectures centred on people, their living environment and care (Figs. 4 and 5). It is considered aesthetically beautiful and brings freedom of form and biophilic properties, which are beneficial in care-related environments. Natural wood has antibacterial properties (Kotradyová and Kaliňáková 2014), which make it a material of choice for the interior design of living spaces for people with high sensitivity (e.g. young children, the elderly, and people with disease). The application of solid wood can contribute to well-being, and is demonstrably suitable for health care, social, and day care facilities in conjunction with appropriate zoning and cleaning (Kotradyová et al. 2019).

Today's discourse in terms of health and well-being frequently emphasises volatile organic compounds (VOCs). VOCs belong to many chemical families of different origins: biogenic VOCs as components of scents and other substances emitted by flowers, fruits, and leaves; but also anthropogenic VOCs as components of solvents, adhesives, and fossil fuels, which can be dangerous pollutants. Both of these classes of VOCs are present in construction, finishing, and renovation materials (Ruiz-Jimenez et al. 2022). Some materials can emit VOCs for several years. In wooden buildings, a number of construction materials, such as reconstituted panels, timber framework, and floors or floor coverings, especially through the impact of additives present in certain wood products, can lead to a degradation of the indoor air quality (e.g. via formaldehyde or acetaldehyde emission) (Adamová et al. 2020). The effects of VOCs on human health can be 'acute' if they are linked to exposure over a short period of time, or 'chronic' if they are linked to continuous exposure over a long period. However, several studies show that the evaluation of building product emissions remains delicate because of the diversity of parameters to check. Therefore, it is imperative to further the knowledge on emission data from wood products (Yrieix et al. 2004; Bluyssen 1997).

With regard to individual perception, research suggests that wood has a positive psychological effect, whether encountered during a walk through the forest or in the interior design of a building. Wood, when left visible, contributes to sensations of warmth (i.e. effusivity) and conviviality (Ibañez et al. 2019). The use of wood as a building material stimulates aesthetic pleasure, enhances the feeling of relaxation, and, put plainly, makes people feel good (Boulet and Achard 2013; Rice et al. 2006).

The benefits of wood on the stimulation of certain senses (e.g. touch, smell, and sight) when used in interior design and structure of buildings can be analysed through the measurement of psychological responses (via surveys) and physiological indicators (i.e. health criteria). Together, the results of several studies (Ikei et al. 2017; Matsubara and Kawai 2014; Akitaka et al. 2011) show that the presence of wood correlates with: (1) positive physiological effects, (2) a lower heart rate and blood pressure leading to reduced fatigue and stress, and (3) bolstering of the immune system. Most survey results indicate a qualitative improvement in states of anxiety, depression, and fatigue in the subjects questioned, and highlight certain qualifying adjectives associated with wood such as 'comfortable', 'relaxing', 'natural', or 'warm'.

5.7 Timber Supply and the Impact on Forest Ecosystems

A pressing question is whether the environmental and experiential gains of building with substantially more wood truly outweigh the apparent environmental harm of cutting down trees. This is a profoundly complex query with nuanced answers beyond the scope of this chapter, with relevant implications also for the regions surrounding BioCities, as analysed in chapter "From BioCities to BioRegions and Back: Transforming Urban-Rural Relationships". However, it is indispensable here to share the following points:

First, the base material, timber, is a renewable resource that can be produced in local, sustainably managed forests, and within short supply chains. By transforming round wood into higher value prefabricated components (e.g. CLT panels) in local factories, timber may increase the assessed value of local forests and therefore incentivise protection or augmentation of forested areas, if appropriate forest and environmental policy measures are adopted as indicated in the New EU Forest Strategy (EEA 2021a). Similarly, job opportunities and economic growth in the forestry and construction sectors could also be spawned, thus mitigating depopulation of the often underdeveloped associated countryside areas.

Second, even though the rate of global deforestation has decreased over the past three decades (UNEP 2020), an estimated 420 million hectares of forest (approaching 10% of the world's total forested area) have been lost to deforestation since 1990 through conversion to other uses, primarily agricultural. This statistic compels caution when proposing a large increase of timber logging and production from existing forests throughout the world. However, drastic differences exist between regions. In developed countries, forest surface and forest productivity is expanding; indeed, within the European Union, more than 40% of land is now covered by forests, with a substantial 0.36 hectares of forest per capita (Eurostat 2018). More significantly, European forests have expanded by about 10 million hectares (Mha), 6% of their present total surface, since 1990 (Ramage et al. 2017). It should also be mentioned that, at global scale, plantation forests are rapidly expanding with an annual surface increase of about 12 Mha in the period from 1990 to 2010, reaching in recent years a total surface of more than 260 Mha (Szuleck et al. 2014). Planted forests also have much greater annual biomass productivity than natural forests (Churkina et al. 2020), capable of providing 40% of the wood harvested globally within a gross surface only 7% of world's cumulative forested area. The greater productivity ratios of planted forests could potentially contribute towards reducing the pressure of future timber production on pristine natural forests, with positive feedback for biodiversity, conservation, and carbon.

Third, the argument that young trees absorb CO_2 at a faster rate than old trees and thus increase carbon sequestration within forests maintained at younger average ages, has been substantially superseded by a more thorough understanding of the value of old forests in terms of biodiversity, ESS, mycorrhizal fungal networks, and the carbon content of healthy soils (Simard et al. 2012). Hence, in the future bioeconomy strategy and management should carefully balance timber production, even for carbon mitigation, with relevant ecological and amenity aspects of forests that play an increasing role for our society (Winkel 2017).

Taking these factors into consideration, detailed projections of potential supply and demand for timber, accommodating the option of increasing wood utilisation to satisfy the future needs of the building industry, have been recently elaborated by different groups of researchers intent on keeping timber supply well below the summary growth increment of world forests (Oliver et al. 2014; Churkina et al.

2020). Two prominent studies concluded that forest planning and governance at global scale have the potential to significantly increase timber production, maintaining sustainable management goals, and meet the growing needs of the wood construction sector by up to three to four times in the next decades, whilst maintaining the carbon stock of managed ecosystems and enhancing the carbon pools stored in cities. However, all modelling exercises must be considered judiciously, as the greatest potentials to sustainably increase wood harvest worldwide rely heavily on tropical forests, which require highly sensitive approaches appropriate to their outstanding biodiversity, and the particular political and economic circumstances of resident societies. In any case, a broad palette of political and technical instruments should be put in place to carefully support planning, monitoring, and enforcement of verifiable sustainable forest management plans around the world, with a bedrock of close cooperation amongst countries and the development of public–private partnerships (PPPs). Also, there remains the clear danger that climate change could increasingly impact forest ecosystems, spreading major disturbances significantly reducing the productivity and health conditions of forests, in different world regions and globally (Seidl et al. 2014).

All things considered, the widespread application of adaptive, climate-smart forest management will be a necessity in the years ahead for many ecosystems, to increase resilience and adaptation. Development and implementation of these plans will enable the collective forest stock to meet global wood demand if managed with 'moderate intensity' by integrating sustainable policies like selective harvesting, strategic replanting, species diversification, and the protection of indigenous communities. Certification schemes, such as those promoted by the Forest Stewardship Council (FSC) and by the Programme for the Endorsement of Forest Certification (PEFC), currently cover about 30% of global forest production and are essential to ensure that long-term sustainability is respected. The 'Think Wood' campaign identifies extensive parallel advantages to active forest management, including the mitigation of fires, replenishment of waterways, habitat expansion, rural job creation, and an overall reduction in carbon emissions (Think Wood 2021). Of course, forest management practices will nonetheless be compelled to adapt to global climate change, which through exacerbation of issues like drought, wildfire, and infestation may result in unpredictable fluctuations of timber availability and cost. Climate change is undeniably relevant to most all aspects of BioCities, and is thus discussed in greater depth in chapter "Towards BioCities: The Pathway to Transition".

Another key issue is the lack of appropriate physical infrastructures for implementing innovative timber architecture *en masse*. This covers both the industrial infrastructures required for the effective, efficient, and locally integrated processing of raw timber into high-performing engineered timber products, as well as infrastructures for collecting, cataloguing, storing, and redistributing timber products capable of cascading through multiple uses upon being decommissioned from any singular use (Oliver et al. 2014).

Although establishing industries and infrastructures is no small feat, mass timber is predisposed to streamlined supply chains (plus erection, demounting, and recycling) thanks to its ability to provide both desirable finished surfaces and the bearing structure of a building in a single combined assembly. Such multi-functionality stands once more in stark contrast to light frame construction, which typically necessitates complicated layerings of gypsum board, structural studs, cavity insulation, and vapour/air membranes to attain the same performance.

6 Decentralisation, Distribution, and Mixed Use

In general terms, innovative construction models for BioCities can be viewed through the dual lenses of decentralisation and distribution, leading to systemic resilience via integration, redundancy, and polycentricity. The following has implications for buildings of all scales, but given the focus of this book on BioCities, urban buildings of three or more storeys are a priority.

The use and contents of architectural and urban constructions, in terms of activities accommodated, must serve the BioCity through decentralisation and distribution. Polycentric urbanisms which optimise the diversity of uses per unit of urban area through the implementation of mixed-use architectures containing multiple functions within the same building, or a dense variety of architectural typologies, engender the development of local character through the creation of vibrant, distinctive neighbourhoods. In combination with high-quality exterior urban spaces, such distribution encourages citizens to prioritise walking, cycling, and gathering outdoors, rather than rely primarily on resource-intensive private or public motor vehicles and conditioned indoor spaces. This concept has recently been popularised as the '15-minute city', a model in which all essential urban amenities are reachable within a 15-minute walk or bicycle ride, and is featured amongst the 10 functional properties of BioCities indicated in chapter "Towards the Development of a Conceptual Framework of BioCities" (Moreno et al. 2021).

6.1 *Bottom-Up Decision-Making*

The principles of decentralisation and distribution ought to be applied not only to the end results or goals of the transition towards BioCities, but also to the processes of transition themselves, beginning with the most basic levels of decision-making. Co-governance methods (discussed in greater detail in chapter "Green Infrastructure and Urban Forests for BioCities: Strategic and Adaptive Management") engage citizens and residents in determining what to build, where to build, and how to build in their respective cities lead to multiple benefits, though they may require intense social and cultural dialogues and occasionally be confronted with conflicting demands. Clearly, the buildings in a BioCity must be optimally adapted to the local conditions, following the same tenet of evolving in response to environmental feedback as biological natural selection. Interestingly, the richest store of locally-specific, place-based knowledge, especially priceless tacit knowledge regarding appropriate technologies or strategies and characteristic qualitative values, is the minds of local citizens. This knowledge can only be accessed and applied through sincere involvement. Furthermore, participatory co-design, co-development, or co-realisation processes engender feelings of ownership, prompting collective stewardship of shared assets and a sense of belonging and civic pride which promotes the longevity of realised designs and improves outcomes (Lachapelle 2008; Waag Society et al. 2017). New approaches to citizen science initiatives (as extensively documented in chapter "Biodiversity and Ecosystem Functions as Pillars of BioCities") empowering the public to participate in urban data gathering and analysis programmes with accessible technologies can have similar effects, but tend to receive disproportionate inputs from younger 'digital natives', risking the exclusion of older persons. Therefore, a mixed-methods approach is likely beneficial (Kullenberg and Kasperowski 2016).

6.2 *Digital Fabrication*

Once decisions regarding the trajectory of the built environment are made, their realisation should likewise be conducted in a decentralised, distributed, and locally integrated manner. As communicated by the German Working Group on Industry 4.0, digital fabrication tools and workflows (including, but not limited to, laser or plasma cutting, CNC machining, 3D printing, or additive manufacturing) improve precision, reduce production time, reduce waste, enable greater design complexity, enable automation, and, in many cases, enable operation by relatively unskilled individuals (BMBF 2021). Summarily, these assets predispose digital fabrication towards spatial distribution so as to optimise the distances between raw materials, manufacturing facilities, and points of end-use. Apart from offering such potential for improved logistical efficiency, distributed production enabled by digital fabrication is a boon for BioCities due to the ability to embed fabrication within the fabric of public life by siting production facilities in mixed urban settings rather than purely industrial districts, and inviting citizens to participate first hand in production processes. These concepts have been successfully demonstrated by the proliferation of the Fab Lab Network, which, since its inception at the Massachusetts Institute of Technology (MIT) in 2001, has established approximately 1500 publicly accessible digital fabrication workshops in over 90 countries, empowering the masses to make 'almost anything', from precision farming and forestry to mechanical equipment, and from fashion to electronics and drones (Fab Foundation 2021). Equally important, the unprecedented precision and customisation afforded by digital fabrication is a fundamental enabler of the aforementioned prefabricated, modular, demountable assemblies which will underpin the built environments and circular bioeconomies of BioCities.

6.3 A Network of Networks to Support Urban Metabolisms

Normally deemed 'utilities', we can think of the energy, water, and waste systems of buildings in BioCities as 'metabolic systems', in that they constitute the processes occurring within the building that maintain life. As with uses, decision-making, and production, metabolic systems should be decentralised and distributed by designing each building with infrastructures contributing to the satisfaction of its own requirements. Energy can be generated and stored by individual buildings via renewable technologies such as photovoltaics and battery banks, as well as optimised via high-efficiency components and intelligent design, mainly with regard to thermodynamic comfort. Known as passive design, optimising building envelopes and sizing, and placing openings and overhangs so as to modulate the amount of solar energy the building's interior or materials are exposed to over the course of a day, can achieve great savings in the energy required for mechanical conditioning (PHI 2015). Individual buildings can also optimise water by collecting rainwater fallen on horizontal building or landscape surfaces, and hierarchically structuring water systems based on contamination levels (i.e. fresh, rain, grey, or black) to maximise the number of times a single litre is used before being permitted to exit the system. For instance, rainwater can be used directly for irrigation or minimally treated to become fresh water; fresh water can be used in showers, baths, or hand sinks; grey water can be reused for flushing toilets or for irrigation, and finally blackwater from kitchen sinks or flushed toilets can be either treated, reused to irrigate specially designed organic ecosystems such as constructed wetlands, or to fuel biogas production whilst organic solids can be removed for composting to produce heat, energy, and fertiliser (Jaeger et al. 2019).

6.4 Information and Control Systems

Decentralised information and control systems at the building scale, or building automation, have been studied in an effort to close the persistent problematic 'performance gap' observed between expected and actual energy performance (Build Up 2020). Often, these discrepancies stem from unanticipated behaviours of the building occupants, unforeseen complications associated with renovations, or an inability to use efficient technologies. Broadly speaking, by incorporating automatic controls for heating, ventilation, and air conditioning (HVAC) systems, as well as for lighting, access control, energy management, fire sensing, and other actuators with a digital building management system (BMS), building automation can deliver critical information on operational performance to inform optimisation. With the integration of basic artificial intelligence (AI), BMS can exceed the mere communication of critical information and take automated actions, thereby enhancing the safety and comfort of the occupants and ameliorating operational inefficiencies. To progress from the smart-building/city paradigm in which building automation gained momentum to the subsequent bio-building/city paradigm, decentralised information, and control systems can be paired with the metabolic systems distributed at the building scale to facilitate the networking of individual buildings. This networking ought not to be thought of only in terms of information, but also in terms of communicating and sharing energy, water, and waste resources. In this way, provided reciprocal infrastructural investments, the BioCity will become a network of networks as envisioned in chapter "Towards the Development of a Conceptual Framework of BioCities", gaining resilience to disruption through redundancy, and the ability to dynamically adjust the flow paths of resources as supplies and demands spontaneously emerge and evolve (Guallart 2014).

6.5 Integrated Green Systems

As reviewed more thoroughly in chapter "Green Infrastructure and Urban Forests for BioCities" and "Mitigation and Adaptation for Climate Change", green and blue infrastructures such as vegetated roofs and walls, permeable pavements, shade trees, and innovative porous and natural materials are necessary urban strategies for BioCities. Specific to buildings, green roofs and walls, paired with appropriate insulation and irrigation technologies, help prevent solar gain, cool the surrounding microclimate through evapotranspiration, and reduce water runoff. Roofs typically make up 20–25% of an average city's surfaces, providing a significant opportunity for cities to retrofit and modify the urban environment (Susca et al. 2011). Green roofs can range from a thin vegetation layer (lawns) to trees and shrubs; therefore, they are suitable in cities with sufficient precipitation and require buildings' structures to support their weight. Properly sized and maintained green roofs can extend the life of the underlying roof and provide significant value to cities struggling with stormwater management (Rosasco and Perini 2019). Moreover, in urban zones with insufficient green spaces, green roofs can supply additional areas usable by the community as gardens or for socialising. Green roofs thus have great potential for deployment in many locations in dense urban environments where land has a high premium value. Alternatively, green walls are vertical systems of plants (hedges and shrubs) applied to a building's external walls. Green walls are less common than other strategies, consequently markets are less developed and prices remain relatively high. Besides being integrated with buildings, green walls can also be constructed on the pillars of viaducts, retaining walls, and other boundary walls, forming a useful component of a city's greening and cooling portfolio.

Green roofs and walls should be prioritised solutions because, when social benefits are accounted for, both prove to be cost-effective (Blackhurst et al. 2010), and help cities to mitigate and adapt to climate change whilst making urban surfaces more liveable, desirable, and comfortable. In addition to planning green roofs and walls for residential and office buildings, BioCities should encourage green roofs and walls for public buildings and infrastructures, such as libraries, city halls, university campuses, recreation centres, transit stations, and public housing developments; thereby allowing all residents and users, especially those not otherwise enfranchised with accessible green and blue spaces, to enjoy the benefits of urban greening and the innovative use of materials.

Complementarily, greenhouses can be designed in a lightweight, modular fashion (using engineered timber systems) for ease of installation on existing urban roofs and other underutilised built surfaces, or integrated into new building designs. Urban greenhouses are strategically positioned to address the intersectional food, water, energy, and ecosystem nexus. Beyond producing food locally, recent studies demonstrate urban greenhouses' potential to generate electrical energy with photovoltaics and recycle/purify wastewater without reducing growing capacities (Ravishankar et al. 2021). Greenhouses, green roofs, and other urban agriculture installations have proven to be exceptional tools for engaging urban citizens with limited access to nature, poor nutrition, a lack of economic opportunities, and low social cohesion; all of which directly impact well-being (Aznar-Sánchez et al. 2020; FAO 2015).

7 Case Studies

1. *Grand Genève Constellation Métropolitaine*

   This case study is a polycentric vision developed between 2018 and 2020 for the Genève metropolitan area based on forestry strategies, timber production, sustainable supply chains, construction processes, and resource cycles (Fig. 6).

   For the Consultation Grand Genève, Stefano Boeri Architetti and seven other teams of interdisciplinary professionals were asked to propose strategies and

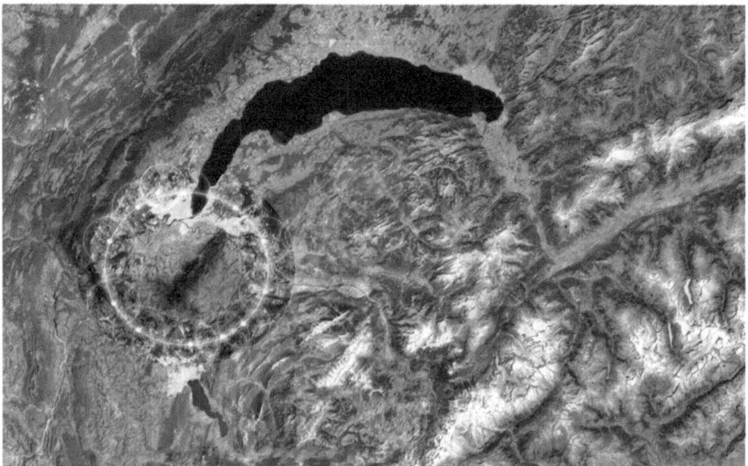

Fig. 6 Grand Genève Constellation Métropolitaine. © Stefano Boeri Architetti (Fondation Braillard Architectes 2020)

solutions for the ecological transition of the transnational area of Geneva, in which 350,000 new inhabitants are expected to live by 2050. These newcomers will need new houses, new schools, new public spaces, and new infrastructure. The associated growth should not waste agricultural soils, compromise the environment, nor increase emissions. The vision relies on the transformation of a centralised urban territory into a metropolitan constellation derived from the concept of a metropolitan archipelago. Eleven different urban nuclei comprise the cities of Geneva and Annecy, leaving the Salève mountain at its core. The new *Constellation Métropolitaine* will have at its heart not a city but a natural formation, the quintessential habitat for both non-domestic (e.g. chamois and wolves) and domestic species (e.g. cattle and sheep). The metropolis of Geneva extends around Salève, alternating urban and natural areas strengthened by urban forestry, afforestation, and agroforestry. It will, therefore, become a manifestation of human/nature coexistence, no longer based on authoritarian and aggressive anthropocentrism.

To meet the housing demand of 350,000 new inhabitants, Grand Genève will require a huge amount of construction timber. At a transnational level, Grand Genève will define agreements with Swiss and French forestry companies to secure the possibility of acquiring all the timber necessary for the realisation of the interventions for the first years of the project. The timber processing will be distributed to the regional level and new sawmills will be placed at the borders of the metropolitan area, in order to establish a short chain between harvesting activities, places of manufacturing, and places of utilisation. Through this strategy, the time and energy needed for transportation will be reduced.

The products of this process are to be used for the construction of buildings within the region or for the energy retrofitting and volumetric expansion of the existing construction stock. At the end of the lifespan of the timber buildings, the deconstructed wooden components will find new reuse and recycling possibilities within Grand Genève allowed by the implementation of digital and physical infrastructures. Following this plan, the geographical constellation of the new Grand Geneva will become an exemplar of a new form of metropolis, ready to face the challenges of the near future.

For more information visit: braillard.ch/consultation-grand-geneve/

2. *Botanica Tower*

The *BOTANICA Tower* is a new model of a garden tower for Milan designed in 2021 and is currently under development, with the aim of improving urban biodiversity by promoting the coexistence and interdependency of flora, fauna, humans, and architecture (Fig. 7). The tower has a wooden structure and a photovoltaic facade. *BOTANICA* is conceived as a self-sufficient and integrated high-rise building; it produces, consumes, recycles, and regenerates primary resources, supported by a continual waste resource stream. The large amount of vegetation that extends over the building's floors allows a significant reduction in temperature on its façades, providing considerable total energy savings. *BOTANICA*'s structure consists of prefabricated mixed wood and steel elements, achieving a substantial reduction in construction times and a significant decrease in CO_2

Fig. 7 BOTANICA Tower. © Stefano Boeri Architetti, Diller Scofidio + Renfro (Domus 2021)

emissions compared to other building material systems. The building applies a sustainable life cycle approach to the entire construction process, from construction to disassembly. *BOTANICA* generates clean renewable energy, thanks to the installation of photovoltaic panels on its façade, providing up to 65% of the building's energy requirements.

For more information visit: domusweb.it/en/architecture/gallery/2021/02/03/the-botanical-tower-by-stefano-boeri-and-diller-scofidio--renfro-for-pirelli-39.html

3. *The Niu Haus and the Voxel: A Quarantine Cabin*

The Niu Haus and *The Voxel: a Quarantine Cabin* represent the collective final projects of the 2018–2019 and 2019–2020 editions of the Institute for Advanced Architecture of Catalonia's (IAAC) Master of Advanced Ecological Buildings and BioCities programme (MAEBB) (Fig. 8). Both are small constructions located in Barcelona's Collserola Natural Park, but conceived as prototypes proposing scalable ideas for urban development.

These buildings demonstrate a comprehensive set of ecological building techniques including: photovoltaic energy systems with battery storage; composting or biogas toilets; passive air-conditioning strategies; reliance on locally sourced and processed natural materials; design for disassembly; and optimisation through parametric design. Taken together, their systems achieve a high standard of fully off-grid living for up to 14 days without resupplying water, and up to 4 days without sun, whilst actively promoting the inhabitants' awareness of their relationship to their greater environment. All timber was selectively

Fig. 8 The Niu Haus and The Voxel. © IAAC (photographs by Adrià Goula) (IAAC 2020)

harvested by students from the surrounding forest in accordance with the area's sustainable forest management plan, and each element is fully traceable from its exact point of origin to its final position in the building. All building components are rigorously quantified in terms of geographic source and embodied carbon, accounting for each fuel or energy input throughout the entire respective life cycle. It is thus possible to evidence that the overall construction of each sequester over 3000 kg of CO_2. Moreover, software has been developed to display this information with interactive graphics easily understood by non-experts, including an augmented reality application powered by a rural 5G network. Such global awareness and hyper-localism are further combined with the reimagination of linear cycles of material waste as circular flows. In the case of *The Voxel*, for instance, repurposing off-cuts from the on-site CLT fabrication as an organic skin of charred slats with naturally formed profiles that blend harmoniously with the landscape and remain fully compostable due to the avoidance of any chemical additives.

For more information visit: valldaura.net/research/self-sufficient-buildings/.

4. *Mjøstårnet*

Mjøstårnet is an 18-storey timber building situated in Brumunddal, Norway, and opened in 2019 (Fig. 9) (Abrahamsen 2018). The building houses offices, apartments, a hotel, a restaurant, conference rooms, and a rooftop terrace. The initiative came from investor Arthur Buchardt, who sought to build the tallest timber building in the world using local resources, suppliers, and competences. The architects of the project are Voll Arkitekter from Trondheim. The Moelven glulam factory (situated 15 kilometres from Mjøstårnet) produced the glulam structures. The highest occupied floor is at 68.2 m and the architectural top (the pergola) reaches 85.4 m. The main load-bearing structure consists of large-scale glulam trusses along the façades as well as internal columns and beams. The trusses handle the global forces in the horizontal and vertical directions and give the building its necessary stiffness (Abrahamsen 2017). CLT walls are used for

Fig. 9 Mjøstårnet. © Voll Arkitekter (photographs by Lone Ross Gobakken) (Voll Arkitekter 2019)

secondary load bearing of three elevators and two staircases, but do not contribute to the building's horizontal stability. Moelven's proprietary floor system (Trä8) uses less wood material compared to CLT decks, and are light and quick to assemble. Large, prefabricated façade panels are attached to the outside of the timber structures to form the envelope of the building. These sandwich-type elements come with insulation and external panels prefixed.

For more information visit: vollark.no/portfolio_page/mjostarnet/.

5. *HoHo Wien*

HoHo Wien, by Rudiger Lainer + Partner Aarchitekten, is an 84-metre-tall wood hybrid high-rise in Austria completed in 2020 (Fig. 10). The structure is 75% composed of timber, including CLT walls, CLT floor slabs, and glulam beams. Assembly of each floor's timber components was completed in only 4 days with pre-manufactured CLT walls partially exposed and internally protected by a UV and water-repellent finish, and supplied with pre-installed windows. Roughly 4350 cubic metres of wood is used in the entire construction. Compared to reinforced concrete construction, the use of wood avoids some 2800 tonnes of CO_2.

For more information visit: www.lainer.at/projekte/hoho-hoho-wien-holzhochhaus-1220-wien-in-bau-2016/

6. *Triodos Bank*

The *Triodos Bank* building, finished in 2019 in Driebergen-Rijsenburg, The Netherlands, by RAU Architects, is made of CLT with sculptural glulam beams and columns, apart from basement areas (due to the presence of groundwater), cores, floors, and roofs, which are made of concrete. CLT was also used for stairs

Fig. 10 HoHo Wien. © Rudiger Lainer + Partner Aarchitekten (Rüdiger Lainer + Partner Aarchitekten 2022)

Fig. 11 Triodos Bank. © RAU Architects (Photographs by Ossip van Duivenbode, Bert Reitberg and Marcel van der Burg) (RAU Architects 2019)

and roof slabs (Fig. 11). All CLT panels and major elements used were considered in terms of design for disassembly and maximum value reuse from the outset. In other words, the building is an example of a material bank. Data from all the key elements of the building have been collected, and every element identified and documented with a material passport. RUA Architects used Madaster Database, a commercial materials passport platform that aims to 'eliminate waste by providing materials with an identity'. Moreover, all structural connections and fittings are made using dry processes in order to allow for maximum flexibility and potential for disassembly and reuse.

For more information visit: www.rau.eu/portfolio/triodos-bank-nederland/

8 Outcomes and Concluding Remarks

The composition of the built environment must reflect the underlying processes and practices leveraged by BioCities to 'promote life'. Whilst wood and engineered timber products from sustainably managed forests should be considered principle materials for the built environments of BioCities, direct 1:1 substitution for conventional materials such as concrete and steel is inadequate. Rather, the built environment of BioCities must be evaluated holistically in terms of LCA/LCC, embodied carbon, integrated green systems, resilience, decentralisation and distribution, and health and well-being.

References

Abrahamsen R (2017) Mjøstårnet – Construction of an 81 m tall timber building. www.forum-holzbau.com/pdf/32_IHF2018_Abrahamsen.pdf

Abrahamsen R (2018) Mjøstårnet – 18 storey timber building completed. www.forum-holzbau.com/pdf/32_IHF2018_Abrahamsen.pdf

Adamová T, Hradecký J, Pánek M (2020) Volatile organic compounds (VOCs) from wood and wood-based panels: methods for evaluation, potential health risks, and mitigation. Polymers 12(10):E2289. https://doi.org/10.3390/polym12102289

Akitaka K, Hiroyuki S, Sei S, Mitsuyoshi Y, Mitsuyoshi Y (2011) Psychological and physiological effects in humans induced by the visual and olfactory stimulations of an interior environment made of hiba (Thujopsis dolabrata) wood. 木材学会誌 57(3):150–159

Architecture 2030 (2018). New buildings: Embodied carbon. architecture2030.org/new-buildings-embodied/

Architecture 2030 (2022) Why the built environment? Architecture 2030

Aznar-Sánchez JA, Velasco-Muñoz JF, López-Felices B, Román-Sánchez IM (2020) An analysis of global research trends on greenhouse technology: towards a sustainable agriculture. Int J Environ Res Public Health 17(2):664. https://doi.org/10.3390/ijerph17020664

BAMB (2019) Materials passports. Buildings as material banks. www.bamb2020.eu/topics/materials-passports/

Bayer C, Michael G, Gentry R, Surabhi J (2010) AIA guide to building life cycle assessment in practice. American Institute of Architects

Bazli M, Heitzmann M, Ashrafi H (2022) Long-span timber flooring systems: a systematic review from structural performance and design considerations to constructability and sustainability aspects. J Build Eng 48:103981. https://doi.org/10.1016/j.jobe.2021.103981

Blackhurst M, Hendrickson C, Matthews HS (2010) Cost-effectiveness of green roofs. J Architect Eng 16(4):136–143. https://doi.org/10.1061/(ASCE)AE.1943-5568.0000022

Blaß HJ, Sandhaas C (2017) Timber engineering—principles for design. https://doi.org/10.5445/KSP/1000069616

Bluyssen P (ed) (1997) Evaluation of VOC emissions from building products: solid flooring materials. Office for Official Publications of the European Communities

BMBF-Internetredaktion (2021) *Industrie 4.0—BMBF*. Bundesministerium für Bildung und Forschung - BMBF. www.bmbf.de/de/zukunftsprojekt-industrie-4-0-848.html

Boulet S, Achard G (2013) Study of the impact of wood material on hygrothermal comfort via an objective and subjective evaluation of indoor environment. 11th REHVA world congress CLIMA, Prague

Buchanan A, Abu AK (2017) Structural design for fire safety, 2nd ed. Wiley. www.wiley.com/en-us/Structural+Design+for+Fire+Safety%2C+2nd+Edition-p-9780470972892

Buck D, Wang X, Hagman O, Gustafsson A (2016) Further development of cross-laminated timber (CLT): Mechanical Tests on 45° Alternating Layers. www.researchgate.net/publication/313192887_Further_Development_of_Cross-Laminated_Timber_CLT_Mechanical_Tests_on_45_Alternating_Layers

Build Up (2020) Overview: building automation as a means for efficient building operation. Build Up. www.buildup.eu/en/news/overview-building-automation-means-efficient-building-operation

Build-in-Wood (2019) Build-in-Wood | EU horizon 2020 project for sustainable construction. Build-in-Wood. www.build-in-wood.eu/

Ceccotti A, Sandhaas C, Okabe M, Yasumura M, Minowa C, Kawai N (2013) SOFIE project – 3D shaking table test on a seven-storey full-scale cross-laminated timber building. Earthquake Eng Struct Dyn 42(13):2003–2021. https://doi.org/10.1002/eqe.2309

CF Møller (2015) Akershus University hospital—new ahus. CF Møller. www.cfmoller.com/p/Akershus-University-Hospital-New-Ahus-i269.html

Churkina G, Organschi A, Reyer CPO, Ruff A, Vinke K, Liu Z, Reck BK, Graedel TE, Schellnhuber HJ (2020) Buildings as a global carbon sink. Nat Sustain 3(4):269–276. https://doi.org/10.1038/s41893-019-0462-4

Craig S, Halepaska A, Ferguson K, Rains P, Elbrecht J, Freear A, Kennedy D, Moe K (2021) The design of mass timber panels as heat-exchangers (dynamic insulation). Front Built Environ 6: 606258. https://doi.org/10.3389/fbuil.2020.606258

Dahl R (2010) Green washing: do you know what you're buying? Environ Health Perspect 118(6): 6. https://doi.org/10.1289/ehp.118-a246

Davies C, Hansen R, Rall E, Pauleit S, Lafortezza R, Bellis Y, Santos A, Tosics I (2015) Green infrastructure planning and implementation: the status of European green space planning and implementation based on an analysis of selected European city-regions. Green Surge. https://doi.org/10.13140/RG.2.1.1723.0888

De Wolf C, Pomponi F, Moncaster A (2017) Measuring embodied carbon dioxide equivalent of buildings: a review and critique of current industry practice. Energy Build 140:68–80. https://doi.org/10.1016/j.enbuild.2017.01.075

Dixit MK, Fernández-Solís JL, Lavy S, Culp CH (2010) Identification of parameters for embodied energy measurement: a literature review. Energy Build 42(8):1238–1247. https://doi.org/10.1016/j.enbuild.2010.02.016

Domus (2021) The botanical tower by Stefano Boeri and Diller Scofidio + Renfro for Pirelli 39. Domus. www.domusweb.it/en/architecture/gallery/2021/02/03/the-botanical-tower-by-stefano-boeri-and-diller-scofidio%2D%2Drenfro-for-pirelli-39.html

EC (2018) Level(s). European Commission. ec.europa.eu/environment/levels_en

EC (2019) A European green deal. European Commission. ec.europa.eu/info/strategy/priorities-2019-2024/european-green-deal_en

EC (2021). Overview of natural and man-made disaster risks the European Union may face. European Commission. ec.europa.eu/echo/sites/default/files/overview_of_natural_and_man-made_disaster_risks_the_european_union_may_face.pdf

EEA (2021a) A new EU Forest strategy. European Environment Agency. www.eea.europa.eu/policy-documents/the-eu-forest-strategy-com

EEA (2021b) Climate change mitigation. European Environment Agency. www.eea.europa.eu/themes/climate

ESEAN (2010) ESEAN Lab—ESEAN - APF Nantes. rehab-lab.org/organizations/esean-lab-nantes

ETH Zurich, Arup, Universiteit Gent, Promat (2021) COST Action FP 1404: Fire Safe Use of Bio-Based Building Products. costfp1404.ethz.ch/

Eurostat (2018) Over 40% of the EU covered with forests. Eurostat. ec.europa.eu/eurostat/web/products-eurostat-news/-/edn-20180321-1

Fab Foundation (2021) The fab Lab network. The Fab Foundation. fabfoundation.org/global-community/

FAO (2015) Growing greener cities: food and nutrition security. Food and Agriculture Organisation of the United Nations. www.fao.org/ag/agp/greenercities/en/whyuph/foodsecurity.html

FAO (2021) World Food and Agriculture – Statistical Yearbook 2021. FAO. https://doi.org/10.4060/cb4477en

Fatima T, Srivastava A, Hanur VS, Rao MS (2018) An effective wood DNA extraction protocol for three economic important timber species of India Am J Plant Sci 9(2):139–149. https://doi.org/10.4236/ajps.2018.92012

Follesa M, Fragiacomo M, Casagrande D, Tomasi R, Piazza M, Vassallo D, Canetti D, Rossi S (2018) The new provisions for the seismic design of timber buildings in Europe. Eng Struct 168: 736–747. https://doi.org/10.1016/j.engstruct.2018.04.090

Fondation Braillard Architectes (2020) Genève: Constellation métropolitaine, Stefano BOERI, Consultation Grand Genève. www.youtube.com/watch?v=BInx-NRw7Oo

Fragiacomo M, Menis A, Clemente I, Bochicchio G, Ceccotti A (2013) Fire resistance of cross-laminated timber panels loaded out of plane. J Struct Eng 139(12):04013018. https://doi.org/10.1061/(ASCE)ST.1943-541X.0000787

Friedlingstein P, Jones MW, O'Sullivan M, Andrew RM, Hauck J, Peters GP, Peters W, Pongratz J, Sitch S, Le Quéré C, Bakker DCE, Canadell JG, Ciais P, Jackson RB, Anthoni P, Barbero L, Bastos A, Bastrikov V, Becker M, Zaehle S (2019) Global carbon budget 2019. Earth System Science Data 11(4):1783–1838. https://doi.org/10.5194/essd-11-1783-2019

FSUW (2008) Fire safe use of wood. fsuw.com/

Guallart V (2014) The self-Sufficient City—internet has changed our lives but it hasn't changed our cities, yet. Actar Publishers, Barcelona. actar.com/product/the-self-sufficient-city/

Hammond GP, Jones CI (2008) Embodied energy and carbon in construction materials. Proc Instit Civil Eng - Energy 161(2):87–98. https://doi.org/10.1680/ener.2008.161.2.87

Hegnes A W, Gobakken LR, Norhagen E (2019) Understanding and practicing wood waste qualities in Norway—a case of adaptation work in circular bioeconomy. PLATE (product lifetimes and the environment), the 3rd international conference, Berlin

Hoennige A (2018) Identification of distinctive features for cascading wood waste in Eastern Norway. Northern European Network for Wood Science and Engineering, National School of Agricultural Sciences and Engineering, Bordeaux Aquitaine

Höglmeier K, Weber-Blaschke G, Richter K (2017) Potentials for cascading of recovered wood from building deconstruction: a case study for south-East Germany. Resour Conserv Recycling 117:304–314. https://doi.org/10.1016/j.resconrec.2015.10.030

IAAC (2020) Self-sufficient buildings – Valldaura. valldaura.net/research/self-sufficient-buildings/

Ibañez D, Hutton JE, Moe K (eds) (2019) Wood urbanism: from the molecular to the territorial. Actar Publishers, Barcelona

IEA (2020) Industry direct CO2 emissions in the sustainable development scenario, 2000–2030 – charts – Data & Statistics. International Energy Agency. www.iea.org/data-and-statistics/charts/industry-direct-co2-emissions-in-the-sustainable-development-scenario-2000-2030

IEA and UNEP (2018) 2018 global status report—towards a zero-emission, efficient and resilient buildings and construction sector. World Green Building Council. www.worldgbc.org/news-media/2018-global-status-report-towards-zero-emission-efficient-and-resilient-buildings-and

Ikei H, Song C, Miyazaki Y (2017) Physiological effects of touching wood. Int J Environ Res Public Health 14(7):801. https://doi.org/10.3390/ijerph14070801

Ingwersen W, Kuczenski B, Mutel C, Srocka M, Scanlon K (2019) Building the LCA data pipeline—automating product system creation. United States Environmental Protection Agency. cfpub.epa.gov/si/si_public_record_report.cfm?dirEntryId=347737&Lab=CESER&showCriteria=2&fed_org_id=111&TIMSType=Presentation&

Jaeger M, Webb C, Muñoz P, Southard E (2019) Water petal permitting guidebook. International Living Building Institute. living-future.org/wp-content/uploads/2019/02/WaterPetal_PermittingGuidebook_FINAL.pdf

Jensen A, Norford L, Grinham J (2020) Mass(ive) timber: examining the thermally massive behavior of mass timber construction. Technology|Architecture + Design 4(2):186–199. https://doi.org/10.1080/24751448.2020.1804763

Joshi SM (2008) The sick building syndrome. Indian J Occup Environ Med 12(2):61–64. https://doi.org/10.4103/0019-5278.43262

Konijnendijk CC, Ricard RM, Kenney A, Randrup TB (2006) Defining urban forestry – a comparative perspective of North America and Europe. Urban Forest Urban Green 4(3–4): 93–103. https://doi.org/10.1016/j.ufug.2005.11.003

Kotradyová V, Kaliňáková B (2014) Wood as material suitable for health care and therapeutic facilities. Adv Mater Res 1041:362–366. https://doi.org/10.4028/www.scientific.net/AMR.1041.362

Kotradyová V, Vavrinsky E, Kalinakova B, Petro D, Jansakova K, Boles M, Svobodova H (2019) Wood and its impact on humans and environment quality in health care facilities. Int J Environ Res Public Health 16(18):E3496. https://doi.org/10.3390/ijerph16183496

Kovacic I, Honic M, Rechberger H (2019) Proof of concept for a BIM-based material passport. Springer Professional. www.springerprofessional.de/en/proof-of-concept-for-a-bim-based-material-passport/16183096

Kullenberg C, Kasperowski D (2016) What is citizen science? A scientometric meta-analysis. PLOS ONE 11(1):e0147152. https://doi.org/10.1371/journal.pone.0147152

Lachapelle P (2008) A sense of ownership in community development: understanding the potential for participation in community planning efforts. Community Dev 39(2):52–59. https://doi.org/10.1080/15575330809489730

Le Duigou A, Castro M, Bevan R, Martin N (2016) 3D printing of wood fibre biocomposites: from mechanical to actuation functionality. Mater Des 96:106–114. https://doi.org/10.1016/j.matdes.2016.02.018

Li J, Greenwood D, Kassem M (2019) Blockchain in the built environment and construction industry: a systematic review, conceptual models and practical use cases. Autom Constr 102: 288–307

Lowe G (2020) Wood, well-being and performance: the human and organisational benefits of wood buildings. Naturally Wood. www.naturallywood.com/resource/wood-well-being-and-performance-the-human-and-organizational-benefits-of-wood-buildings/

Malmqvist T, Glaumann M, Scarpellini S, Zabalza I, Aranda A, Llera E, Díaz S (2011) Life cycle assessment in buildings: the ENSLIC simplified method and guidelines. Energy 36(4): 1900–1907. https://doi.org/10.1016/j.energy.2010.03.026

Malone D (2016) Military giants: cross-laminated timber construction gets a salute from the Army. Building Design + Construction. www.bdcnetwork.com/military-giants-cross-laminated-timber-construction-gets-salute-army

Markström E, Bystedt A, Fredriksson M, Sandberg D (2016) Use of bio-based building materials: perceptions of Swedish architects and contractors. Forest Products Society International Convention: 26/06/2016 - 29/06/2016. urn.kb.se/resolve?urn=urn:nbn:se:ltu:diva-28406

Matsubara E, Kawai S (2014) VOCs emitted from Japanese cedar (Cryptomeria japonica) interior walls induce physiological relaxation. Build Environ 72:125–130. https://doi.org/10.1016/j.buildenv.2013.10.023

Moreno C, Allam Z, Chabaud D, Gall C, Pratlong F (2021) Introducing the "15-Minute City": sustainability, resilience and place identity in future post-pandemic cities. Smart Cities 4(1): 93–111. https://doi.org/10.3390/smartcities4010006

Oliver CD, Nassar NT, Lippke BR, McCarter JB (2014) Carbon, fossil fuel, and biodiversity mitigation with wood and forests. J Sustain Forestry 33(3):248–275. https://doi.org/10.1080/10549811.2013.839386

PACE (2021) The Circularity Gap Summary 2021. https://www.circularity-gap.world/2021

Padilla-Rivera A, Amor B, Blanchet P (2018). Evaluating the link between low carbon reductions strategies and its performance in the context of climate change: a carbon footprint of a wood-frame residential building in Quebec, Canada. Sustainability 10(8):2715. https://doi.org/10.3390/su10082715

PHI (2015) Passivhaus Institut. Passive House Institute. passivehouse.com/

Praeger T (2019) Understanding the acoustical challenges of mass timber buildings. Construction Canada. www.constructioncanada.net/understanding-the-acoustical-challenges-of-mass-timber-buildings/

Preservation Green Lab, Cascadia Green Building Council, Green Building Services, Skansa, Quantis (2011) The greenest building: quantifying the environmental value of building reuse. National Trust for Historic Preservation

Privett I (2020) Demountable building: don't think straight, think circular. WORKTECH Academy. www.worktechacademy.com/demountable-building-dont-think-straight-think-circular/

Ramage MH, Burridge H, Busse-Wicher M, Fereday G, Reynolds T, Shah DU, Wu G, Yu L, Fleming P, Densley-Tingley D, Allwood J, Dupree P, Linden PF, Scherman O (2017) The wood from the trees: the use of timber in construction. Renew Sustain Energy Rev 68:333–359. https://doi.org/10.1016/j.rser.2016.09.107

RAU Architects (2019) Triodos Bank Nederland. RAU Architects. www.rau.eu/portfolio/triodos-bank-nederland/

Ravishankar E, Charles M, Xiong Y, Henry R, Swift J, Rech J, Calero J, Cho S, Booth RE, Kim T, Balzer AH, Qin Y, Hoi Yi Ho C, So F, Stingelin N, Amassian A, Saravitz C, You W, Ade H, O'Connor BT (2021) Balancing crop production and energy harvesting in organic solar-powered greenhouses. Cell Rep Phys Sci 2(3):100381. https://doi.org/10.1016/j.xcrp.2021.100381

Rice J, Kozak RA, Meitner MJ, Cohen DH (2006) Appearance wood products and psychological well-being. Wood Fiber Sci 38(4):644–659

Rosasco P, Perini K (2019) Selection of (green) roof systems: a sustainability-based multi-criteria analysis. Buildings 9:134. https://doi.org/10.3390/buildings9050134

Rüdiger Lainer + Partner Aarchitekten (2022) *HOHO, HOHO Wien - Holzhochhaus, 1220 Wien, in Bau 2016*—Rüdiger Lainer + Partner. www.lainer.at/projekte/hoho-hoho-wien-holzhochhaus-1220-wien-in-bau-2016/

Ruiz-Jimenez J, Heiskanen I, Tanskanen V, Hartonen K, Riekkola ML (2022) Analysis of indoor air emissions: from building materials to biogenic and anthropogenic activities. J Chromatograph Open 2:100041. https://doi.org/10.1016/j.jcoa.2022.100041

Saling P, ISO Technical Committee 207/SC 5 (2006) Environmental management—life cycle assessment—principles and framework (ISO 14040:2006). International organisation for standardisation. www.iso.org/standard/37456.html

Sartori T, Drogemuller R, Omrani S, Lamari F (2021) A schematic framework for life cycle assessment (LCA) and green building rating system (GBRS) J Build Eng 38:102180. https://doi.org/10.1016/j.jobe.2021.102180

Sciomenta M, Spera L, Bedon C, Rinaldi V, Fragiacomo M, Romagnoli M (2021) Mechanical characterisation of novel homogeneous beech and hybrid beech-Corsican Pine thin cross-laminated timber panels. Construct Build Mater 271:121589. https://doi.org/10.1016/j.conbuildmat.2020.121589

Seidl R, Schelhaas MJ, Rammer W, Verkerk PJ (2014) Increasing forest disturbances in Europe and their impact on carbon storage. Nat Climate Change 4(10):930. https://doi.org/10.1038/nclimate2393

Simard SW, Beiler KJ, Bingham MA, Deslippe JR, Philip LJ, Teste FP (2012) Mycorrhizal networks: mechanisms, ecology and modelling. Fung Biol Rev 26(1):39–60. https://doi.org/10.1016/j.fbr.2012.01.001

Singh R, Gehlot A, Vaseem Akram S, Kumar Thakur A, Buddhi D, Kumar Das P (2021) Forest 4.0: digitalisation of forest using the Internet of Things (IoT). J King Saud Univ - Computer and Information Sciences. https://doi.org/10.1016/j.jksuci.2021.02.009

Susca T, Gaffin SR, Dell'Osso GR (2011) Positive effects of vegetation: urban heat Island and green roofs. Environ Pollut 159(8):2119–2126. https://doi.org/10.1016/j.envpol.2011.03.007

Sutton A, Black D, Walker P (2011) Natural fibre insulation: an introduction to low-impact building materials. Information paper, 18/11. IHS-BRE Press. www.google.com/url?sa=t&rct=j&q=&esrc=s&source=web&cd=&cad=rja&uact=8&ved=2ahUKEwjDntm2lLL6AhUBaMAKHa4uD9sQFnoECAgQAQ&url=https%3A%2F%2Fwww.bre.co.uk%2Ffilelibrary%2Fpdf%2Fprojects%2Flow_impact_materials%2FIP18_11.pdf&usg=AOvVaw31PIETD0oSgwzvzzxgE0ns

Szuleck J, Pretzsch J, Secco L (2014) Paradigms in tropical forest plantations: a critical reflection on historical shifts in plantation approaches. Int Forestry Rev 16(2):128–143

Think Wood (2021) Building with Wood. Think Wood. www.thinkwood.com/

UNECE, FAO (2014) Forest Products Annual Market Review 2013–2014. United Nations Economic Commission for Europe. unece.org/forests/publications/forest-products-annual-market-review-2013-2014

UNEP (2020) The state of the world's forests: forests, Biodiversity and People. UN Environment Programme. www.unep.org/resources/state-worlds-forests-forests-biodiversity-and-people

UNFCCC (2015) Paris Agreement to the United Nations framework convention on climate change (no. 16–1104). ec.europa.eu/clima/eu-action/international-action-climate-change/climate-negotiations/paris-agreement_en

US GSA (2019) 1.8 Life Cycle Costing. United States General Services Administration. www.gsa.gov/node/81412

Vassallo D, Follesa M, Fragiacomo M (2018) Seismic design of a six-storey CLT building in Italy. Eng Struct 175:322–338. https://doi.org/10.1016/j.engstruct.2018.08.025

Voll Arkitekter (2019) Mjøstårnet—Verdens høyeste trehus. Voll Arkitekter. vollark.no/portfolio_page/mjostarnet/

von der Leyen U (2020) State of the Union 2020. European Commission. ec.europa.eu/commission/presscorner/detail/en/SPEECH_20_1655

Waag Society, IAAC, University of Dundee, Peer Educators Network, Joint Research Centre, Fab Lab Network (2017) Making sense. Making Sense. making-sense.eu/

Williams RS, Jourdain C, Daisey GI, Springate RW (2000). Wood properties affecting finish service life. J Coatings Technol 72(3):35–42. https://doi.org/10.1007/BF02698003

Winkel G (ed) (2017) Towards a sustainable European forest-based bioeconomy – assessment and the way forward (no. 8; what science can tell us). European Forest Institute. efi.int/publications-bank/towards-sustainable-european-forest-based-bioeconomy-assessment-and-way-forward

WoodWorks (2021) Mass timber building insurance. WoodWorks Wood Products Council. www.woodworks.org/mass-timber-building-insurance/

Yates M, Linegar M, Dujic B (2008) Design of an 8-story residential tower from KLH cross laminated solid timber panels. In: Proceedings of the Wood is good; properties, technology, valorisation, application, proceedings of the 19th international scientific conference, Zagreb, Croatia

Yrieix C, Maupetit F, Ramalho O (2004) Determination of VOC emissions from French wood products. Archive Ouverte HAL. 4th European wood-based panel symposium, Hanover. hal.archives-ouvertes.fr/hal-00688527

Zikeli F, Vinciguerra V, Sennato S, Scarascia Mugnozza G, Romagnoli M (2020) Preparation of lignin nanoparticles with entrapped essential oil as a bio-based biocide delivery system. ACS Omega *5*(1):358–368. https://doi.org/10.1021/acsomega.9b02793

Zikeli F, Romagnoli M, Mugnozza GS (2022) Lignin nanoparticles in coatings for wood preservation. In: Puglia D, Santulli C, Sarasini F (eds) Micro and Nanolignin in aqueous dispersions and polymers. Elsevier, pp 357–384. https://doi.org/10.1016/B978-0-12-823702-1.00014-1

Zimmer K, Flindall O, Gobakken LR, Nygaard M (2020). Weathering of unpainted wooden façades—experience and examples 2020. In *64*. NIBIO. nibio.brage.unit.no/nibio-xmlui/handle/11250/2638753

The Social Environment of BioCities

Giovanna Ottaviani Aalmo, Silvija Krajter Ostoic, Divina Gracia P. Rodriguez, Liz O'Brien, and Constanza Parra

1 Introduction

An increasing number of cities are becoming a striking illustration of the maldistribution of resources. These resources, which are both physical and societal, lead to inequalities which are at the root of issues such as societal tensions, poverty, alienation, and marginalization of particular groups from the public discourse (Cassiers and Kesteloot 2012). The interrelationships between the urban social environment and urban environmental conditions, alongside political and economic structures, define the distribution and access to the benefits and services that are linked to nature in the cities (O'Brien et al. 2017a, b).

BioCities include the traditional components of the social environment (e.g. families, associations, neighbourhoods, institutions, and norms) that are interconnected by communication and collaboration amongst stakeholders, resulting in a widespread awareness of the mutual responsibility for all processes. Co-creation processes, including planners and diverse stakeholder groups, can avoid the current exclusion of specific societal strata and clusters, and ensure that society at large benefits from all of the advantages of future urban settings in the BioCities model. Involvement in urban policy, planning, and management processes has the potential to create jobs for urban dwellers, and their inclusion can secure greater benefits in

G. O. Aalmo (✉) · D. G. P. Rodriguez
Norwegian Institute of Bioeconomy Research (NIBIO), Ås, Norway
e-mail: giovanna.ottaviani.aalmo@nibio.no

S. Krajter Ostoic
Croatian Forest Research Institute (CFRI), Jastrebarsko, Croatia

L. O'Brien
Society and Environment Research Group, Forest Research, Surrey, UK

C. Parra
KU Leuven, Leuven, Belgium

G. E. Scarascia-Mugnozza et al. (eds.), *Transforming Biocities*, Future City 20,
https://doi.org/10.1007/978-3-031-29466-2_9

terms of ecosystem services, from better air quality to recreational activities and access to green spaces.

BioCities also provide opportunities to lead healthy and empowered lives, as increased connection and understanding of nature and its processes give the opportunity to appreciate it, care more for it, and be at one with nature, resulting in positive impacts on people's brains, bodies, feelings, thought processes, and social interactions (see chapter "BioCities as Promotors of Health and Wellbeing"). Moreover, BioCities function as a getaway from the stress and hectic lifestyle typical of urban context. BioCities would also be able to better include all groups, in all steps of the decision-making processes, starting from the planning stages, as elaborated in chapter "Urban Sustainable Futures: Concepts and Policies Leading to BioCities". Ownership of the processes and the results of planning (implementation and management) can therefore become more inclusive, by following processes of co-creation (Basnou et al. 2020).

The problem is to really create 'true involvement' from the different groups, however, as their engagement can be diverse given different educational backgrounds, motivations, and time availability. Nevertheless, the rewards from bringing people together can be great in terms of increasing all capitals (e.g. natural, human, social, and financial) whilst ensuring environmental capital benefits the most. For the concept of BioCities to be successful, it becomes paramount to accurately define the society that will live in the BioCities, the characteristics that the communities within the BioCities need to have, and to clearly outline any known hindrances related to achieve our BioCities concept, whilst describing the potential of overcoming them through adopting a more sustainable and inclusive way of living.

2 Human–Nature Relationship and Their Impacts on the Urban Environment

Human relationships with urban nature are complex. BioCities should explore and address this complexity in pursuit of social justice and equity, hence building a nature–human nexus in the BioCity.

2.1 Importance of Studying Human–Nature Relationships

Humans and nature are part of the urban socio-ecological system (Seymour 2016). Humans have certain perceptions, preferences, attitudes, and values with regard to urban nature (De Vreese et al. 2016). They may or may not use different types of urban nature in different ways. The importance of including human perspectives of urban nature is recognised in the European Landscape Convention that puts human perception at the centre by defining a landscape as 'an area, as perceived by people, whose character is the result of the action and interaction of natural and/or human factors' (Europe 2000). Human relationships with urban nature are complex and need continuous exploration. The main reasons for including human perspectives are to:

- Avoid potential conflicts of various community values and views of urban green space planning and management.
- Monitor and understand any evolving human preferences over time.
- Study the effect of various policies (e.g. health policies) targeting the human uses of urban green spaces or how they encourage active lifestyles (or pro-environmental behaviour).

2.2 *Overview of Theories Addressing Human–Nature Relationship*

Tveit et al. (2018) explain human landscape/environmental preferences as either innate (evolutionary) or learnt (cultural). In the context of evolutionary theories, E.O. Wilson's 'biophilia hypothesis' (1984) states that people have an innate affinity towards nature, whilst the 'habitat theory' claims that people have innate preference for savannah-like landscapes (Orians 1980). Prospect-refuge theory suggests that humans prefer landscapes where they have the possibility to observe and hide without being seen (Ruddell and Hammitt 1987). And the 'Preference Matrix' (Kaplan and Kaplan 1989) specifies that the human urge for exploration and understanding influence their landscape perceptions.

On the other hand, cultural theories claim that landscape preferences are influenced by socio-cultural and personal characteristics. The most popular theories are 'topophilia theory' (Tuan 1990) and 'ecological aesthetic theory' (Nassauer 1992; Gobster 1999; Carlson 2009). The former infers that people prefer locations that they are familiar with, whilst the latter sees knowledge as the key to understanding human preferences. Amongst other theories, 'aesthetics of care theory' puts an emphasis on the importance of human perceptions of landscape patterns and processes (Nassauer 1995, 1997).

Some human preferences may be considered as universal, such as human preference for water elements, whilst others may vary amongst groups and cultures. New theories integrating both evolutionary and cultural theories are therefore needed for studying human preferences (Tveit et al. 2018).

2.3 State of the Art of Scientific Literature on Human–Nature Relationship

The number of studies of human–nature interactions have been growing over the last couple of decades. A recent systematic review of scientific papers dealing with urban green spaces identified seven main topics of importance (Kabisch et al. 2015):

- Conceptual focus on tool and theory development.
- Development, planning, and management of Urban Green Spaces (UGS).
- Satisfaction and preference of park characteristics.
- Social and environmental justice, including equal provision and access to UGS by different social groups.
- Social cohesion, particularly focusing on methods used to gauge community participation.
- Direct and indirect health effects of UGS.
- Economic value analyses.

Some topics, such as development, planning, and management; satisfaction and preference; and social and environmental justice, received more attention than others. The same paper discussed research gaps and suggestions for future research. It was found that most of the studies analysed were mainly conducted in only one city and on one site, hence they were cultural specific. Only rarely were studies transnational, including people with very different age groups, or covering different types of UGS in a single study.

European cities have been increasingly multicultural, and this most likely applies to BioCities as well. As population grows, cities of the future must take into account not only the economic, environmental, societal, and technological futures, but also the cultural identity, as the key building blocks of city design. Different countries, regions, and cultural identities generate different design propositions in terms of urban systems, architectural forms, and use of materials. Hence, learning about cross-cultural similarities and differences, in terms of perceptions, preferences, and values people associate with UGS, as well as how they use UGS, is critically important. Since human perceptions of UGS are not always positive, in BioCities both positive and negative perceptions/experiences (also called ecosystem services and disservices) should be addressed in terms of a balanced management approach (Skår 2010; Lyytimäki 2014).

Urban nature encompasses various types of UGS and people may perceive and use these types differently (O'Brien et al. 2017a, b; Krajter Ostoić et al. 2020). Some types of green space, such as urban forests and parks, are studied more than others (O'Brien et al. 2017a, b). Hence, there is still a need to understand how citizens of BioCities perceive and use different types of UGS.

2.4 Methods Used for Studying Human–Nature Relationship

Various methods and tools are available for studying human perceptions, preferences, and understanding human behaviour with regard to UGS. Indeed, questionnaire surveys are the most common method used for studying human-nature relationships (Ostoić et al. 2017; Madureira et al. 2018). There are many other methods available, however, such as individual or (focus) group interviews (Žlender and Thompson 2017; Krajter Ostoić et al. 2020) and (non-) participatory observations (such as studying how people use UGS) (Goličnik and Thompson 2010; Adinolfi et al. 2014). Recently, there has been a technological development that includes the use of computers for creating virtual environments for studying human preferences (Gao et al. 2019), GPS tracking of human movements in UGS (Korpilo 2018), application of eye tracking (Li et al. 2020), and online PPGIS (which combines questionnaire surveys with spatial data) (Rall et al. 2017). User-generated spatial data (using of GPS tracking or PPGIS and collected from social media) are useful for studying the dynamic use of UGS, but can come with certain limitations. Social media data, such as photographs, are more likely to depict large scenic UGS rather than smaller scale UGS that are used as part of everyday life (Heikinheimo et al. 2020). When selecting the most appropriate method(s) for the research question, researchers should bear in mind the advantages and limitations of each method.

2.5 Human–Nature Relationship in Urban and Green Space Planning and Management

Urban planners, green space planners, and managers should take into consideration that people may perceive and use different types of UGS in various ways. A network of accessible high-quality green spaces comprising various types and sizes of UGS should be prioritized over favouring only certain types of UGS. A recent and useful 'rule of thumb' for promoting health and well-being of BioCitizens is the '3–30–300 Rule' for urban and green space planning (Konijnendijk van den Bosch 2021). According to this rule, people should be able to see at least three trees from their homes, live in a neighbourhood with at least 30% canopy cover, and have access to quality green space for recreation not more than 300 m from their home.

Ideally, results of human–nature relationship studies should be used more regularly in urban and green space planning and management than currently (see chapter "Green Infrastructure and Urban Forests for BioCities: Strategic and Adaptive Management"). For the best/most comprehensive results and whenever possible, a representative sample of citizens should be included in community planning and consultation, covering various ages, genders, and ethnicities. In this way, equitable participation and representation of different groups are enabled. Citizens of BioCities should be involved at the very beginning of the planning process. An example is provided below for the case of Helsinki (Kahila-Tani et al. 2016), on how public participation GIS (PPGIS) tools can be used to support the creation of a master plan, and have the potential to evolve into a more comprehensive participatory planning support system (Fig. 1).

2.6 Environmental Justice

Environmental justice addresses different aspects of human–nature interaction including (Brulle and Pellow 2006; Miranda et al. 2011):

- Climate change
- Energy
- Air and water pollution (including toxins and pesticides)
- The location of industrial plants
- Waste disposal, including recycling and toxic waste
- Worker and community health and safety
- Distribution, accessibility, and management of green areas and public parks
- Wildlife conservation and protection

Environmental justice should be fully integrated in the governance of BioCities. The concept of environmental justice presupposes that (1) there is a just relationship between people and nature, and (2) that there is equal right to human access to natural capital (de Oliveira Finger and Zorzi 2013). The unequal distribution of environmental benefits and burdens has been studied since the late 1970s and has mainly aimed at addressing (1) toxic waste pollution, and (2) environmental racism, primarily in the USA (Perez et al. 2015). This was highlighted by activists' who mobilised to provide empirical evidence of environmental inequalities. Since then, the concept of environmental justice has undergone a process of evolution and significant change from being solely about where adverse environmental impacts are distributed, to include and emphasise the fundamental inequalities in decision-making (Cousins 2021). Environmental justice has therefore become a paradigm which, whilst identifying the problem (i.e. the existence of environmental inequalities), also provides the keys for solving it (Hafner 2020). Though because of the wide array of different interpretations of what justice refers to (whether just equity, equality, or more), environmental justice must be integrated with the notions of community recognition and political participation (Schlosberg 2003, 2004, 2009).

The distribution of impacts issue is closely linked to the lack of recognition of the groups against which such injustices are committed (Walker 2009). In turn, the lack of recognition causes the exclusion of (discriminated) groups from participating in decision-making processes, making distribution, recognition, and participation the three dimensions of justice that must be included in the notion of environmental justice (Schlosberg 2009; Walker and Day 2012). These dimensions are interpreted as inequality in the distribution of environmental goods and environmental risks, as a failure to recognize the most vulnerable social groups, and as exclusion from

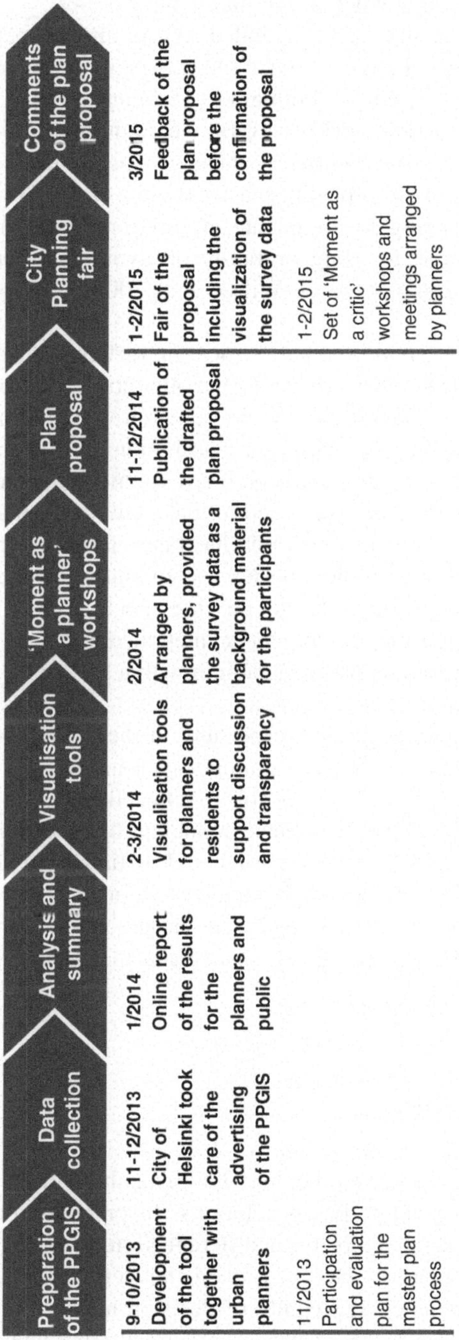

Fig. 1 The process of integrating the PPGIS tool into the master plan process of Helsinki (Kahila-Tani et al. 2016)

political decision-making. In this way, environmental (in-)justice is a single term that envelops economic, social, gender, and political inequalities.

The groups that tend not to be included in the environmental discourse, in general, are also typically the most vulnerable (i.e. people of colour, ethnic minorities, indigenous people, minors, immigrants, women, people with disabilities, people with low or no income, and the LGBT+ community) (Wilson 2009; Mohai et al. 2009). The injustices undergone by these groups are difficult to address and mitigate as they become environmental legacies (Grove et al. 2018). Urban planning, which has been contributing to the debate on environmental justice, has a more forward-looking character, focusing primarily on avoiding future injustice rather than on solving current and past issues (Wilson et al. 2008; Certomà and Martellozzo 2019).

Research related to environmental justice in different disciplines has been increasing in the past few years, thanks to the attention raised by different movements at different levels (Das 2021). Whilst research will continue addressing the issues related to human health, pollution, distribution, and access to greenspaces (Bowen 2002), there are emerging aspects of environmental justice that are gaining attention and need to be further researched to make environmental justice solutions sustainable in the long run. Little research has been done so far on the impact of environmental exposure on people's health and mortality in relation to racial and socio-economic disparities. In order to provide much-needed sustainability, the question of how these emerging topics can be integrated in the environmental justice discourse must be answered by a new paradigm with a broader sense of distribution (Mohai et al. 2009).

To increase environmental justice, especially in the urban context, its different dimensions must be addressed at the planning stages and beyond (Haaland and van Den Bosch 2015). BioCities will then be able to reflect the commitment to leaving no one behind, and therefore strive to achieve equal and fair distribution and access to greenspaces and green jobs through inclusive and participatory urban planning and management. At the same time, the highly transdisciplinary essence of BioCities will allow society to address current and ongoing environmental justice issues caused by the fragmentation of the existing green spaces and their related different benefits further mitigating the injustices.

3 Inclusive BioCities

The creation of BioCities has the potential to mitigate the impacts of climate change and advance the use of nature-based solutions. In practice, the BioCity will be governed by a network of actors from different sectors that have partly aligned and partly conflicting interests. Hence, it is crucial to understand power dynamics and enhance the transparency and equity of decision-making in any project (Reed et al. 2009). The transition to a BioCity will require altering many of the social and biological systems, affecting spatial, economic, and social relations. Human and social well-being will depend on whether BioCities can create an inclusive culture, where 'no one should be left behind', as stated in the UN Sustainable Development Goals. As outlined in chapter "Towards the Development of a Conceptual Framework of BioCities", an inclusive BioCity can then be defined as a city where all stakeholder groups, including people of all functional levels, ages, genders, and ethnicities (i.e. all who can benefit from and contribute to the building of a BioCity) are taken into account in terms of:

1. *Spatial Inclusion*

   In the classic sense, spatial inclusion means that all stakeholder groups in the city are provided with affordable basic necessities (e.g. housing, clean water, and sanitation) and have access to essential infrastructure and services (e.g. green spaces and leisure areas) (World Bank 2015). Failed spatial inclusion can mean (1) residential segregation, ghettos, and excessive gentrification, (2) unequal access to institutions and services, (3) land use policies that are unresponsive to distinct residential, recreational, religious, and cultural needs, and (4) recurring spatial reminders of advantage for some, and deprivation for others (Shah et al. 2015; Siemiatycki 2021). How the urban forest and green spaces are filled with meaningful and efficient elements will be a question for the BioCity, so that ecosystem services and co-benefits are realised to their full potential.
2. *Economic Inclusion*

   As economic growth does not always translate to a common good, there is a need to ensure that all stakeholder groups, in particular, the vulnerable groups and those most affected by BioCity-related changes, are provided with secure and dignified employment and opportunities to enjoy the benefits of the BioCity. To create an inclusive economy in the context of a BioCity, local government and planners must provide opportunities to its residents to adapt their capacity, resources, and skills to the changing demand of the new city. For example, local government can provide training/learning programmes on the use and development of local bio-based and recycled materials to manufacture the products required for the functions of a BioCity. They can also assist businesses/firms to launch, scale, and innovate goods and services associated with the functions of a BioCity by, for example providing economic incentives for investing in new digital and nature-based solutions. Special attention should be paid to the ecosystem services that cannot be easily valued in monetary terms (Bockarjova et al. 2020). There will be new job opportunities to establish, implement, and manage ecosystem service accounting.
3. *Social Inclusion*

   It is important to ensure that all stakeholder groups, especially the vulnerable, marginalised, and under-represented, have a representative voice in strategy development, planning, and implementation to realise the benefits of a BioCity through processes of co-governance. From a normative perspective, it is crucial to involve a wide range of stakeholders in the BioCity in all stages of decision-making because it is their 'right' to participate regardless of their socio-economic status, age, gender, ethnicity etc. It is also the moral duty of city planners and decision makers to involve all people in decisions that affect their lives. In the era of COVID-19, with its multi-faced health, economic, and social deprivations, the needs and wants of all groups of citizens, as well as the different capabilities, capacities, and constraints of people to benefit from goods and services, have to be considered (Sullivan III 1994; Matsuoka and Kaplan 2008; Costanza-van den Belt et al. 2021). During the COVID-19 pandemic, for example vulnerable citizens have been asked to stay at their home, limiting their ability to procure goods from common retail channels. Because of their age and/or limited access to the Internet, basic goods and services became limited. For other societal categories, such as refugees, asylum seekers, stateless persons, and migrant workers, the constraints were hindering access to basic human needs such as nutritious food, sufficient water, sanitation, secure housing, and electricity. BioCities will protect the right to health and other economic, social, and cultural rights, of members of marginalized groups by including them in the decision-making process and giving them a voice.

4 Role of Green Space in Community Building

BioCities can foster community building through green space planning and management. Conventional cities in the Global North and Global South are currently busy with the design, planning, and implementation of a wide variety of greening projects and initiatives. Various types of urban greening interventions, leading to the creation of new green spaces in cities, are considered to be a crucial step in enhancing climate adaptive capacity (Lehmann 2021). At the same time, urban greening is thought of in connection with both mental and physical health of human beings, as thoroughly elaborated in chapter "BioCities as Promotors of Health and Wellbeing". In recent years, several studies have indicated to the benefits of urban greenery in terms of socio-psychological well-being, relaxation and stress alleviation, social cohesion, reduced impact from pollution and noise, amongst others (Söderlund 2019).

Green space contained in a BioCity has a dynamic hybrid nature, involving interconnecting biophysical and social features. On one hand, green space, including natural areas and cultivated greenery, consists of unsealed, porous, and soft surfaces such as grass, shrubs, trees, parks, residential greenery, allotment gardens, amongst others (Swanwick et al. 2003; Spijker and Parra 2018). On the other hand, green space holds an intrinsic social dimension that has thus far received less attention, and which refers more specifically to the social, cultural, economic, and political dynamics underlying the different uses, transformations, and decisions over green space in cities.

Unpacking the meaning of the social aspect entails interpreting green space from at least three interrelated perspectives. First, a socio-cultural perspective refers to the individual and collective perceptions, emotions, values, meanings, and forms of attachment revolving around green space (Nieto-Romero et al. 2019). Second, a socio-political perspective highlighting the collective action underlying the social leading up to the creation of these spaces and their different uses, and the societal drivers keeping these places alive and appreciated by the individuals and groups around them (Spijker and Parra 2018). This second perspective also comprises the actions, programmes, and policies of governments and other authorities regulating green space and the entire urban area, as well as the various community initiatives imagining, creating, caring, and keeping in good condition the green commons of our cities (Parra 2013). Third, a socio-economic perspective which has to do with equity, justice, and power dynamics within and across the different social groups striving and negotiating the access, control, and distribution of these green spaces and their benefits (Anguelovski 2013). This third perspective calls attention to the need to consider green spaces, not as green islands within the BioCity, but as spaces anchored in a larger socio-spatial plexus involving multiple interconnected spatial scales (Ignatieva 2021).

The evolution of the social communities supporting green space will inevitably face the challenge of implementing and maintaining urban green space in an inclusive and equitable manner. Spijker and Parra (2018) question how we can germinate and cultivate green space governance capable of acting as a socially innovative catalyst for stimulating (1) relations between human beings and nature that are grounded in the values of care, responsibility, and respect for more than humans; and (2) social relations and modes of governance in which bottom-up community initiatives are better connected to formal spatial planning, management, and policies regulating the city as a whole. This would entail bottom-linked governance dynamics and processes (Miquel et al. 2013) through which the various involved parties are mutually inspired, equipped, and empowered within their different place-making and place-keeping efforts. Spijker and Parra (2018) state that bottom-linked governance, defined as a multi-level middle ground, where actors from various political levels, geographical scales, and industry sectors come together to share decision-making (Castro-Arce and Vanclay 2020), and that place-keeping is a collaborative and inclusive activity. The actors have connected roles to play in the long-term governance and sustainability of green spaces. Individuals and municipalities might lack the capacity or resources to engender the interest and compliance of communities for successful and sustainable long-term collaborative processes. Adopting bottom-linked practices can contribute to long-term place-keeping, allowing for bottom-up initiatives developed at the community level to flourish with the support from public institutions (i.e. policy making and other facilitations).

5 Case Study for Sustainable Place-Keeping: The Geogarden

The Geogarden is a bottom-up community project led by the Department of Earth and Environmental Sciences at the University of Leuven, Belgium. Established in 2018 with the aim to grow food in a sustainable manner, help to make the Arenberg campus flourish, and stimulate socio-ecological experimentation (Fig. 2). Advancing different types of sustainability, enhancing connection with nature, learning, sharing, caring, and relaxing are the major drivers underpinning this initiative. After more than 3 years of action, the Geogarden continues developing with fruit trees, vegetables, berry bushes, edible flowers, and other connected projects. There are beehives for the local species of 'dark bees', which are producing honey in partnership with a local NGO (Zwartebeij.org). The Geogarden has a composting system set up with the support of the Municipality of Leuven, and a weather station installed as part of a project investigating the urban heat island effect (Leuven Cool 2022). In a relatively short period of time, the Geogarden has become an important meeting place for students and staff, including gatherings, social activities, small parties, and daily lunch breaks taking place in the garden. The garden also hosts visitors, such as academics from different parts of the world, school children coming to discover the university, and public interested in gardening, amongst many others. In October 2021, Brazilian artist Julia Mota Alburquerque contributed to the Geogarden community with a mural entitled 'Connecting Nature & People' (GeoGarden 2022).

Fig. 2 From place-making to place-keeping (photos by C. Parra)

6 Public Participation/Stakeholder Engagement

BioCities should foster stakeholder involvement and co-creation of knowledge in addressing urban challenges. The idea of participation and involving diverse groups of citizens and stakeholders is not new and it is often called for, however, it is not always practised. Within BioCities, public participation should be integrated in decision-making at both the city-wide and local neighbourhood levels. The BioCity approach should test different approaches to participation and involvement as socio-economic, cultural, and institutional factors can all influence the outcomes of participation. It is still important to understand, however, the definition of relevant terms begin to understand how participation might work in the BioCity.

6.1 Definitions

The 'public' (sometimes called citizen) participation approach involves people in decision-making process regarding policies, plans, or programmes in which they have an interest. 'Community engagement' is also a term that is frequently used, which is based on outreach to communities of interest in which there is collaborative groups of people working in particular places or with specific interests or facing similar issues. 'Stakeholder engagement' involves people active in decision-making and the design of programmes that have direct interest or a stake in the project (Reed et al. 2018). The term 'stakeholder' is broad and can range from organisations, decision makers, policy makers, and practitioners from public, private, or non-governmental organisations. Bell and Reed (2021) provide a new model for inclusive decision-making, highlighting the importance of creating a safe space for deliberation, having an inclusive process, and removing barriers to involvement.

All of these approaches and processes have a focus on improving democratic participation, so that those affected by decisions can have an input and the ability to influence a decision or outcome. For planners and organisational representatives, engagement and participation can often be a requirement of their role (i.e. a job). For the public/community/stakeholders, however, participation is mostly voluntary (Salbitano et al. 2016), which can lead to issues of time, power, and finding the right moment and right setup for the activities.

The classic ladder of participation outlined by Arnstein (1969) helped people to understand progress from agency control to community control. It has been adapted and updated over the years, but the key issue remains of who gets to participate, who is the focus of consulting or empowering people to get involved, and where does the power lie in the participation process.

6.2 Why Participation?

Participation in environmental decision-making became a right in 1998 under the Aarhus Convention (McAllister 1999), and Principle 10 of the Rio Declaration (UN 1992) outlines that environmental issues should involve the participation of all concerned publics (UNEP 2011). Participation and engagement of citizens and stakeholders are crucial for BioCities for a range of reasons including the following:

- Can encourage co-design and co-creation of solutions to address the challenges faced by BioCities (Arlati et al. 2021).
- Gain input from local (indigenous) knowledge.
- Improve cooperation and collaboration.
- Has the potential to improve decision-making.
- Can help avoidance of conflict.
- Gain ideas and perspectives that can enrich a design, project, or nature-based solution.
- Can build skills and capacity of those involved.
- Can improve a sense of ownership and empowerment of a project, plan, or policy.
- Can contribute to knowledge sharing, collective learning, and social cohesion (Ferreira 20).
- Builds trust and improves transparency.
- Mobilise knowledge exchange between publics and institutions.

There is potential for new technology, including artificial intelligence, the Internet of Things (IoT), and e-democracy to contribute to making smart cities more liveable and resilient by prioritising public participation and investment in human and social capital (Bricout et al. 2021). Smart cities should be part of a deliberative planning process in which communities with shared interests will engage in grassroots planning and design. Smart cities focus strongly on technology and the use of data to gain insights that can feed into management, whereas BioCities are a broader more holistic concept that includes nature-based solutions and a circular bioeconomy, as well as the inclusion of diverse peoples in decision-making.

6.3 Core Values of Participation

The International Association for Public Participation outlines key values for participation (IAP^2 2022):

- People affected by a decision have a right to involvement in decision-making.
- The contribution of people will influence the decision.
- Information is provided that enables people to participate in a meaningful way.
- People are informed about how their input has affected the decision that was made.

It is critical that the process of participation should be fair and equitable. Tools, methods, and guidelines are needed to support this, as well as new ways of working to encourage transformative change and adaptive capacity. Deliberation (i.e. debate) that is considered, evaluated, and appraised (Kenter et al. 2016) can support social learning and inform decisions.

6.4 When to Do It?

The FAO calls for community engagement as a paradigm in the governance of cities and urban landscapes (Salbitano et al. 2016), which encompasses ideas of engagement being embedded across all areas of decision-making in cities. Early engagement of people is particularly important when co-creation/design is the approach being taken. A stakeholder/public analysis can help to identify those who have a stake in a specific decision and help to categorise stakeholders (Reed et al. 2009). Consideration needs to be given as to whether participation should be ongoing or a one-off approach specific to a particular decision.

6.5 Who to Engage and Include?

The process needs to be fair and equitable and should involve those with an interest or stake in the decision-making process. Representation of appropriate groups increases legitimacy of the process and it is important to take into account the power of different people/stakeholders (Buizer et al. 2015). There are sections of society, that because of their skills, knowledge, attitudes towards participation, social accountability, commitment, and position in the society, are often left out of decision-making. These groups can be, but they are not limited to, children and young people, indigenous groups, migrants and refugees, diverse ethnic groups, the disabled, and deprived groups (Arlati et al. 2021). Good representation and inclusion need to be considered in public/stakeholder participation. The COVID-19 pandemic has highlighted many existing inequalities within European society, making inclusive approaches more important than ever (e.g. high-density, low-income urban areas typically lacking green infrastructures indicated higher rates of infection when compared to lower density wealthier areas). Including city dwellers residing in all of these areas is important to gather feedback on their needs to address their issues properly. Potential stakeholders of BioCities could be academia and research institutions, experts and scientists, local and regional administration, financial suppliers/investors, citizens, government, property developers, Civil Society Organisations (CSOs), NGOs, planners, policy makers, media, energy suppliers, and political institutions.

6.6 Challenges of Participation and Engagement

There are a range of challenges for publics and stakeholder participation including a perception by authorities that it can take too much time, that it is too costly, and there may be a lack of political support (Ferreira et al. 2020). Costs associated with public engagement depend on the intensity and timescale needed to gain meaningful engagement. Ferreira et al. (2020) suggest that a key challenge is overcoming poor social mobilisation when urban residents perceive management and stewardship of nature and green infrastructure as the responsibility of government and not their own. Conflicts and tensions may arise from participation, which could benefit from conflict management or mediation. A lack of awareness and knowledge about environmental issues and problems, and a lack of support and guidance can also hamper participation. Organising a quality process is more important than focusing on the number of people involved.

7 Case Studies of Stakeholder and Public Participation

7.1 Stakeholder Engagement for Nature, Liveability, and Sustainability

Engagement and participation of stakeholders are taking place through networks of cities and towns focused on nature and sustainability. These networks involve stakeholder engagement often with a focus on specific projects, campaigns, and programmes.

- The Biophilic Cities Network aims to connect cities, advocates, and scholars through an understanding of how nature contributes to the lives and liveability of cities (Biophilic Cities 2022).
- The Local Governments for Sustainability Network includes 1500 cities, towns, and regions globally and has a focus on sustainable, low carbon, resilient, and biodiverse economies (ICLEI 2022).
- The European Commission has a Smart Cities marketplace platform to bring together cities, industries, businesses, researchers, and investors (EC 2022).
- The Under2 Coalition is a global community of regional and state governments focused on climate action (Under2 Coalition 2022).

7.2 Public Participation for Participatory Democracy

Participation of the public is taking place through participatory democracy, deliberative democracy, and e-democracy. Citizens mobilise into groups or forums often when dissatisfied with government/organisation decisions. Public dialogue, face-to-face and online engagement, and special events are being used as tools to debate key issues.

- In 2016, the Government of Ireland created a Citizens' Assembly to explore how to make Ireland a leader in tackling climate change (Citizens' Assembly 2022). Ninety-nine Irish citizens were chosen at random to represent the Irish people in an 'Assembly'. The Assembly made 13 recommendations to the State, although they were non-binding. Another Citizens' Assembly was created in 2022 to address biodiversity loss.
- In England, there is currently a programme of citizen engagement to improve understanding of what people value about the environment and their priorities in relation to it (NCCPE 2022).
- The City of Utrecht, in the Netherlands, facilitated public participation in green infrastructure (GI) development with the aim of creating a bottom-up process of working with the public through neighbourhood councils, and grassroots and civil society organisations. Over 140 GI projects were eventually implemented (Ambrose-Oji et al. 2017).

8 Conclusions

To be able to realise the BioCities concept and to build a strong and cohesive BioCity society, diverse stakeholder views of must be incorporated in all aspects of planning and management of BioCities, such as infrastructure, development, and public health. Additionally, to be able to sustain the BioCity concept over time and maintain it as up-to-date, the preferences of city dwellers must be continuously monitored. BioCity researchers should focus on studying the effects of policies targeting humans in the context of urban green space (e.g. health policies) and encouraging active lifestyles (or pro-environmental behaviour).

Furthermore, the outcomes of the studies on the relationship between humans and nature should be used in planning and management of urban and green space, making sure that participation of all diverse groups is enabled. By doing so, it will address the different aspects of human–nature interactions thereby achieving greater environmental justice. BioCities will reflect a commitment to leaving no one behind, and hence taking a step towards equal and fair access to ecosystem services within cities.

Any method of stakeholder inclusion should include spatial, economic, and social aspects. The identification of all stakeholders is vital for the success of the BioCities. BioCities can become a dynamic hybrid that interconnects biophysical and social features and ensures inclusivity and diversity, reduced inequality, embedded engagement and participation in decision-making, access to nature for all, and liveable cities for both people and wildlife. With the ever-increasing population and rate of urbanisation, additional research should target knowledge gaps regarding (i) issues of urban scale and the important role it plays in the future urban sustainability and (ii) the specific and general benefits and advantages of BioCities.

References

Adinolfi C, Suárez-Cáceres GP, Carinanos P (2014) Relation between visitors' behaviour and characteristics of green spaces in the city of Granada, south-eastern Spain. Urban Forestry Urban Green 13(3):534–542

Ambrose-Oji B, Buijs A, Gerőházi E, Mattijssen T, Száraz L, Van der Jagt A, Hansen R, Rall E, Andersson E, Kronenberg J, Rolf W (2017) Innovative governance for urban green infrastructure: a guide for practitioners. Green surge project deliverable 6.3, University of Copenhagen, Copenhagen

Anguelovski I (2013) New directions in urban environmental justice: rebuilding community, addressing trauma, and remaking place. J Plan Educ Res 33(2):160–175

Arlati A, Rödl A, Kanjaria-Christian S, Knieling J (2021) Stakeholder participation in the planning and design of nature-based solutions. Insights from CLEVER cities project in Hamburg. Sustainability 13(5):2572

Arnstein SR (1969) A ladder of citizen participation. J Am Inst Plann 35(4):216–224

Basnou C, Baró F, Langemeyer J, Castell C, Dalmases C, Pino J (2020) Advancing the green infrastructure approach in the province of Barcelona: integrating biodiversity, ecosystem functions and services into landscape planning. Urban Forestry Urban Green 55:126797

Bell K, Reed M (2021) The tree of participation: a new model for inclusive decision-making. Commun Dev J 57(4):595–614

Biophilic Cities (2022) Accessed December 1. Connecting cities with nature. https://www.biophiliccities.org

Bockarjova M, Botzen WJ, Koetse MJ (2020) Economic valuation of green and blue nature in cities: a meta-analysis. Ecol Econ 169:106480

Bowen W (2002) An analytical review of environmental justice research: what do we really know? Environ Manag 29(1):3–15

Bricout J, Baker PM, Moon NW, Sharma B (2021) Exploring the smart future of participation: community, inclusivity, and people with disabilities. Int J E-Plann Res 10(2):94–108

Brulle RJ, Pellow DN (2006) Environmental justice: human health and environmental inequalities. Annu Rev Public Health 27:103–124

Buizer I, Elands BH, Mattijssen T, Jagt A, Ambrose-Oji B, Gerohazi E, Santos E (2015) The governance of urban green spaces in selected EU-cities: policies, practices, actors, topics. European Union Green Surge Project

Carlson A (2009) Nature and landscape: an introduction to environmental aesthetics. Columbia University Press

Cassiers T, Kesteloot C (2012) Socio-spatial inequalities and social cohesion in European cities. Urban Stud 49(9):1909–1924

Castro-Arce K, Vanclay F (2020) Transformative social innovation for sustainable rural development: an analytical framework to assist community-based initiatives. J Rural Stud 74:45–54

Certomà C, Martellozzo F (2019) Cultivating urban justice? A spatial exploration of urban gardening crossing spatial and environmental injustice conditions. Appl Geogr 106:60–70

Citizens' Assembly (2022) Accessed December 1. https://www.citizensassembly.ie/en/

Costanza-van den Belt M, O'Donnell T, Webb R, Robson E, Costanza R, Ling J, Crowe S, Han H (2021) Community preferences for urban systems transformation in Australia. Sustainability 13(9):4749

Council of Europe (2000) European Landscape Convention. European Treaty Series #176

Cousins JJ (2021) Justice in nature-based solutions: research and pathways. Ecol Econ 180:106874

Das U (2021) Environmental justice research – limitations and future directions using qualitative research methods. Qual Res J 21:469

de Oliveira Finger M, Zorzi FB (2013) Environmental justice. UFRGS Model United Nations J 1: 222–243

De Vreese R, Leys M, Fontaine C, Dendoncker N (2016) Social mapping of perceived ecosystem services supply – the role of social landscape metrics and social hotspots for integrated ecosystem services assessment, landscape planning and management. Ecol Indic 66:517–533

EC (2022) Accessed December 1. Smart cities marketplace. European Commission. https://smart-cities-marketplace.ec.europa.eu

Ferreira V, Barreira AP, Loures L, Antunes D, Panagopoulos T (2020) Stakeholders' engagement on nature-based solutions: a systematic literature review. Sustainability 12(2):640

Gao T, Liang H, Chen Y, Qiu L (2019) Comparisons of landscape preferences through three different perceptual approaches. Int J Environ Res Public Health 16(23):4754

GeoGarden (2022) Accessed November 29. KU Leuven, Belgium. https://geogardenleuven.wordpress.com/2021/10/13/27-mural-opening/

Gobster PH (1999) An ecological aesthetic for forest landscape management. Landsc J 18(1):54–64

Goličnik B, Thompson CW (2010) Emerging relationships between design and use of urban park spaces. Landsc Urban Plan 94(1):38–53

Grove M, Ogden L, Pickett S, Boone C, Buckley G, Locke DH, Lord C, Hall B (2018) The legacy effect: understanding how segregation and environmental injustice unfold over time in Baltimore. Ann Am Assoc Geogr 108(2):524–537

Haaland C, van Den Bosch CK (2015) Challenges and strategies for urban green-space planning in cities undergoing densification: a review. Urban Forestry Urban Green 14(4):760–771

Hafner R (2020) Environmental justice Incommensurabilities framework: monitoring and evaluating environmental justice concepts, thought styles and human-environment relations. DIE ERDE–J Geograph Soc Berlin 151(2–3):67–76

Heikinheimo V, Tenkanen H, Bergroth C, Järv O, Hiippala T, Toivonen T (2020) Understanding the use of urban green spaces from user-generated geographic information. Landsc Urban Plan 201:103845

IAP^2 (2022) Accessed November 30. IAP^2 core values. International Association for Public Participation. https://www.iap2.org/page/corevalues

ICLEI (2022) Accessed December 1. ICLEI cities biodiversity center. https://cbc.iclei.org

Ignatieva M (2021) Evolution of the approaches to planting design of parks and gardens as main greenspaces of green infrastructure. In: Catalano C (ed) Urban services to ecosystems, Springer, pp 435–452

Kabisch N, Qureshi S, Haase D (2015) Human–environment interactions in urban green spaces—a systematic review of contemporary issues and prospects for future research. Environ Impact Assess Rev 50:25–34

Kahila-Tani M, Broberg A, Kyttä M, Tyger T (2016) Let the citizens map—public participation GIS as a planning support system in the Helsinki master plan process. Plan Pract Res 31(2): 195–214

Kaplan R, Kaplan S (1989) The experience of nature: a psychological perspective. Cambridge University Press

Kenter JO, Reed MS, Fazey I (2016) The deliberative value formation model. Ecosyst Serv 21:194–207

Konijnendijk van den Bosch C (2021) Promoting health and wellbeing through urban forests – introducing the 3-30-300 rule. https://www.linkedin.com/pulse/promoting-health-wellbeing-through-urban-forests-rule-cecil/?trackingId=

Korpilo S (2018) An integrative perspective on visitor spatial behaviour in urban green spaces: linking movement, motivations, values and biodiversity for participatory planning and management. Dissertation, Faculty of Biological and Environmental Sciences, University of Helsinki, 38 p

Krajter Ostoić S, Marin AM, Kičić M, Vuletić D (2020) Qualitative exploration of perception and use of cultural ecosystem services from tree-based urban green space in the city of Zagreb (Croatia). Forests 11(8):876

Lehmann S (2021) Growing biodiverse urban futures: renaturalization and rewilding as strategies to strengthen urban resilience. Sustainability 13(5):2932

Leuven Cool (2022) Accessed November 29. https://www.leuven.cool

Li J, Zhang Z, Jing F, Gao J, Ma J, Shao G, Noel S (2020) An evaluation of urban green space in Shanghai, China, using eye tracking. Urban Forestry Urban Green 56:126903

Lyytimäki J (2014) Bad nature: newspaper representations of ecosystem disservices. Urban Forestry Urban Green 13(3):418–424

Madureira H, Nunes F, Oliveira JV, Madureira T (2018) Preferences for urban green space characteristics: a comparative study in three Portuguese cities. Environments 5(2):23

Matsuoka RH, Kaplan R (2008) People needs in the urban landscape: analysis of landscape and urban planning contributions. Landsc Urban Plan 84(1):7–19

McAllister ST (1999) The convention on access to information, public participation in decision-making, and access to justice in environmental matters. Colorado J Int Environ Law Policy 10: 187

Miquel MP, Cabeza MG, Anglada SE (2013) Theorizing multi-level governance in social innovation dynamics. In: Moulaert F, McCallum D, Mehmood A, Hamdouch A (eds) International handbook on social innovation: Collective action, social learning and transdisciplinary research. Edward Elgar Publishing Limited

Miranda ML, Hastings DA, Aldy JE, Schlesinger WH (2011) The environmental justice dimensions of climate change. Environ Justice 4(1):17–25

Mohai P, Pellow D, Roberts JT (2009) Environmental justice. Annu Rev Environ Resour 34:405–430

Nassauer JI (1992) The appearance of ecological systems as a matter of policy. Landsc Ecol 6(4): 239–250

Nassauer JI (1995) Culture and changing landscape structure. Landsc Ecol 10(4):229–237

Nassauer JI (1997) Cultural sustainability: aligning aesthetics and ecology. Island Press

NCCPE (2022) Accessed December 1. Citizen engagement on the environment. National Co-ordinating Centre for Public Engagement. https://www.publicengagement.ac.uk/nccpe-projects-and-services/nccpe-projects/citizen-engagement-environment

Nieto-Romero M, Valente S, Figueiredo E, Parra C (2019) Historical commons as sites of transformation. A critical research agenda to study human and more-than-human communities. Geoforum 107:113–123

O'Brien L, De Vreese R, Atmiş E, Olafsson AS, Sievänen T, Brennan M, Sanchez M, Panagopoulos T, de Vries S, Kern M (2017a) Social and environmental justice: diversity in access to and benefits from urban green infrastructure–examples from Europe. In: Pearlmutter D, Calfapietra C, Samson R, O'Brien L, Krajter Ostoić S, Sanesi G, Alonso del Amo R (eds) The urban forest, Springer, pp 153–190

O'Brien L, De Vreese R, Kern M, Sievänen T, Stojanova B, Atmiş E (2017b) Cultural ecosystem benefits of urban and peri-urban green infrastructure across different European countries. Urban Forestry Urban Green 24:236–248

Orians GH (1980) Habitat selection: general theory and applications to human behavior. In: Lockard JS (ed) The evolution of human social behavior. Elsevier, Amsterdam, pp 49–63

Ostoić SK, van den Bosch CCK, Vuletić D, Stevanov M, Živojinović I, Mutabdžija-Bećirović S, Lazarević J, Stojanova B, Blagojević D, Stojanovska M (2017). Citizens' perception of and satisfaction with urban forests and green space: results from selected southeast European cities. Urban Forestry Urban Green, 23: 93–103

Parra C (2013) Social sustainability: a competing concept to social innovation? In: Moulaert F, McCallum D, Mehmood A, Hamdouch A (eds) International handbook on social innovation: collective action, social learning and transdisciplinary research. Edward Elgar Publishers, pp 142–154

Perez AC, Grafton B, Mohai P, Hardin R, Hintzen K, Orvis S (2015) Evolution of the environmental justice movement: activism, formalization and differentiation. Environ Res Lett 10(10): 105002
Rall E, Bieling C, Zytynska S, Haase D (2017) Exploring city-wide patterns of cultural ecosystem service perceptions and use. Ecol Indic 77:80–95
Reed MS, Graves A, Dandy N, Posthumus H, Hubacek K, Morris J, Prell C, Quinn CH, Stringer LC (2009) Who's in and why? A typology of stakeholder analysis methods for natural resource management. J Environ Manag 90(5):1933–1949
Reed MS, Vella S, Challies E, De Vente J, Frewer L, Hohenwallner-Ries D, Huber T, Neumann RK, Oughton EA, Sidoli del Ceno J (2018) A theory of participation: what makes stakeholder and public engagement in environmental management work? Restor Ecol 26:S7–S17
Ruddell EJ, Hammitt WE (1987) Prospect refuge theory: a psychological orientation for edge effect in recreation environments. J Leis Res 19(4):249–260
Salbitano F, Borelli S, Conigliaro M, Chen Y (2016) Guidelines on urban and peri-urban forestry. FAO Forestry Paper 178
Schlosberg D (2003) The justice of environmental justice: reconciling equity, recognition, and participation in a political movement. Moral Polit Reason Environ Pract 77:106
Schlosberg D (2004) Reconceiving environmental justice: global movements and political theories. Environ Politics 13(3):517–540
Schlosberg D (2009) Defining environmental justice: theories, movements, and nature. Oxford University Press
Seymour V (2016) The human–nature relationship and its impact on health: a critical review. Front Public Health 4:260
Shah P, Hamilton E, Armendaris F, Lee H (2015) World inclusive cities approach paper. The World Bank. Report no. AUS8539
Siemiatycki M (2021) Developing homeless shelters through public–private partnerships: the case of the red door family shelter in Toronto. J Urban Aff 43(2):236–250
Skår M (2010) Forest dear and forest fear: dwellers' relationships to their neighbourhood forest. Landsc Urban Plan 98(2):110–116
Söderlund J (2019) The emergence of biophilic design. Springer
Spijker SN, Parra C (2018) Knitting green spaces with the threads of social innovation in Groningen and London. J Environ Plan Manag 61(5–6):1011–1032
Sullivan W III (1994) Perceptions of the rural-urban fringe: citizen preferences for natural and developed settings. Landsc Urban Plan 29(2–3):85–101
Swanwick C, Dunnett N, Woolley H (2003) Nature, role and value of green space in towns and cities: an overview. Built Environ (1978-) 94–106
Tuan YF (1990) Topophilia: a study of environmental perception, attitudes, and values. Columbia University Press
Tveit MS, Ode Sang Å, Hagerhall CM (2018) Scenic beauty: visual landscape assessment and human landscape perception. In: Steg L, de Groot JIM (eds) Environmental psychology: an introduction, 2nd ed. Wiley, pp 45–54
UN (1992) The Rio declaration on environment and development. United Nations
Under2 Coalition (2022) Accessed December 1. Climate Group. https://www.theclimategroup.org/under2-coalition
UNEP (2011) Guidelines for the development of National Legislation on access to information, public participation and access to justice in environmental matters. United Nations Environment Programme
Walker G (2009) Beyond distribution and proximity: exploring the multiple spatialities of environmental justice. Antipode 41(4):614–636
Walker G, Day R (2012) Fuel poverty as injustice: integrating distribution, recognition and procedure in the struggle for affordable warmth. Energy Policy 49:69–75
Wilson EO (1984) Biophilia. Harvard University Press, Cambridge

Wilson SM (2009) An ecologic framework to study and address environmental justice and community health issues. Environ Justice 2(1):15–24

Wilson S, Hutson M, Mujahid M (2008) How planning and zoning contribute to inequitable development, neighborhood health, and environmental injustice. Environ Justice 1(4):211–216

World Bank (2015). Everyone counts making the cities of tomorrow more inclusive. World Bank. https://www.worldbank.org/en/news/feature/2015/10/29/a-new-approach-to-cities-including-inclusion

Žlender V, Thompson CW (2017) Accessibility and use of peri-urban green space for inner-city dwellers: a comparative study. Landsc Urban Plan 165:193–205

From BioCities to BioRegions and Back: Transforming Urban–Rural Relationships

Bart Muys, Eirini Skrimizea, Pieter Van den Broeck, Constanza Parra, Roberto Tognetti, David W. Shanafelt, Ben Somers, Koenraad Van Meerbeek, and Ivana Živojinović

1 Introduction

Cities are hubs of money, power, and information. Characterised by high population density, numerous built structures, extensive impervious surfaces, decreased vegetative cover, and highly modified ecosystem services, cities or urban areas are surrounded by less-densely populated areas with less built-up space, referred to as rural areas (Wu 2014). Rural areas are perceived as a mosaic of land uses with various types of human intervention and productivity, including various degrees of naturalness. In Europe, one of the most intensely anthropised areas of the world, very few natural areas have been left untouched, and thus the degree of naturalness of rural areas is relatively low. Notwithstanding certain benefits that rural areas gain from cities, such as market access, investment inputs, or employment opportunities (Gebre and Gebremedhin 2019), cities have generally developed an extractive relationship with those areas. The countryside is perceived as a source of food, water, materials, and energy to serve the needs of cities, which behave as accumulative economic nodes (McHale et al. 2015). Widespread policy attention on cities

B. Muys (✉) · E. Skrimizea · P. Van den Broeck · C. Parra · B. Somers · K. Van Meerbeek
KU Leuven, Leuven, Belgium
e-mail: bart.muys@kuleuven.be

R. Tognetti
University of Molise, Campobasso, Italy

D. W. Shanafelt
Université de Lorraine, Université de Strasbourg, AgroParis Tech, Centre National de la Recherche Scientifique (CNRS), Institut National de Recherche pour l'Agriculture, l'Alimentation et l'Environnement (INRAE), Bureau d'Economie Théorique et Appliquée (BETA), Strasbourg, France

I. Živojinović
University of Natural Resources and Life Sciences, Vienna (BOKU), Vienna, Austria

G. E. Scarascia-Mugnozza et al. (eds.), *Transforming Biocities*, Future City 20,
https://doi.org/10.1007/978-3-031-29466-2_10

has marginalised rural areas from territorial development planning, forgetting to account for the urban and the rural on equal terms (Urso 2020). Urban dwellers, particularly in middle to high-income countries, are physically and mentally disconnected from the rural world, at most knowing it as a space for recreation and distraction (Roberts and Hall 2001). This disconnection could hamper awareness about overconsumption of natural resources and environmental degradation happening in the rural world (Church 2013), even though environmental driven protest against impacts of climate change and natural resource extraction is often originating from urban citizens (Scheidel et al. 2020).

The extractive relationship between cities and rural areas has often resulted in socio-economic and environmental decay of the latter, but has drawbacks for the city too. This is illustrated by the trends in urban–rural relationships in Europe since the nineteenth century, especially after World War II. The contemporary agribusiness model is based on mechanisation, use of biocides and fertilisers, increased productivity, and decreased consumer prices for agricultural commodities. It has come at a high cost of biodiversity loss and strongly simplified landscapes (Vanbergen et al. 2020). It has also triggered massive job losses in rural areas, migration of the rural population to cities, and land abandonment in marginal areas (Lasanta et al. 2017; MacDonald et al. 2000).

Cities have grown and expanded at the expense of rural land. Development of large-scale industrial areas, housing developments, urban sprawl, and touristic exploitation has occupied significant areas of often very fertile land (Antrop 2004; Hennig et al. 2015; Zambon et al. 2019). This process has included land grabbing (Van der Ploeg et al. 2015), privatisation and urbanisation of agricultural land, forested areas, lakes and riversides, and coastlines and mountain villages (Antrop 2004; Jiménez et al. 2019), with a serious loss of unique and fragile ecosystems. Overall, these trends have led to an alteration of the social fabric in rural areas (Vanbergen et al. 2020), diminishing the socio-cultural identity of rural communities, and the deterioration of cultural and aesthetic values of rural landscapes (Chaudhary et al. 2018; Leal Filho et al. 2016). They have also led to drawbacks for the urban and peri-urban populations themselves (Seifollahi-Aghmiuni et al. 2022), including the loss of confidence in the quality and sustainability of the food provided by the industrial food chain, slow or problematic integration of large groups of rural newcomers in the social fabric of the city, touristic saturation and loss of touristic potential of the countryside, and increased risk of life-threatening megafires as a consequence of land abandonment and urban sprawl.

In the context of planetary urbanisation (Brenner and Schmid 2014), a complex and dynamic web of spatial and functional interdependencies is shaping the fortunes of cities and countryside alike (Davoudi and Stead 2002). Citing Merrifield (2011), 'the urban unfolds into the countryside just as the countryside folds back into the city'. Acknowledging this complexity and in line with the UN Sustainable Development Goal 11 (Sustainable Cities and Communities), this chapter introduces the concept of 'BioRegion' to complement the BioCity. The challenge of this chapter is to document the potential for more sustainable links between urban and rural areas with the following three specific objectives: (1) conceptualise BioCities and

BioRegions as an integrated, complex social-ecological system; (2) apply this improved system's understanding as a means to develop more sustainable cities in a harmonious co-evolution with their rural areas; and (3) give illustrative examples of cities, which connect with the rural realm in a fair and sustainable way.

2 State of the Art and Trends

2.1 BioCity and BioRegion as a Complex Social-Ecological System

Complex living systems are thermodynamically open systems, which receive *exergy*, another word for useful energy, from external sources (e.g. solar energy and fossil energy), and use this energy to build up a certain level of organisation and stability (Muys 2013). To maintain order, they continue to metabolise exergy, resulting in an increase of entropy of their environment (Pelorosso et al. 2017). Complex living systems have several properties that allow them to create and keep order, even under changing conditions or when facing internal or external factors of disturbance or destabilisation, as described in chapter "Towards the Development of a Conceptual Framework of BioCities" introducing the BioCity concept. For this reason, they are also named Complex Adaptive Systems (CAS).

Two crucial properties of CAS are the *ability to self-organise* and the *ability to learn*. The ability of self-organisation strongly depends on a continuous exergy flow. It allows complex systems to grow and become more efficient and stable by increasing connectivity for energy or information flows between system components. The ability to learn allows the system to memorise and further improve the successful pathways of building order and connectivity. In ecosystems, the evolutionary processes captured in the genetic code of the biodiversity are essential to this idea. In cities, it directly connects to humans' cognitive and communication capacities (Muys 2013; Pelorosso et al. 2017; Skrimizea et al. 2019; Wu 2014), but also to patterns of human social processes as embedded in socio-economic, socio-political, and socio-cultural institutions, their continuities, discontinuities, and transformations (Moulaert et al. 2016; Servillo and Van Den Broeck 2012).

As a result of the exceptional learning ability of human beings, the human–nature relationship and its thermodynamics have changed drastically over time. Over the last 10,000 years, human–nature relationships changed from a social-ecological system of hunter–gatherers via an agrarian system to a predominantly urban-industrial system (Fig. 1a–c). As human societies mobilised their ability to learn and develop a global urban-industrial society, the exergy and resources needed to sustain societies have grown tremendously. Cities require a large amount of matter and energy from their surroundings to self-organise and function (Giampietro 2019). Cities feed on limited natural resources and on resources external to the biosphere (e.g. oil, coal, natural gas, and uranium from the geosphere), and increase the entropy

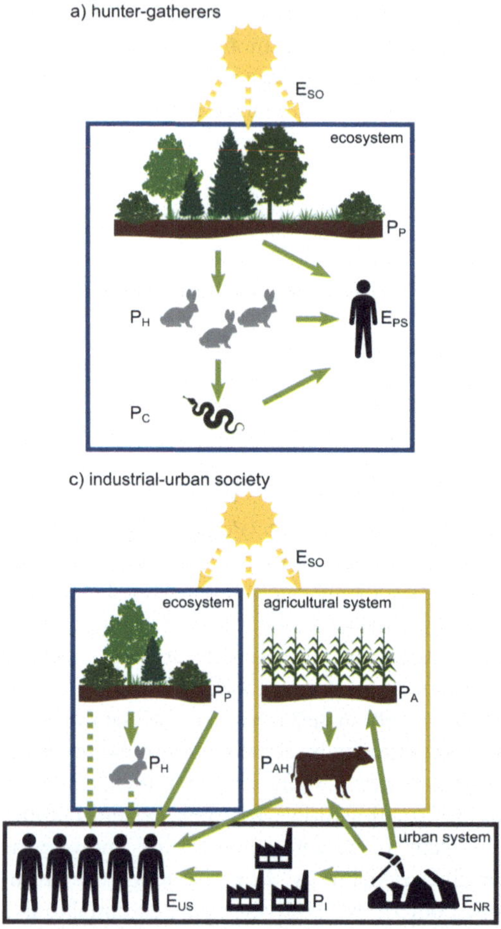

Fig. 1 A simplified representation of the energetic relationships between humankind and nature in (**a**) a hunter-gatherer society; (**b**) an agrarian society; (**c**) an industrial-urban society; and (**d**) a circular BioSociety composed of BioCities and their BioRegions (modified after Muys 2013, using icons by Ola Möller, Joel McKinney, and Andrejs Kirma through the Noun Project). Legend of symbols: ES_O = incoming solar energy; P_P = primary production of plant biomass in the ecosystem; P_H = production of herbivores in the ecosystem; P_C = production of carnivores in the ecosystem; E_{PS} = energy needs of the primitive society; P_A = primary production in the agricultural ecosystem; P_{AH} = production of herbivores in the agricultural ecosystem; E_{AS} = energy needs of the agrarian society; EN_R = non-renewable energy sources; P_I = industrial production; E_{US} = energy needs of the urban society; P_R = renewable energy production. Dashed, resp. dotted lines indicate fluxes of relatively decreasing importance, which in absolute terms may be increasing. Note that through the evolution from hunter–gatherer over agrarian to industrial-urban society the human population increases, the area of (semi-) natural systems decrease in favour of agricultural and urban land; wildlife decreases, and large predators become extinct. In the urban-industrial society, societal metabolism thrives mainly on fossil energy, rather than solar. In the circular bioeconomy of the future, the human population stagnates or gradually decreases, the system adapts to climate change, the agricultural production becomes more predominantly vegan, and the industrial production is emission-free and based on renewable energy and circularity

of their environment, exemplified by resource consumption and depletion, land and water pollution, toxic or climate forcing emissions, waste production, and biodiversity loss and degradation of the biosphere on which cities depend (Muys 2013; Pelorosso et al. 2017). Through this process, human societies have accumulated wealth and shaped human–nature relations in various ways, some more sustainable than others. Some cities have developed as spatial expressions of highly unequal societies structured around excessive power differences, driving extractivist and accumulative tendencies, leaving behind vulnerable groups and contributing to environmental and climate injustice (Cole et al. 2017). Others have developed ecologically more sustainable and socially more just political mechanisms, economies, and agricultural systems.

Sustainable development (i.e. restoring the balance between the global urban–industrial system and the biosphere on which it depends) is a difficult challenge, and requires systemic changes throughout the complex social-ecological system. The BioCity, as an upgraded complex social-ecological system, represents a more sustainable future alternative based on a low-entropy release system (Pelorosso et al. 2017), evolving in harmony with the BioRegion (Fig. 1d). Building on Goh (2020) and Thackara (2019), the BioRegion refers to a refined urban regional scale that includes the BioCity and its surroundings and is formulated and defined by natural and social interconnections (e.g. watersheds, foodsheds, and food systems) rather than administrative and economic boundaries. This means that the term BioRegion is more consistent with a 'network' rather than with a spatially compact region. In the circular BioSociety, formulated by the BioCity and its BioRegion together, the use of resources and the economy around it is more cyclical. The impact on nature and the dependence on energy sources, such as fossil fuels, is minimised, and climate change and adaptation to global and regional environmental change are integral to its functions. The links of the BioCity with its surroundings are less extractive, and socially, economically, and ecologically more sustainable than in present-day unsustainable cities. Multi-level governance safeguards the BioSocieties human well-being and social equity.

The social-ecological system of a BioCity and its BioRegion comprises a variety of organisms, compartments, and processes, with distinct traits and interdependent flows of ecosystem functions and services across a blue and green network within the landscape (Fig. 2). As will become clear from the rest of this chapter, these natural elements will serve as essential components of sustainable BioCities, where they will integrate harmoniously with human communities and human build-up structures.

2.2 Dominant Processes: Urbanising the Rural

Quoting Neil Brenner (2019), 'Nothing escapes urbanism'. The twentieth century has been characterised by an unprecedented accumulation of people and resources in cities, accompanied by an intensification of rural land through agribusiness and of

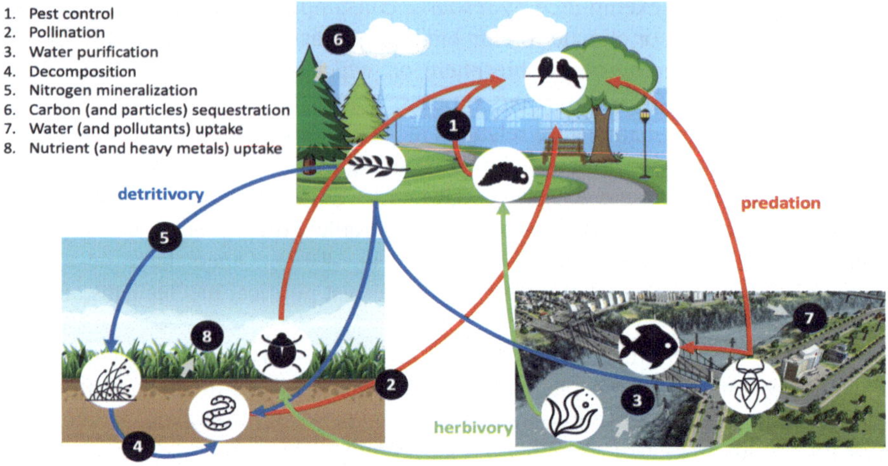

Fig. 2 Arrows indicate directional fluxes (energy, matter) amongst functional groups, highlighting how ecological processes permeate urban boundaries (vegetation, soils, and freshwater). These ecosystem fluxes are analogous to respective ecosystem functions, many of which can be directly translated into ecosystem services

forests through conversion of natural forests to agricultural land, tree plantations of fast-growing trees. and recreational resorts. These often-unsustainable pathways of twentieth century urbanisation and globalisation have strong effects on cities and rural areas. These include urban sprawl, agricultural and forestry intensification, and land abandonment. The latter being a side effect of urbanisation and agricultural intensification, which is very widespread over large parts of Europe, with complex positive and negative consequences.

2.2.1 Urban Sprawl

Urban sprawl (i.e. the rapid expansion of the geographic extent of cities and towns), often characterised by low-density residential housing, is a key driver in the loss of open space worldwide (Lichtenberg 2011). The Organisation for Economic Co-operation and Development (OECD) monitored urban sprawl between 1990 and 2014 across 1100 urban areas spread over 29 countries, and found that over 60% of urban space is actually sparsely populated (OECD 2018). Sprawl is the result of a complex set of interrelated socio-economic, socio-political, and cultural processes (Brody 2013). These include a growth-oriented economic system, increasing consumption norms triggered by individualisation of rights and duties, rising incomes, systemic land speculation, and preference for living in low-density areas (Schrank et al. 2012). The OECD also points to the policy as an important driver of urban sprawl: ‘Maximum density restrictions, specific zoning regulations, tax systems that are misaligned with the social cost of low-density development, the under-pricing of car use externalities and the massive investment in road infrastructure contribute to this phenomenon' (OECD, 2018, pp. 10). The literature reports many environmental and social consequences of urban sprawl. From an environmental perspective, urban sprawl is related to higher air pollution, habitat loss, degradation, and fragmentation. From a social perspective, urban sprawl leads to increased traffic and traffic jams, car dependency, spatial segregation between residential and commercial areas, and negative impact on lifestyles and human health (e.g. obesity and stress) (Hasse and Lathrop 2003; Johnson 2001; Power 2010). Urban sprawl further drastically increases the cost of public services (e.g. waste collection, water treatment, road infrastructure, Internet and telecommunication infrastructure, and public transport) compared to more dense development strategies (Carruthers and Ulfarsson 2003).

2.2.2 Agricultural and Forestry Intensification

Since the middle of the twentieth century, the agricultural sector in Europe experienced a rapid process of intensification driven by mechanisation and increased application of chemicals (pesticides and fertilisers). This led to a fourfold increase in yields between 1945 and 2000 (Robinson and Sutherland 2002). Intensification of land-use practices has reduced farmland biodiversity across different species groups such as plants (Storkey et al. 2012), birds (Donald et al. 2006), insects (Hallmann et al. 2017), and soil organisms (Tsiafouli et al. 2015), leading to a multitrophic homogenisation of the agricultural landscape (Gossner et al. 2016). The effort to spare land for biodiversity conservation has only partly offset the negative impact of intensification on biodiversity (Reidsma et al. 2006).

Afforestation efforts, together with land abandonment, have increased the forested area in Europe by about 25% since the 1950s (Fuchs et al. 2013), to 35% of land cover today (Forest Europe 2020). This increase in forest cover, which has not been even across the biogeographical regions of Europe, mainly reflects the expansion of forest plantations (+14.5% to 8.1 million ha) and spontaneous forest regrowth (+13.1% to 199.6 million ha) over the last three decades (Forest Europe 2020). The area of undisturbed forests also slightly increased to 4.7 million ha (2.2% of the forested area). However, when looking at longer time trends (since 1750), the area of unmanaged forests declined drastically, together with the share of coppice and broadleaved species (Naudts et al. 2016). All these changes in the forest landscape led to biodiversity loss, but less drastically than in the agricultural land.

2.2.3 Land Abandonment

Urbanisation and agricultural intensification are leading to abandonment of areas less suitable for agriculture, further driven by a declining profitability of farming on marginal lands (e.g. mountain systems) as well as rural-to-urban migration (e.g. younger generations moving to cities) (Navarro and Pereira 2012). Farmland abandonment continues to be an important land-use change process up until today, mainly concentrated in marginal lands of eastern and southern Europe (Estel et al. 2015; Kuemmerle et al. 2016). However, farmland abandonment has also occurred in the urban–rural fringe in rapidly industrialising and urbanising regions (Zhou et al. 2020). This process may provide opportunities for rewilding (i.e. restoration aiming to (partially), restore self-sustaining and complex ecosystems via restoring natural ecological processes whilst minimising human interventions (Perino et al. 2019) on abandoned landscapes (Pereira and Navarro 2015). But it can also lead to the loss of species that are adapted to open cultural landscapes (Van Meerbeek et al. 2019). In close proximity to urban areas, agricultural land and associated production costs have become increasingly expensive. In peri-urban areas, landowners are often waiting for opportunities to convert farmland into built-up land, speculating on rising land prices. The conversion of productive agricultural land and semi-natural ecosystems into built-up areas has led to habitat degradation and fragmentation. More recently, new ideas regarding the conservation of peri-urban agricultural areas have emerged, pushing conversion plans of cropland into urbanised areas to consider social and ecological issues.

2.3 New Processes Ruralising the Urban

Important trends during recent decades have appeared to boost the transformation of BioCities and BioRegions, in the sense of creating social-ecological networks and flows cutting across artificial urban–rural divisions, through community agriculture, city greening, and joint land-sparing/sharing approaches.

2.3.1 Community-Based Agriculture

Novel forms of food production and acquisition can be observed in urban contexts, through various forms of direct citizen engagement in allotment, urban gardening, and wild food foraging, or more passive engagement in local food systems such as community supported agriculture. Although having older foundations, allotment and community gardens grew in extent, especially during the nineteenth and early twentieth centuries in industrial European countries as a response to unhealthy living conditions and poverty in workers' settlements following rapid urbanisation (Steel 2013; Keshavarz and Bell 2016). Different urban gardening forms appeared across Europe, from legally designed allotment gardening to illegal occupation in newly constructed modernist neighbourhoods (Čepić et al. 2020). Nowadays, we witness an increase in non-governmental organisations forming urban gardening spaces across cities to achieve environmental, social, economic, and health benefits. Wild food foraging by the urban population occurs often in sub-urban and nearby rural areas, and reflects many changes in nature–society relationships. Foraging provides material benefits to urban dwellers, and it has the potential to enhance human health and well-being and can be a useful means of education about biodiversity conservation, food, and nutrition (Schunko et al. 2021).

Community Supported Agriculture (CSA) assumes establishment of direct relationships between consumers (often of urban character) and producers (often based in sub-urban and rural areas) based on values like solidarity, risk-sharing, community building, respect for the environment, and strong regional anchoring. The CSA consumer agrees to pay the producer in advance for a harvest share whilst the producer is committed to supply fresh, local, and good quality products (Egartner et al. 2020; Hinrichs 2000; Wellner and Theuvsen 2016).

2.3.2 City Greening Ideas

Forest gardens and so-called 'tiny forests' are strategies to implement green infrastructure in dense cities that lack urban greenery. Forest gardening is an approach to gardening that constructs a garden to resemble the structure of a natural forest (or modify an already existing forest with desirable plants), planting or arranging fruits and vegetables, herbs, fungi, or any other food, fuel, or fibre, based on their requirements for light, nutrients, and water at each level of the forest (Bukowski and Munsell 2018; Hart 1996). Forest gardens complement traditional agriculture and provide ecosystem services such as biodiversity refuges (Eyzaguirre and Linares 2010; McConnell 2017). The concept can be implemented on small scales, such as in backyards or gardens, or larger-scale properties like community gardens or woodlands (e.g. the Beacon Hill 'Food Forest' in Seattle, Washington, USA). In a similar vein, 'tiny' forests represent a way of replanting species-rich, native small forest patches in urban landscapes. Originally conceived in Japan by botanist Akira Miyawaki, the tiny forest concept in Europe was first implemented in The Netherlands in 2015, with the planting of over 100 tennis court-sized degraded patches with indigenous trees (Bleichrodt et al. 2018; Ottburg et al. 2018). The patches, dispersed throughout the country, were initially planted with trees close together to quickly regenerate forests, and could be thinned later as growing trees competed with each other for resources. Tiny forests aim to increase urban biodiversity, emphasising engagement with local urban communities whilst fostering education and awareness (Ottburg et al. 2018).

The planning and creation of parks, allotment gardens, urban forests, and other sustainable spatial planning tools, should not be dissociated from the imperative of securing enhanced equity and social sustainability within and between different living environments (Gibbs et al. 2013; Krueger and Gibbs 2007; Parra 2013). Anguelovski et al. (2019a, b), Cole et al. (2017), and other scholars have questioned the assumption that urban greening acts as a public good for everyone. By examining the relationship between greening and gentrification, it is argued that the expected benefits of urban greening are often only reaching the elite and leading to the creation of ghettos of environmental privilege and green gentrification (Anguelovski et al. 2019b). Under such a scenario, these green benefits are not only leaving behind vulnerable groups and lower-income places in the city, but also exacerbating actual burdens and vulnerabilities. By applying the concepts of environmental and climate justice, claims are made to pay attention to issues of green/climate-led displacement, dispossession, increased property prices, green accumulation, and inequalities between nature-rich versus nature-poor areas within and between cities (Anguelovski et al. 2019b).

2.3.3 Land Sparing, Land Sharing, and Rewilding

Land sparing and land sharing are two different land-use strategies to address the effects of human land uses on biodiversity (Lin and Fuller 2013). Land-sparing approaches look for spatial separation between conservation areas and intensive forms of human land use. Land sharing aims at integrating conservation targets into sustainable types of human land use (e.g. organic agriculture and close-to-nature forestry) (Phalan et al. 2011).

In an urban context, land sparing and densification of residential areas may enable the establishment and/or conservation of green spaces in an urban matrix. Strong environmental governance is needed, however, to avoid the expansion of residential areas into these green spaces under increasing urbanisation pressure (Ceddia et al. 2014). On the other hand, land sharing implies residential areas and rural spaces may form a fragmented landscape, including smaller settlements and less intensively developed urban and peri-urban areas. This means that green spaces may be closer to the residences of urban dwellers (e.g. private gardens), facilitating a flow of ecosystem services, but requiring a larger land area is needed (Soga et al. 2014).

The relative conservation benefits of land sharing and land sparing depend on the level of urbanisation and the quality of governance (Sushinsky et al. 2013). In intensively urbanised contexts, the adoption of land-sparing approaches that maintains a binary landscape (i.e. close-to-nature blocks outside urban boundaries segregated from developed regions) is often implemented. With this strategy, spatial optimisation is obtained at the expense of socio-ecological equilibrium. Also, such strong land use control often discards participatory approaches. The concept of urban green infrastructure, on the other hand, strives to integrate and promote ecosystem services provided by urban greenery in urban planning and policy, and is related to the land-sharing approach. BioCity development will need to recognise the trade-offs, but also the synergies between land sharing and land sparing, and will require an integrated approach that considers both rewilding processes (related to minimising human intervention) and socio-economic factors (related to societal development) in order to reconcile the embedded tensions (Dennis et al. 2019) (Fig. 3).

Rewilding is a land-use strategy focussing on restoring self-regulating ecosystems and phasing out human intervention, and thus ultimately aiming at a land-sparing strategy (Van Meerbeek et al. 2019). In this sense, rewilding links to the novel urban green space emerging from vacant lots, and the natural regeneration of abandoned urban-industrial sites and brown fields, the so-called 'nature of the fourth kind' proposed by Kowarik since the 1990s (Kowarik 2013). Rewilding in urban

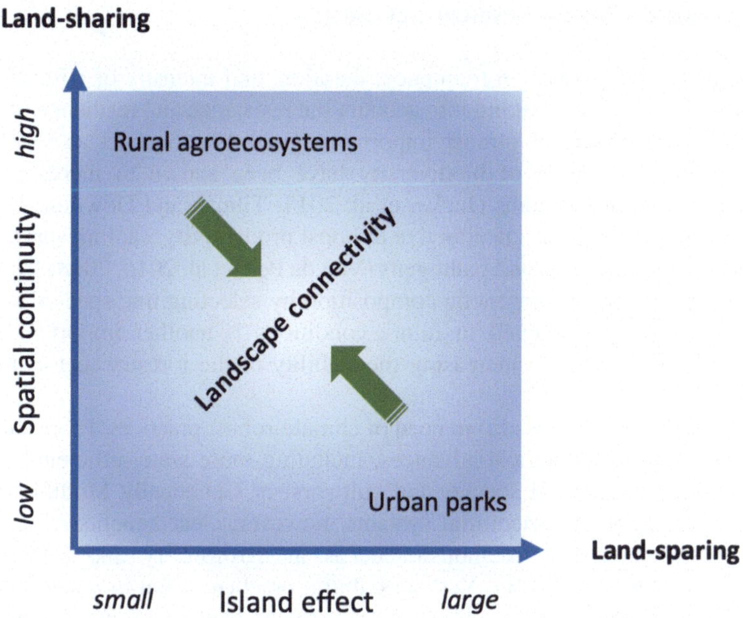

Fig. 3 Land-sharing and land-sparing measures cover multiple spatial scales and fall along a sharing–sparing continuum. Their combination in land-sharing/-sparing landscapes promotes connectivity for both the maintenance of ecological processes and the provisioning of ecosystem services. High connectivity across the urban–rural landscape matrix is needed for land-sharing and -sparing to be successful. The connectivity matrix ensures spill over from agroecosystems as well as from green spaces

contexts entails many challenges (Lehmann 2021). It is a matter of regaining ecological integrity of altered environmental conditions, meeting the needs of local communities and city dwellers, with the aim of gradual restoration of ecological processes and social benefits (Clancy and Ward 2020). Although re-establishing original ecosystem processes can be impractical or even undesirable, supporting the development and interconnection of green patches inside cities and the countryside with urban and farming systems has enormous potential for biodiversity conservation (Aronson et al. 2014). To create citizen ownership of such developments and ensure the equality of the many benefits people gain from green spaces, city dwellers should be granted certain levels of access to these green spaces. Many studies have pointed out that access to urban green spaces varies amongst social groups based on a wide variety of socio-economic factors, from demographical and ethnic factors, to income, education levels, migration status etc. (Wu and Kim 2021). To address this social justice issue, which has important implications for people's health, well-being, and social participation, efforts to tackle social-ecological resilience, health, spatial, and environmental inequalities should be fully addressed in the nature-based solutions proposed here.

2.3.4 Creating Climate-Resilient Landscapes

With the predicted increase in frequency, duration, and intensity of climate-related disturbances in the future, taking into account the resistance and resilience of natural and rural landscapes is of utmost importance (Van Meerbeek et al. 2021). The protection and restoration of biodiversity have been shown to increase overall stability in natural ecosystems (Jucker et al. 2014; Tilman and Downing 1994). In forestry, mixing tree species is effective to boost productivity, sapling survival, and the resistance against pests and pathogens (Van de Peer et al. 2016, 2018; Jactel et al. 2021). Adapting species or genetic composition by selecting tree species or provenances that are more adapted to future conditions is another important climate change adaptation measure increasing the stability of the forestry sector (Fremout et al. 2021).

The agricultural sector is also in need of climate-robust practices. Proposals so far have focused on technological advances, including more water-efficient irrigation practices, drought-resistant species and cultivars, or Genetically Modified Organisms (GMOs). In rural agricultural systems, however, clear benefits of increasing crop diversity have been demonstrated, as in agroforestry and intercropping (Tornaghi and Dehaene 2021). Yet, agricultural practices seem to reside in a lock-in of business-as-usual reinforced by over investment, bank loans, insurances, public subsidies, and disaster funds, which impede necessary transformations in the sector. In peri-urban agriculture, emerging trends like food forests and agro-ecological urbanism, aim to improve climate robustness by betting on biodiversity and sustainable agricultural practices.

3 Restoring Urban–Rural Relationships: From Concepts to Good Practices

BioCities require a paradigm shift towards a more holistic understanding of society–nature relationships. Contemporary urban–rural relationships are unsustainable and based on the dualistic opposition between nature and culture that has dominated the Western view of humanity's place in nature (Haila 2000). Systemic changes can ultimately eliminate the nature–culture dualism and pave the way for a novel socio-ecological contract of BioCities in a sustainable relationship with their surrounding BioRegions.

3.1 Reconfiguring the Nature–Culture Nexus

Reconnecting urban environments to natural landscapes is more than creating open areas or taking care of green spaces. Collectively, creating the BioCity and the

BioRegion involves much more than planting trees, ruralising, implementing green technology, and/or building infrastructure for increasing cities' capacity to adapt to climate change. There is a need for a new nature–culture nexus that addresses and renegotiates entangled ecological and socio-economic realms and processes. At the core of this is a new social-ecological contract in urban–rural relationships.

3.1.1 Towards a New Social-Ecological Perspective for Urban–Rural Relationships

Starting from a social-ecological understanding of urban–rural relationships, we see cities and villages as habitats, humans as nature, and nature as embedded in complex social-ecological processes (Huntjens 2021). Urban and rural needs should be considered as connected through both ecological and social metabolic processes (Swyngedouw 2006). Both the urban and the rural are ecological expressions, and both are part of economic, political, cultural, and governance manifestations. Changing cities, in such a way that they become less extractive, more ecologically sound, more liveable, and more just, requires a reinvention of the urban, but also the rural and especially the relations between the two (Gebre and Gebremedhin 2019). More than the spatial segregation between cities and their surroundings, their socio-spatial relations should be reconstructed and their ecosystem services need to be linked (Gebre and Gebremedhin 2019).

A social-ecological perspective can build on the fundamental rights of humans and non-humans in liveable cities and rural areas such as (Moulaert et al. 2012):

- The right to clean air, which requires the drastic reduction of sources of pollution, expanding natural networks, changing mobility systems, and changing consumption practices and norms.
- The right to living water and soil requires programmes of unsealing, soil protection and improvement, stimulating agro-ecological food production, reduction of water and soil pollution, and restoration of soils.
- The right to ecologically and socially sustainable human settlements requires circular economies that are also just. This may include: reconceptualisation of infrastructures; social housing programmes; rethinking of mobility systems; stimulating locally embedded economies; green and edible cities; collectivisation of services, housing systems, and spaces; and harnessing cultural and natural legacies to support a sense of place.

The social-ecological perspective and the fundamental rights are the non-negotiable foundations of an otherwise contested and always evolving social-ecological contract of BioCities and BioRegions.

3.1.2 Co-production and Governance of Transformed Urban–Rural Relationships

To implement the right to clean air, to living water and soil, and to ecologically and socially sustainable settlements, current extractive urban–rural relationships need to be transformed into considerate and restoring endeavours. To achieve this, it is necessary to tackle the drivers of urbanisation through diverse ecological restoration policies such as implementing agro-ecological urbanism (Tornaghi and Dehaene 2021), promoting rewilding and restoration of biodiversity and ecosystem functioning (Navarro and Pereira 2012), and creating edible cities (Sartison and Artmann 2020). Re-introducing social redistributive mechanisms (fair fiscal policies, protection of labour and wages, accessible public services, and egalitarian consumption norms), limiting the production of social inequalities (softening the unequal accumulation of wealth), de-commodifying the foundational economy (water, energy, education, transport, health care, elderly care), and stimulating circular and regenerative cities (control of extraction, material and waste policies) is also necessary (Huntjens 2021).

This social-ecological transition also needs social value systems, institutions, and governance systems to be able to formulate novel modes to organise societal responses for the way ecosystems are managed, stewarded, and transformed (Frantzeskaki et al. 2021). In this respect, the BioCity will depend on the inclusive engagement of citizens and the day-to-day practices that keep the BioCity alive and flourishing. Knowledge and learning to accommodate nature in daily life become very relevant. Thus, co-imagining and co-producing novel urban–rural relationships will demand not only paid labour, but also the voluntary engagement of people (i.e. active citizenship) rooted in environmental stewardship that goes beyond immediate personal benefit and reflects wider social-ecological values (Buijs et al. 2016). It will also demand an enabling and stimulating governance approach that harnesses the transformative potential of active citizenship and adopts, scales out, and multiplies alternative practices, including bottom-up and local innovation initiatives (Buijs et al. 2016).

A main challenge is how to jointly consider social and ecological matters to prevent the BioCity from becoming a fancy but ineffective eco-effort. There is already evidence that the benefits of urban greening are only reaching an elite, leading to green gentrification (Anguelovski et al. 2019a). Hence, there has to be careful consideration of a BioCity model that accounts for environmental and climate justice, putting in place mechanisms that promote economic development in harmony with the biosphere whilst supporting social sustainability.

3.1.3 The Crucial Role of Co-learning

Collective learning (or co-learning) is a crucial component in the journey towards a sustainable reconnection between the city and the rural world. Learning is place-based and time-sensitive (Bakema et al. 2017), and it occurs through storing knowledge of past events in the memories of people and communities. Individual and collective learning is both a process and an outcome, which is mediated and reshaped by pre-existing individual and social values (Tidball et al. 2010). From a social-ecological systems perspective, learning and the combination of different types of knowledge are crucial components for sustainability and resilience.

In the case of BioCities operating under the ambition to stimulate a more harmonious coexistence with rural areas, connecting learning to experimentation and social innovation can respond to uncertain and unstable futures (Bakema et al. 2017; Pahl-Wostl 2009). Learning, being it at the individual, community, institutional, or systemic level, requires self-reflexivity and exchanges between different types of knowledge, including different bodies of relevant legitimated facts, beliefs, and perceptions, which are usually informed by multiple values and ethical choices (Skrimizea et al. 2019). In this context, the EU Covenant of Mayors for Climate and Energy can be considered a successful example of a socially innovative bottom-up co-learning platform It gathers thousands of local governments in Europe, and is led by public authorities acting, since 2008, within a voluntary, multi-actor, self-reflective, and context-sensitive approach to co-learning in the face of climate change.

The involvement of educational institutions of different levels (from primary to higher education) is fundamental for the promotion of collective learning. Focusing on universities, the provision of technical skills in ecology, biodiversity, urban planning etc., and the promotion of critical and systemic thinking on sustainability should be accompanied by the inclusion of curricula such as ethics and culture. Combined with instilling values such as mutual help, solidarity, and compassion, this will enable learning and education for sustainability through emerging issues (Leal Filho et al. 2018). Moreover, overcoming disciplinary boundaries and adopting transdisciplinary approaches is encouraged to address societally relevant challenges, such as creating BioCities and BioRegions (Colucci-Gray et al. 2013; Leal Filho et al. 2018; Lehtonen et al. 2018). In this sense, universities are more than teaching institutions that transfer knowledge, they foster mutual learning. Non-academic knowledge and expertise are treated as equal and complementary to scientific knowledge, with citizens and practitioners being knowledgeable, respected, and active agents that cooperate with scientists in a co-learning process. Developing diverse partnerships and strategies aimed at communities and stakeholders' engagement and participation is then essential for embedding sustainability in higher education institutions (Leal Filho et al. 2018), bringing together different 'agents of change' who co-construct the BioCity and BioRegion.

4 Case Studies: Successful Transformational Processes in Urban–Rural Relationships

There are already examples of transformational processes in urban–rural relationships that contain aspects of the BioCities and BioRegions vision. These examples come from Valencia (Spain), Rome (Italy), and Flanders (Belgium).

1. *Valencia, Spain—Urban–Rural Interfaces Building Forest Fire Resilience*

   For several decades, the rural areas of Valencia, as with many other regions in the Mediterranean, have been suffering from depopulation, economic decline, and large wildfires (Fig. 4). How to build social-ecological resilience given these conditions, and having as a backdrop the big wildfires in Valencia at the beginning of the 1990s, has become a major challenge for practitioners, policy makers, rural inhabitants, and the academic world (Rodriguez Fernandez-Blanco et al. 2022). The Valencian *'A paso lento'* initiative is a source of ideas for BioCities, BioRegions, and their multiple sustainability ambitions. *A paso lento* is a cooperative project gathering shepherds from Alcublas (La Serranía, Valencia), the local municipality, and Som Alimentació, the first cooperative supermarket in the city of Valencia. The extensive livestock production system combines improved grazing infrastructure in Alcublas and the setting up of a new channel for the commercialisation of goat and lamb. In addition, it provides wildfire prevention for a municipality that was devastated in 2012 by a very large wildfire. Through a mix of social economy and agroforestry, pastoralism becomes a venue to keep the rural world alive and dynamic from a socio-cultural angle, strengthening the rural–urban ties, whilst generating economic revenues. Food is produced according to the local resources and conditions, food miles are reduced, and food production and commercialisation are guided by sustainability principles. From an ecological perspective, grazing systems based on grasslands, shrubs, and forest vegetation reduce the risk of wildfire spread.
2. *Rome, Italy—One of the Largest Rural Communities*

   Rome has the largest municipal area in Europe (about 1500 km^2 including the municipality of Fiumicino), with an inhabited surface area only slightly smaller

Fig. 4 Fire-prone Mediterranean landscape at Alcublas, Valencia. Photo Enric Díaz, permission granted

than that of Greater London, and almost double that of the inner Paris suburbs (the Petite Couronne). Rome is also the largest rural municipality in Italy including areas with high environmental, historic, or cultural value. As such, the rural area of Rome is engaged in a New Master Plan that promotes multifunctional and high-quality agriculture within Natura 2000 sites, and other areas of high natural value, incorporating green connectivity and ecological corridors to re-connect natural and cultural capital (Marcelloni 2003). The Master Plan aims at strengthening urban–rural linkages that allow a more enabling market environment for smallholders, whilst preserving cultural traditions and natural resources (Perrin et al. 2018). The urban regeneration axes outline several rural-based initiatives that have the potential to boost rural business, support local services, and build upon good practices whilst exploring new opportunities for sustainable land management. The sustainable rural regeneration of Rome's large peri-urban area will depend on its reintegration into the continuum of the municipality in social, cultural, and environmental terms (di Zio et al. 2018). In this context, the conservation and restoration of natural and forested areas play a central role.

3. *Flanders, Belgium—Treescape Design Remediating Diffuse Urbanisation*

   Flanders, the northern part of Belgium including the historical cities of Bruges, Ghent, Antwerp, and Leuven, is one of the least forested areas in Europe and has not succeeded in increasing its forest cover. This urbanised region is characterised by extensive urban sprawl, intensive industrial agriculture, and a heterogeneous landscape, where nearly every square metre of land is intensively occupied. Though there is available space that trees and forests can grow, the space often remains 'invisible' because it is related to other types of land use or is 'untouchable' because of sectoral claims. There is a need to find a new spatial paradigm to introduce more trees and forests within the fabric of an urbanised territory. The Treescape Research Project (Carron et al. 2021) aims to explore new strategies and test concepts to intertwine trees and forests in unique configurations with other types of urban and peri-urban land uses. Examples of such configurations are woodland gardens, residential forest allotments, agroforestry and food forests, forest business sites, and road infrastructure. By means of a design-driven research approach, possible tree and forest configurations are explored in a Treescape Catalogue and Treescape Atlas within the central Flanders study area (Fig. 5).

5 Outcomes and Concluding Remarks

The BioCity concept and vision can be intertwined with the larger landscape focus of the BioRegion to consider urban and rural communities together in pursuit of establishing more sustainable linkages. Starting from the premise that cities have generally developed an extractive relationship with the countryside, we have reconceptualised the BioCity and the BioRegion as a complex social-ecological,

Fig. 5 Example of heavily sealed peri-urban area with large Treescape potential in Flanders. From the Treescape Atlas by Bjoke Carron (permission granted)

low-entropy release system. Restored and new urban–rural relationships enable the restructuring and embedding of ecological and social metabolic processes, from 'accumulative and extractive' to 'circular and regenerative'. Urban sprawl, agricultural and forestry intensification, and land abandonment, are major examples of extractive urbanisation, in contrast to processes that ruralise the city and boost the transition to the BioCity (i.e. community agriculture, alternative city greening, and joint land-sharing/-sparing approaches). To increase the effectiveness and longevity of the latter (and similar efforts), an in-depth reconfiguration of our understanding and governance of the nature–culture nexus is fundamental. What is needed is a new social-ecological contract for BioCities, planned and implemented through the inclusive engagement of citizens in governance processes.

A major question at this stage is 'what policies and programmes could constitute part of a social-ecological planning and governance toolkit for BioCities and BioRegions to support the sustainable urban–rural relationships and the new nature-culture nexus?' At the EU level, the European Green Deal (European Commission 2019) seems a promising policy framework, enabling a socially and territorially fair transition to a resource-efficient circular economy through the EU Forest Strategy, the EU Biodiversity Strategy, the Farm to Fork Strategy (European Commission 2019), and the new EU Strategy on Adaptation to Climate Change. All of these programmes have the ambition of creating a more sustainable reconnection between society and nature.

Beneath the state level, BioCities and BioRegions need to be intentionally created through enabling policies and governance frameworks at the city-region level. This is needed because local and regional authorities are ideal partners to stimulate and foster citizen and civil society initiatives. In this respect, we support municipal or regional-level policies that enable a broad range of synergies and partnerships, where active citizens and civil society organisations (i.e. grassroots and community-based organisations, NGOs, coalitions, and other organisational forms) complement policy makers, scientists, and other actors, by providing local knowledge and innovative ideas, citizen science, and engagement programmes.

In view of the 15th Conference of the Parties to the Convention for Biological Diversity in October 2021, cities are expected to play an increasingly significant role in global biodiversity enhancement and conservation. To strengthen this role of cities, policy frameworks that incorporate biodiversity objectives in urban planning using nature-based solutions are of crucial importance. This needs to be connected to climate action through approaches that adapt urban nature to the changing climate.

City-level responsibilities around biodiversity and nature-centred policies should go hand-in-hand with policy mechanisms and urban planning considerations that prevent different types of risks, such as green gentrification whilst stimulating a circular economy that interacts with other progressive economic forms such as social, sharing, and collaborating economies. In this sense, policies on biodiversity restoration and green infrastructure support the transition to BioCities/BioRegions, especially when they intersect with regulatory strategies (e.g. affordable housing), public investments (e.g. land banks), and collective ownership models (e.g. community land trusts and cooperatives).

Finally, research needs to foster the connection between BioCities and BioRegions, including: (1) better understanding of (the drivers of) extractive mechanisms; (2) the role of humans in both stressing and producing nature; (3) the interactive processes of humans and non-humans in urbanisation dynamics; and (4) the creation of community-developed alternatives and their interaction with local, regional, and state governments.

References

Anguelovski I, Connolly JJ, Pearsall H, Shokry G, Checker M, Maantay J, Gould K, Lewis T, Maroko A, Roberts JT (2019a) Why green "climate gentrification" threatens poor and vulnerable populations. Proc Natl Acad Sci 116(52):26139–26143

Anguelovski I, Connolly JJ, Garcia-Lamarca M, Cole H, Pearsall H (2019b) New scholarly pathways on green gentrification: what does the urban 'green turn' mean and where is it going? Prog Hum Geogr 43(6):1064–1086

Antrop M (2004) Landscape change and the urbanisation process in Europe. Landsc Urban Plan 67: 9–26

Aronson MF, La Sorte FA, Nilon CH, Katti M, Goddard MA, Lepczyk CA, Warren PS, Williams NSG, Cilliers S, Clarkson B, Dobbs C, Dolan R, Hedblom M, Klotz S, Kooijmans JL, Kühn I, MacGregor-Fors I, McDonnell M, Mörtberg U, Pyšek P, Siebert S, Sushinsky J, Werner P, Winter M (2014) A global analysis of the impacts of urbanisation on bird and plant diversity reveals key anthropogenic drivers. Proc R Soc B Biol Sci 281(1780):20133330

Bakema MM, Parra C, McCann P, Dalziel P, Saunders C (2017) Governance in shaky societies: experiences and lessons from Christchurch after the earthquakes. Environ Policy Gov 27(4): 365–377

Bleichrodt D, Bruns M, Teunissen W, Laine E (2018) Tiny forest handbook. IVN Natuur Educatie, Utrecht

Brenner N (2019) New urban spaces: urban theory and the scale question. Oxford University Press, London

Brenner N, Schmid C (2014) The urban age in question. In: Brenner N (ed) Implosions/explosions: towards a study of planetary urbanisation. Jovis, pp 310–337

Brody S (2013) The characteristics, causes, and consequences of sprawling development patterns in the United States. Nat Educ Knowl 4(5):2

Buijs AE, Mattijssen TJ, Van der Jagt AP, Ambrose-Oji B, Andersson E, Elands BH, Møller MS (2016) Active citizenship for urban green infrastructure: fostering the diversity and dynamics of citizen contributions through mosaic governance. Curr Opin Environ Sustain 22:1–6

Bukowski C, Munsell J (2018) The community food forest handbook: how to plan, organize, and nurture edible gathering places. Chelsea Green Publishing, Vermont

Carron B, Muys B, Van Orshoven J, Leinfelder H (2021) Landscape design to meet the societal demand for ecosystem services: a perspective. Chall Sustain 9:28–44

Carruthers JI, Ulfarsson GF (2003) Urban sprawl and the cost of public services. Environ Plann B: Plann Des 30(4):503–522

Ceddia MG, Bardsley NO, Gomez-y-Paloma S, Sedlacek S (2014) Governance, agricultural intensification, and land sparing in tropical South America. Proc Natl Acad Sci 111:7242–7247

Čepić S, Dubljević-Tomićević J, Živojinović I (2020) Is there a demand for collective urban gardens? Needs and motivations of potential gardeners in Belgrade. Urban Forestry Urban Green 26716

Chaudhary S, Wang Y, Khanal NR, Xu P, Fu B, Dixit AM, Yan K, Liu Q, Lu Y (2018) Social impact of farmland abandonment and its eco-environmental vulnerability in the high mountain region of Nepal: a case study of Dordi River Basin. Sustainability 10(7):2331

Church SP (2013) Urban dwellers and neighbourhood nature: exploring urban residents' connection to place, community, and environment. PhD Dissertation, University of British Columbia

Clancy C, Ward K (2020) Auto-rewilding in post-industrial cities: the case of inland cormorants in urban Britain. Conserv Soc 18(2):126–136

Cole HV, Lamarca MG, Connolly JJ, Anguelovski I (2017) Are green cities healthy and equitable? Unpacking the relationship between health, green space and gentrification. J Epidemiol Commun Health 71(11):1118–1121

Colucci-Gray L, Perazzone A, Dodman M, Camino E (2013) Science education for sustainability, epistemological reflections and educational practices: from natural sciences to trans-disciplinarity. Cult Stud Sci Educ 8:127–183

Davoudi S, Stead D (2002) Urban-rural relationships: an introduction and a brief history. Built Environ 28(4):269–277

Dennis M, Scaletta KL, James P (2019) Evaluating urban environmental and ecological landscape characteristics as a function of land-sharing-sparing, urbanity and scale. PLoS One 14(7): e0215796

Di Zio S, Montanari A, Staniscia B (2018) Simulation of urban development in the City of Rome. J Transp Land Use 3:85–105

Donald PF, Sanderson FJ, Burfield IJ, Van Bommel FP (2006) Further evidence of continent-wide impacts of agricultural intensification on European farmland birds, 1990–2000. Agric Ecosyst Environ 116(3–4):189–196

Egartner S, Ayrault J, Niedermayr J (2020) Development of community supported agriculture in Austria and France: a comparative analysis in the context of social innovation. Austr J Agric Econ Rural Stud 29:28

Estel S, Kuemmerle T, Alcántara C, Levers C, Prishchepov A, Hostert P (2015) Mapping farmland abandonment and recultivation across Europe using MODIS NDVI time series. Remote Sens Environ 163:312–325

European Commission (2019) The European Green Deal. Communication from The Commission to The European Parliament, The European Council, The European Economic and Social Committee, and The Committee Of The Regions

Eyzaguirre PB, Linares OF (eds) (2010) Home gardens and agrobiodiversity. Smithsonian Books, Washington, DC

Forest Europe (2020). State of Europe's forests 2020. Ministerial Conference on the Protection of Forests in Europe

Frantzeskaki N, McPhearson T, Kabisch N (2021) Urban sustainability science: prospects for innovations through a system's perspective, relational and transformations' approaches. Ambio 50:1650–1658

Fremout T, Thomas E, Bocanegra-González KT, Aguirre-Morales CA, Morillo-Paz AT, Atkinson R, Muys B (2021) Dynamic seed zones to guide climate-smart seed sourcing for tropical dry forest restoration in Colombia. For Ecol Manag 490:119127

Fuchs R, Herold M, Verburg PH, Clevers JGPW (2013) A high-resolution and harmonised model approach for reconstructing and analysing historic land changes in Europe. Biogeosciences 10: 1543–1559

Gebre T, Gebremedhin B (2019) The mutual benefits of promoting rural-urban interdependence through linked ecosystem services. Glob Ecol Conserv 20:e00707

Giampietro M (2019) On the circular bioeconomy and decoupling: implications for sustainable growth. Ecol Econ 162:143–156

Gibbs D, Krueger R, MacLeod G (2013) Grappling with smart city politics in an era of market triumphalism. Urban Stud 50:2151–2157

Goh K (2020) Planning the green new deal: climate justice and the politics of sites and scales. J Am Plan Assoc 86:188–195

Gossner MM, Lewinsohn TM, Kahl T, Grassein F, Boch S, Prati D, Allan E et al (2016) Land-use intensification causes multitrophic homogenisation of grassland communities. Nature 540:266–269

Haila Y (2000) Beyond the nature-culture dualism. Biol Philos 15:155–175

Hallmann CA, Sorg M, Jongejans E, Siepel H, Hofland N, Schwan H, Stenmans W, Müller A, Sumser H, Hörren T, Goolson D, de Kroon H (2017) More than 75 percent decline over 27 years in total flying insect biomass in protected areas. PLoS One 12(10):e0185809

Hart R (1996) Forest gardening: cultivating an edible landscape. Chelsea Green Publishing Company, White River Junction, Vermont

Hasse JE, Lathrop RG (2003) Land resource impact indicators of urban sprawl. Appl Geogr 23: 159–175

Hennig EI, Schwick C, Soukup T, Orlitová E, Kienast F, Jaeger JA (2015) Multi-scale analysis of urban sprawl in Europe: towards a European de-sprawling strategy. Land Use Policy 49:483–498

Hinrichs CC (2000) Embeddedness and local food systems: note on two types of direct agricultural market. J Rural Stud 16:295–303

Huntjens P (2021) Towards a natural social contract: transformative social-ecological innovation for a sustainable, healthy and just society. Springer Nature

Jactel H, Moreira X, Castagneyrol B (2021) Tree diversity and forest resistance to insect pests: patterns, mechanisms, and prospects. Annu Rev Entomol 66:277–296

Jiménez V, Sánchez JM, Rengifo JI (2019) A new residential role for the rural environment in Extremadura, Spain. Sustainability 11(2):435

Johnson MP (2001) Environmental impacts of urban sprawl: a survey of the literature and proposed research agenda. Environ Plann A: Econ Space 33(4):717–735

Jucker T, Bouriaud O, Avacaritei D, Coomes DA (2014) Stabilising effects of diversity on aboveground wood production in forest ecosystems: linking patterns and processes. Ecol Lett 17(12):1560–1569

Keshavarz N, Bell S (2016) A history of urban gardens in Europe. In: Bell S, Fox-Kaemper R, Keshavarz N, Benson M, Caputo S, Noori S, Voigt A (eds) Urban allotment gardens Europe. Routledge, New York, NY, pp 9–32

Kowarik I (2013) Cities and wilderness. Int J Wilderness 19(3)

Krueger R, Gibbs D (eds) (2007) The sustainable development paradox: urban political economy in the United States and Europe. Guilford Press

Kuemmerle T, Levers C, Erb K, Estel S, Jepsen MR, Müller D, Plutzar C, Stürck J, Verkerk PJ, Verburg PH, Reenberg A (2016) Hotspots of land use change in Europe. Environ Res Lett 11(6):064020

Lasanta T, Arnáez J, Pascual N, Ruiz-Flaño P, Errea MP, Lana-Renault N (2017) Space–time process and drivers of land abandonment in Europe. Catena 149:810–823

Leal Filho W, Mandel M, Al-Amin AQ, Feher A, Chiappetta Jabbour CJ (2016) An assessment of the causes and consequences of agricultural land abandonment in Europe. Int J Sust Dev World 24(6):554–560

Leal Filho W, Raath S, Lazzarini B, Vargas VR, de Souza L, Anholon R, Quelhas OLG, Haddad R, Klavins M, Orlovic VL (2018) The role of transformation in learning and education for sustainability. J Clean Prod 199:286–295

Lehmann S (2021) Growing biodiverse urban futures: renaturalisation and rewilding as strategies to strengthen urban resilience. Sustainability 13:2932

Lehtonen A, Salonen A, Cantell H, Riuttanen L (2018) A pedagogy of interconnectedness for encountering climate change as a wicked sustainability problem. J Clean Prod 199:860–867

Lichtenberg E (2011) Open space and urban sprawl: the effects of zoning and forest conservation regulations in Maryland. Agric Resour Econ Rev Northeast Agric Resour Econ Assoc 40(3): 1–12

Lin BB, Fuller RA (2013) Sharing or sparing? How should we grow the world's cities? J Appl Ecol 50:1161–1168

MacDonald D, Crabtree JR, Wiesinger G, Dax T, Stamou N, Fleury P, Gutierrez Lazpita J, Gibon A (2000) Agricultural abandonment in mountain areas of Europe: environmental consequences and policy response. J Environ Manag 59(1):47–69
Marcelloni M (2003) Pensare la Città Contemporanea: Il Nuovo Piano Regolatore di Roma. Laterza, Rome
McConnell DJ (2017) The forest farms of Kandy and other gardens of complete design. Routledge
McHale MR, Pickett ST, Barbosa O, Bunn DN, Cadenasso ML, Childers DL, Gartin M, Hess GR, Iwaniec DM, McPhearson T, Peterson N, Poole AK, Rivers L, Shutters ST, Zhou W (2015) The new global urban realm: complex, connected, diffuse, and diverse social-ecological systems. Sustainability 7(5):5211–5240
Merrifield A (2011) The right to the city and beyond. City 15(3/4):468–476
Moulaert F, van den Broeck P, van Dyck B (2012) 'Een groen ruimtelijk beleid' reflectienota voor de Vlaamse groenen. KU Leuven
Moulaert F, Jessop B, Mehmood A (2016) Agency, structure, institutions, discourse (ASID) in urban and regional development. Int J Urban Sci 20(2):167–187
Muys B (2013) Sustainable development within planetary boundaries: a functional revision of the definition based on the thermodynamics of complex social-ecological systems. Chall Sustain 1(1):41–52
Naudts K, Chen Y, McGrath MJ, Ryder J, Valade A, Otto J, Luyssaert S (2016) Europe's forest management did not mitigate climate warming. Science 351(6273):597–600
Navarro L, Pereira H (2012) Rewilding abandoned landscapes in Europe. Ecosystems 15:900–912
OECD (2018) Rethinking urban sprawl: moving towards sustainable cities. OECD Publishing, Paris
Ottburg FGWA, Lammertsma DR, Bloem J, Dimmers WJ, Jansman HAH, Wegman RMA (2018) Tiny Forest Zaanstad: citizen science and determining biodiversity in tiny Forest Zaanstad. Wageningen Environmental Research Report 2882
Pahl-Wostl C (2009) A conceptual framework for analysing adaptive capacity and multi-level learning processes in resource governance regimes. Glob Environ Chang 19(3):354–365
Parra C (2013) Social sustainability, a competitive concept for social innovation? In: Moulaert F, McCallum D, Mehmood A, Hamdouch A (eds) International handbook on social innovation. Edward Elgar Publishers, pp 142–154
Pelorosso R, Gobattoni F, Leone A (2017) The low-entropy city: a thermodynamic approach to reconnect urban systems with nature. Landsc Urban Plan 168:22–30
Pereira HM, Navarro LM (2015) Rewilding European landscapes. Springer Nature
Perino A, Pereira HM, Navarro LM, Fernández N, Bullock JM, Ceaușu S, Wheeler HC et al (2019) Rewilding complex ecosystems. Science 364(6438)
Perrin C, Nougarèdes B, Sini L, Branduini P, Salvati L (2018) Governance changes in peri-urban farmland protection following decentralisation: a comparison between Montpellier (France) and Rome (Italy). Land Use Policy 70:535–546
Phalan B, Onial M, Balmford A, Green RE (2011) Reconciling food production and biodiversity conservation: land sharing and land sparing compared. Science 333(6047):1289–1291
Power A (2010) Social exclusion and urban sprawl: is the rescue of cities possible? Reg Stud 35: 731–742
Reidsma P, Tekelenburg T, Van den Berg M, Alkemade R (2006) Impacts of land-use change on biodiversity: an assessment of agricultural biodiversity in the European Union. Agric Ecosyst Environ 114(1):86–102
Roberts L, Hall D (2001) Rural tourism and recreation: principles to practice. Cabi Publishing
Robinson RA, Sutherland WJ (2002) Post-war changes in arable farming and biodiversity in Great Britain. J Appl Ecol 39:157–176
Rodriguez Fernandez-Blanco C, Gorriz Mifsud E, Prokofieva I, Muys B, Parra C (2022) Blazing the trail: social innovation supporting wildfire-resilient territories in Catalonia (Spain). Forest Policy Econ 138(6):102719

Sartison K, Artmann M (2020) Edible cities: an innovative nature-based solution for urban sustainability transformation? An explorative study of urban food production in German cities. Urban Forestry Urban Green 49:126604

Scheidel A, Del Bene D, Liu J, Navas G, Mingorria S, Demaria F, Avila S, Roy B, Ertor I, Temper L, Martinez-Alier J (2020) Environmental conflicts and defenders: a global overview. Glob Environ Chang 63:102104

Schrank D, Eisele B, Lomax T (2012) Urban mobility report. Texas A&M Transportation Institute, The Texas A&M University System, College Station, TX

Schunko C, Wild AS, Brandner A (2021) Exploring and limiting the ecological impacts of urban wild food foraging in Vienna, Austria. Urban Forestry Urban Green 127164:127164

Seifollahi-Aghmiuni S, Kalantari Z, Egidi G, Gaburova L, Salvati L (2022) Urbanisation-driven land degradation and socioeconomic challenges in peri-urban areas: insights from southern Europe. Ambio 51:1446–1458

Servillo LA, Van Den Broeck P (2012) The social construction of planning systems: a strategic-relational institutionalist approach. Plan Pract Res 27(1):41–61

Skrimizea E, Haniotou H, Parra C (2019) On the 'complexity turn' in planning: an adaptive rationale to navigate spaces and times of uncertainty. Plan Theory 18(1):122–142

Soga YY, Yamaura Y, Koike S, Gaston KG (2014) Land sharing vs. land sparing: does the compact city reconcile urban development and biodiversity conservation? J Appl Ecol 51:1378–1386

Steel C (2013) Hungry City: how food shapes our lives. Random house

Storkey J, Meyer S, Still KS, Leuschner C (2012) The impact of agricultural intensification and land-use change on the European arable flora. Proc R Soc B Biol Sci 279(1732):1421–1429

Sushinsky JR, Rhodes JR, Possingham HP, Gill TK, Fuller RA (2013) How should we grow cities to minimise their biodiversity impacts? Glob Chang Biol 19:401–410

Swyngedouw E (2006) Circulations and metabolisms: (hybrid) natures and (cyborg) cities. Sci Cult 15(2):105–121

Thackara J (2019) Bioregioning: pathways to urban-rural reconnection. She Ji: J Des Econ Innov 5: 15–28

Tidball KG, Krasny E, Svendsen E, Campbell L, Helphand K (2010) Stewardship, learning, and memory in disaster resilience. Environ Educ Res 16(5–6):591–609

Tilman D, Downing JA (1994) Biodiversity and stability in grasslands. Nature 367(6461):363–365

Tornaghi C, Dehaene M (eds) (2021) Resourcing an agroecological urbanism: political, transformational and territorial dimensions. Routledge

Tsiafouli MA, Thébault E, Sgardelis SP, De Ruiter PC, Van Der Putten WH, Birkhofer K, Hedlund K et al (2015) Intensive agriculture reduces soil biodiversity across Europe. Glob Chang Biol 21(2):973–985

Urso G (2020) Metropolisation and the challenge of rural-urban dichotomies. Urban Geography, pp 1–21

Van de Peer T, Verheyen K, Baeten L, Ponette Q, Muys B (2016) Biodiversity as insurance for sapling survival in experimental tree plantations. J Appl Ecol 53(6):1777–1786

Van de Peer T, Verheyen K, Ponette Q, Setiawan NN, Muys B (2018) Overyielding in young tree plantations is driven by local complementarity and selection effects related to shade tolerance. J Ecol 106:1096–1105

Van der Ploeg JD, Franco JC, Borras SM (2015) Land concentration and land grabbing in Europe: a preliminary analysis. Can J Dev Stud 36:147–162

Van Meerbeek K, Muys B, Schowanek SD, Svenning JC (2019) Reconciling conflicting paradigms of biodiversity conservation: human intervention and rewilding. Bioscience 69(12):997–1007

Van Meerbeek K, Jucker T, Svenning JC (2021) Unifying the concepts of stability and resilience in ecology. J Ecol

Vanbergen AJ, Aizen MA, Cordeau S, Garibaldi LA, Garratt MP, Kovács-Hostyánszki A, Lecuyer L, Ngo HT, Potts SG, Settele J, Skrimizea E, Young JC (2020) Transformation of agricultural landscapes in the anthropocene: nature's contributions to people, agriculture and food security. Adv Ecol Res 63:193–253

Wellner M, Theuvsen L (2016) Community Supported Agriculture (CSA): eine vergleichende analyse für Deutschland und Österreich. Yearb Austrian Soc Agric Econ 25:65–74
Wu J (2014) Urban ecology and sustainability: the state-of-the-science and future directions. Landsc Urban Plan 125:209–221
Wu L, Kim SK (2021) Exploring the equality of accessing urban green spaces: a comparative study of 341 Chinese cities. Ecol Indic 121:107080
Zambon I, Cerdà A, Gambella F, Egidi G, Salvati L (2019) Industrial sprawl and residential housing: exploring the interplay between local development and land-use change in the Valencian community, Spain. Land 8(10):143
Zhou T, Koomen E, Ke X (2020) Determinants of farmland abandonment on the urban–rural fringe. Environ Manag 65:369–384

The Enabling Environment for BioCities

Michael Salka, Vicente Guallart, Daniel Ibañez, Divina Garcia P. Rodriguez, Nicolas Picard, Jerylee Wilkes-Allemann, Evelyn Coleman Brantschen, Stefano Boeri, Livia Shamir, Lucrezia De Marco, Sofia Paoli, Maria Chiara Pastore, and Ivana Živojinović

1 Introduction

The *Governing Missions* and *Mission-Oriented Research and Innovation in the European Union* guidelines promoted by the European Commission (EC) are helpful as a starting place for creating the enabling environment for BioCities which follow the principles of natural ecosystems to promote life (Mazzucato 2018, 2019). The strength of mission-oriented policies, defined as systemic public policies that draw on frontier knowledge to attain specific goals, is the empowerment of emergent solutions achieved by: (1) being bold and inspirational with wide social relevance; (2) having a clear direction with targeted, measurable, and time-bound metrics; (3) being ambitious but realistic; (4) being cross-disciplinary and cross-sectoral; and (5) driving multiple bottom-up solutions (Ergas 1987).

M. Salka (✉) · V. Guallart · D. Ibañez
Institute for Advanced Architecture of Catalonia (IAAC), Barcelona, Spain
e-mail: michael.salka@iaac.net

D. G. P. Rodriguez
Norwegian Institute of Bioeconomy Research (NIBIO), Ås, Norway

N. Picard
GIP Ecofor – Ecosystems Forestiers (ECOFOR), Paris, France

J. Wilkes-Allemann · E. C. Brantschen
Bern University of Applied Sciences (BFH), Bern, Switzerland

S. Boeri · L. Shamir · L. De Marco · S. Paoli
Stefano Boeri Architetti (SBA), Milan, Italy

M. C. Pastore
Politecnico of Milan (PoliMi), Milan, Italy

I. Živojinović
University of Natural Resources and Life Sciences, Vienna (BOKU), Vienna, Austria

G. E. Scarascia-Mugnozza et al. (eds.), *Transforming Biocities*, Future City 20,
https://doi.org/10.1007/978-3-031-29466-2_11

Such missions operate at a level of resolution between encompassing challenges like the United Nations' (UN) *Sustainable Development Goals* (SDGs) or *Horizon 2020* (H2020) *Societal Challenges* and *Focus Areas* (UN DESA 2015; Mazzucato 2018). These challenges are useful to ensure focus but too broad to be actionable. In contrast, individual research and innovation projects, which have clear objectives and are actionable, remain isolated in their impacts if not clearly linked to greater challenges (UN DESA 2015). Occupying the critical middle ground, 'missions' set the direction for solutions without prescribing particular methods for achieving success. Rather, missions stimulate experimentation to develop a range of different solutions to achieve the objective, which together makes a significant and concrete contribution towards meeting an SDG or Societal Challenge.

Within the *Horizon Europe* (HEU) funding mission area *Climate-Neutral and Smart Cities*, for example, the EC has proposed a mission called *100 Climate-Neutral Cities by 2030—By and For the Citizens* (EC 2020c). This geo-political and technological mission has a distinct objective and concrete timeline within its very name. Yet, the successive steps necessary for its realisation are left to be co-determined by innovators across many sectors. By virtue of this fact, regardless of whether the primary mandate of the mission is accomplished, a new era of cross-cutting innovations will be enabled and, likely, prove priceless due to the resulting spin-off successes underpinned by countless novel opportunities for experimentation and risk-taking. Of course, some will fail, yet these are learning grounds too as without failure the boundaries of any approach cannot be tested.

To support technical advances and technical concerns, implementing 'missions' requires comprehensive citizen engagement, appropriate public sector capacities, and galvanising finance and funding. In this way, realising a widespread transition to BioCities will exceed the development and application of technologies and technical solutions and require the fostering of legal, policy, cultural, and behavioural changes. The EC has articulated these as 'social innovations'; new ideas that meet social needs, create social relationships, and form new collaborations (EC 2013a, b). These innovations can be products, services, or models addressing unmet needs more effectively, with the objective of encouraging market uptake of innovative solutions and stimulating employment. Reconfigurations of existing practices are fundamental in terms of innovation in processes and outputs (Ludvig et al. 2018).

2 Key Issues and Solutions

Expanding upon the enabling environment outlined in the Food and Agriculture Organisation of the United Nations (FAO) *Guidelines on Urban and Peri-Urban Forestry*, the elaboration for BioCities comprises five broad strategies: (1) governance; (2) policies and legal framework; (3) investment, collaboration, and partnership; (4) social inclusion and participation; and (5) risks and their management (Salbitano et al. 2016). Proposing a coherent enabling environment for BioCities along such lines is considered critical, as so far these topics have often been overlooked in research work on urban green infrastructures (GI) (Krajter Ostoić et al. 2020).

3 Enabling Governance

BioCities will be governed by a network of actors from different sectors and at different levels who lend diverse knowledge, understanding, perceptions, and interests in the BioCity, as introduced via the 'mosaic government approach' in chapter "Green Infrastructure and Urban Forests for BioCities: Strategic and Adaptive Management". Governance is also covered more comprehensively in chapter "Urban Sustainable Futures: Concepts and Policies Leading to BioCities". The following section will briefly review high-level aspects of BioCity governance as they relate to the enabling environment.

Many issues, trade-offs, synergies, and decisions must be defined or made throughout BioCity design, development, and implementation. For example, there may be trade-offs between the co-benefits of nature-based solutions (NBS) and co-harms (Campbell-Lendrum et al. 2015) [e.g. where exposure to infectious diseases linked to wildlife or arthropod vectors is increased (Dick et al. 2020)]. Constraints on land availability in urban areas, especially those that are densely built or populated, may require stronger central governance to deal with trade-offs between different land uses than in areas where land is more readily available. The Gordian Knot in these cases often is governance rather than funding, with the coordination of stakeholders and the support of the public possibly being more important than big budgets (Lawrence et al. 2013; Ordonez et al. 2020). The creation of a BioCity will then have inter-connected effects on human health and well-being, and infrastructure and governance systems (Díaz et al. 2019; Elmqvist et al. 2019).

The development of BioCities depends not only on the technological feasibility and economic viability of selected innovation pathways, but also on the management of a multi-stakeholder system spanning various tiers, sectors, and stakeholder-led pilot implementations. These can be broadly described as applied research, working with and for public, private, third party, or non-governmental organisations (NGOs). Assorted governance challenges will occur (e.g. fostering collaboration amongst different stakeholders) that may require participatory governance to improve BioCity outcomes. Participatory governance can be defined as the 'processes and structures of public decision making that engage actors from the private sector, civil society, and/or the public at large, with varying degrees of communication, collaboration, and delegation of decision power to participants' (Newig et al. 2018). Reaching consensus on tangible solutions satisfying most stakeholders' needs will require bringing diverse multi-stakeholders together to negotiate and compromise (Rodriguez and Prestvik 2020). Engaging stakeholders in the governance and policy processes offer an opportunity to examine and establish optimal, locally attuned pathways that may help achieve successful BioCities by taking into account the following: (1) a more equal distribution of political power; (2) fairer distribution of resources; (3) decentralisation of decision-making processes; (4) the development of a wide and transparent exchange of knowledge and information; (5) the collaborative establishment of progressive partnerships; (6) an emphasis on inter-institutional dialogue; and (7) greater accountability (Fischer 2017).

Attaining consensus across sectors, levels, interest groups, administrative organisations, and scales, multi-stakeholder participatory processes can also mitigate the fragmentation of governance. This can help align decision-making with the natural boundaries of urban ecosystems that support BioCities, as opposed to ecologically arbitrary bureaucratic boundaries determined by top-down governance, in the same spirit as the Paris Agreement (UNFCCC 2015). Multi-stakeholder participation should be leveraged to integrate locally adapted, multiscalar systems of governance. It is important to note that this characterisation runs contrary to many governance models existing today at the city level, thus there is a need for education, raising awareness, and increasing readiness of those actors involved in city governance to pursue new solutions, especially through practicing co-governance where power is shared directly with citizens. That said, several cities are already leading the charge. For example, Milan's ambitious *Forestami* urban afforestation programme should be regarded as a lighthouse from which to glean successful practices, along with learning from failures.

4 Enabling Policies and Legal Frameworks

It will be difficult to promote and implement BioCities as an integral part of the development and planning goals of urban areas without direct policy and institutional support. Policymakers should base their decisions on good quantifiable information from experts about the benefits, costs, and risks associated with the adoption and implementation of BioCity-related initiatives (Bockarjova et al. 2020). BioCity policies are covered more comprehensively in chapter "Urban Sustainable Futures: Concepts and Policies Leading to BioCities". The following section will briefly review high-level aspects of BioCity policies as they relate to the enabling environment.

Intersectorial Nature-Based Solutions (NBS) complemented by sectoral policies for BioCity solutions (e.g. championing social innovation, entrepreneurship, and participatory processes) can help cities to address challenges arising from climate change and urbanisation. Their applicability in practice, however, is often hampered by the lack of supportive policy and legal frameworks (Sarabi et al. 2019). With regard to NBS in particular, the value of the goods and services derived from nature can prove difficult to quantify and thus are not priced in some existing markets (Kotchen and Powers 2006). The same can be said of so-called 'negative externalities', or the costs of causing damage to ecosystems by a producer or consumer but not financially accrued to them (Helbling 2017). An improved understanding of the economic benefits of nature in cities, based on Natural Capital Accounting (NCA) methodologies, can be useful in assessing whether these benefits outweigh the costs of implementing NBS (UN 2017). In theory, NCA can allow local and national policymakers and stakeholders to make better-informed decisions about appropriate strategies and their anticipated impacts on society. That said, it must be noted NCA has been criticised by some for reductively 'putting a price' on 'invaluable' nature we depend on in manners too complex to express in purely quantitative terms, and thus must be treated with caution so as not to 'crowd out', or draw attention away from other conservation efforts that may provide stricter protections (van der Schalk 2018).

NBS, and complementary practices in discrete sectors, may require changes or updates to regulations and/or policies to become legally feasible. The necessary revisions will be both specific to local conditions, as well as entail common frameworks. One identified example of a lacking common framework is a comprehensive, EU-wide set of regulations, policies, and other legal instruments for the certification of bio-based fertiliser products to facilitate, drive, and sustain their production and use (EC 2016). Timber, in terms of its use as a structural material for urban-scale buildings, serves as another primary example of bio-based materials in need of EU-wide regulation (Ludvig et al. 2021; Build-in-Wood 2020). There is also a need for proactive administrative practices enabling the targeted incentivisation of BioCity development with flexible tools, like the transfer of development rights (TDR) (Nelson et al. 2011).

A systematic review of policies, including existing laws and regulations that can aid the integration of NBS and cross-sectoral BioCity solutions in urban areas into city planning, is likewise necessary. This will help to: (1) identify the ways that NBS/BioCity planning is embedded in government structures to bolster its design and implementation; (2) contextualise the institutional instruments that frame the development and implementation of NBS/BioCity solutions; (3) understand the conduciveness of these instruments at all governance levels, hence creating an enabling environment; (4) design operational aspects of policy implementation; and (5) arrange effective adjustments and corrective adaptations in policies and regulations, where necessary, which might facilitate successful integration of NBS/BioCity solutions.

5 Investment, Collaboration, and Partnership

BioCity-related initiatives and activities are most often linked to the public sector, meaning that the research, development, and innovations in BioCities are mainly understood as funded by public sector investments. This is partial because infrastructures and innovations in a BioCity often require high upfront capital costs and offer long-term pay-offs. For example, out of $133 billion worth of investments in NBS, $113 billion is carried out by domestic government bodies whilst the private sector only contributes about $18 billion per year (UNEP 2021). Thus, incentives in the form of publicly-funded research, development, and innovation (R&D + I)

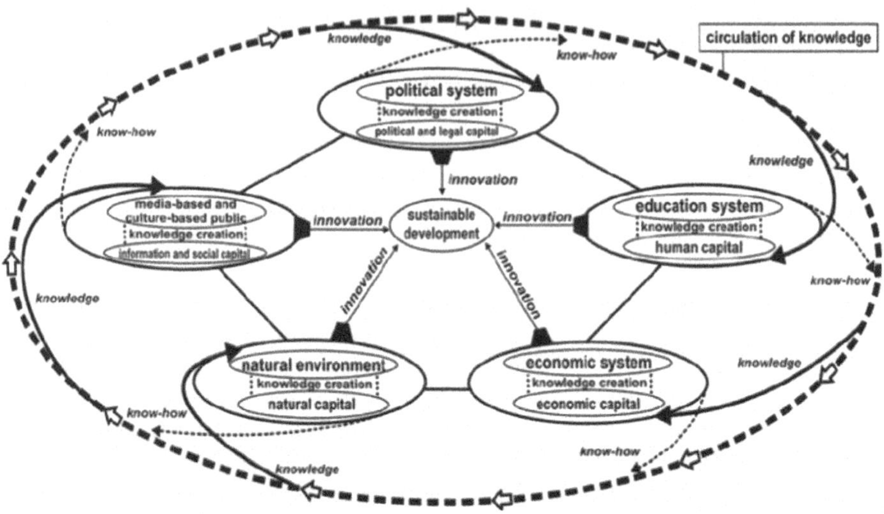

Fig. 1 Quintuple helix model of innovation. © Liyanage and Netswera (Liyanage and Netswera 2021)

tenders, subsidies, tax breaks, and/or other kickbacks (as well as regulatory restrictions of counterproductive activities) are crucial. Public sector investments/spending alone, however, may not be sufficient to stimulate a paradigm shift. This is partly due to fragmentation in public financing schemes along administrative thresholds resulting in the misalignment of funds with overarching or transversal needs, which can be ameliorated through integration in comprehensive, international programmes like H2020 or HEU (EC 2018). Nevertheless, a key element of any BioCity development process is to co-create solutions with relevant stakeholders by involving them from the early stages of the planning process through implementation and monitoring. The EC urges that design and implementation of NBS ought to be co-produced through multi-stakeholder engagement, and lessons learnt should be shared with others (Krull et al. 2015). Still, in many cases, the need for NBS is ignored by the general public (Lorenzoni et al. 2007). Indicatively, 'crowding in' private sector investment via mandates for multi-actor applications to public tenders and co-funding is an explicit priority of HEU (Mazzucato 2019).

Collaborative investments and partnerships between public entities (e.g. governmental institutions); private actors (e.g. small and medium enterprises, consumers, citizens); research bodies (e.g. the socio-scientific community); and NGOs, are vital in effectively developing, implementing, and sustaining a BioCity. Management frameworks organising the communication, knowledge, and innovation pathways amongst these diverse networks of stakeholders—like the quintuple helix model (Fig. 1) (Grundel and Dahlström 2016)—are of the utmost importance for understanding and effectively structuring such complex collaborations. The quintuple helix model was originally developed for application to the redesign of universities, but similar approaches should be extended to innovative activities on

Baugruppen Risk-Reward Table

Baugruppen developments	Traditional development
Buyers own land	Developers own land
Buyers commission architects/builders	Developers commission architects/builders
Buyers get strata title	Buyers get strata title
Buyers get the same tax benefits ad buyer of house-and-land packages	Buyers pay GST and stamp duty
Buyers influence building design	Buyers do not influence building design
Buyers' deposit is 'at risk' and spent on construction	Buyers' depost is held in trust
Buyers recoup saving of up to 30% on market price	Buyers pay market price
Buyers are typically restricted in profiting from immediate resale	Buyers are unrestricted in resale
Buyers can control sustainability features	Buyers cannot control sustainability features
Buyers deposit reduces the cost of investor funding	Buyers deposit does not reduce the cost of investor funding
Buyers emotionally committed to settlement, reducing risk	Buyers, usually investors, are less motionally committed to settlement, increasing risk
Buyers do not pay for the profits of developer and real estate agents	Buyers pay for the profits of developer and real estate agents
Buyers spend time and effort in the developing apartment design	Buyers spend no time or effort in developing their apartment
Buyers live in their apartments because they love them.	Buyers are mostly investors who rent their apartments to others.

Fig. 2 Baugruppen Risk-Reward Table. © Kath Walters (Walters 2021)

the city level, including ecological, environmental, and natural resource elements alongside the social and technical.

It is critical to mobilise private sector investments to fill funding gaps for cost-efficient green infrastructure and innovations in BioCities requiring integration with existing infrastructures. Private sector involvement in BioCities can take the form of public–private partnerships (PPPs) to: (1) ensure collaboration of stakeholders to encourage successful market integration of BioCity-related goods and services; (2) encourage establishment of innovation hubs/centres for ensuring learning from best practices; and (3) provide substantial support for innovative business clusters and networks (Rodriguez and Prestvik 2020). In Japan, for instance, PPPs have been promoted to handle the operation and management of public facilities in the energy, water, and waste sectors, as well as cultural centres and medical facilities (David and Anbumozhi 2018).

Proper legislation can surmount difficulties such as the lack of accountability for negative externalities (valued according to the NCA protocols described above), which influence private profit motives, along with restrictive protections of intellectual property rights (IPR) (Walsh et al. 2021). A potential vehicle for mitigating conflicts between for-profit development models and BioCity ambitions is to enable collective funding, as has been successful in the German Baugruppen model for cohousing (Fig. 2) (Figueira and Trevisan 2019).

6 Social Inclusion and Participation

Planning processes, design, and implementations of BioCities should always have a focus on 'leaving no-one-behind', a central aspiration underpinning the 17 UN SDGs (UN DESA 2015). An inclusive BioCity is an urban place where all stakeholder groups (i.e. all who can benefit from and contribute to the building of a BioCity) are ensured of the following: (1) affordable basic necessities and access to essential infrastructure and services; (2) secure and dignified employment and other economic opportunities; and (3) a representative voice in all stages of decision-making regardless of their socio-economic status, age, gender, or ethnicity, amongst other demographic categories.

One particular challenge is the marginalisation associated with deprivation in a BioCity. 'Green gentrification' might emerge where low-income communities, people of colour, and migrant communities experience residential and social displacement from green and blue climate infrastructure (Shokry et al. 2019; Anguelovski et al. 2016; Gould and Lewis 2018). This can also lead to 'elite capture' where the relatively wealthy have the capacity to manoeuvre the benefits from a BioCity. To avoid green gentrification and elite capture, it is important to identify specific pathways to mainstream BioCity solutions for social inclusion. There are numerous social innovation initiatives developed precisely to respond to the challenge of inclusion of vulnerable groups. Some of these include various care programmes, such as *Green Care* (EU-wide), or more specifically, *Green Care Forest* (in Austria) (Ludvig et al. 2018). These feature therapeutic, social, or educational interventions involving farming, farm animals, gardening, or general contact with nature to help contend with psychological disorders and related physical disabilities in children, adolescents, and adults (Artz and Bitler Davis 2017). Other exemplary initiatives comprise voluntary cooperation for joint goals, such as those regarding mountain bike trails in Switzerland or the woods near Vienna (Wilkes-Allemann et al. 2020), or communal engagement for woodland management with social, cultural, and economic benefits as in Wales (e.g. Woodland Skills Centre, Coppice wood College) (Ludvig et al. 2021).

A distinct benefit of participatory processes is the cultivation of a sense of shared ownership in community development, which can lead to shared stewardship (Lachapelle 2008). 'Citizen science' initiatives empowering the public to participate in urban data gathering, and analysis programmes with accessible technologies can have similar effects (Kullenberg and Kasperowski 2016; Waag Society et al. 2017). Even if ownership, whether felt or documented, is a key determinant for the management of green and blue infrastructures in the BioCity, stewardship, and caretaking by land managers who are not owners should not be neglected (Johnson et al. 2020). This can lead to win-win situations, with both greater social inclusion and a better management of infrastructures. For example, the City of Paris delivers greening permits to citizens to manage micro-gardens on any plot of urban land (i.e. under trees and on rooftops). Similarly, in Vienna, urban renewal offices are working with various city districts, via PPPs, to administer inclusion and participation of local citizens in numerous greening initiatives, such as designing and maintaining micro-gardens under street trees (GB 2021a), or supporting and promoting urban garden projects around the city (GB 2021b). Participatory approaches with micro-forests in cities, like those popularised by the Japanese botanist Miyawaki (i.e. dense copses of fast-growing, biodiverse species that can thrive in areas as small as a tennis court, also known as 'Tiny Forests'), are another example of societal participation in environmental stewardship that can, in the end, provide greater social benefits than those delivered by the green infrastructure itself (Fisher et al. 2015).

7 Risks and Their Management

BioCities must respond and adapt to uncertainty whilst monitoring and managing risks in which changes in knowledge, technologies, and the institutional environment are considered. The type of risks that may occur in a BioCity vary widely within the spectrum of mitigating and adapting to climate change, protecting biodiversity, and ensuring human, social, cultural, financial, and physical well-being. The biggest challenge (and accordingly the highest risk of failure) is in managing a sustainable transition (e.g. ecological, economic, and social) to the BioCity paradigm in a sufficiently short time scale to contend with the rapid changes resulting from climate change. Meeting this challenge will require routine monitoring of relevant, quantifiable metrics encapsulated by progressive certification schemes with hierarchical levels for certifying stepwise achievement of BioCity goals. Importantly, the progressive milestones of these certifications should be made adjustable to remain inclusive of areas with comparably fewer resources.

In designing and implementing certification schemes to avert risk, BioCities would do well to learn from the shortcomings of the Smart City paradigm, summarised as follows (Boorsma 2017):

1. BioCities should prioritise clear understandings and well-defined objectives, outcomes, metrics, and methodologies instead of branding via vague name recognition.
2. BioCities should retain a holistic focus on underlying social, cultural, and behavioural factors, in addition to technical solutions, and avoid technological myopia.
3. BioCities should pursue solutions as means to achieve a desired outcome, rather than for their own sake as demonstrations.
4. BioCities should treat the public sector as one client amongst many, not the predominant or sole customer, and seek to advance PPPs.
5. BioCities should promote interdisciplinary collaboration, and strive for open, non-proprietary sharing of tools and knowledge, whilst planning ahead for replication or scalability.

6. BioCities should be wary of digital divides, engage a variety of citizens through a mixture of strategies, and facilitate integrated decision-making ecosystems over binary top-down versus bottom-up approaches.
7. BioCities should maintain attention on underlying infrastructures, and not exclusively on end-solutions, such as apps.
8. BioCities should keep in mind the far more numerous, and more quickly growing, small and mid-size urban settlements, and foster solutions as appropriate to them, as well as to large cities.

That is not to say technologies including but not limited to, remote sensing networks, automated control systems, digital twins, artificial intelligence (AI), machine learning, and decision support system tools (DSSTs) will not help to moderate the risks of BioCity developments by facilitating the operation of monitoring, modelling, prediction, and response mechanisms. On the contrary, they will likely prove indispensable. Still, such technical solutions should not be allowed to monopolise resources and energies which must also be invested in intermediary participatory practices. Cultural programmes for incorporating knowledge from communities beyond technology, design, and administration are also central considerations. In any case, appropriate key performance indicators (KPIs) for evidencing progress or shortcomings, and solution development and delivery to mitigate unexpected ramifications are indispensable.

Risks related to natural hazards associated with the onset of climate change, such as flooding, drought, fire, and sea-level rise, present a significant challenge for BioCities (EC 2020b). Besides, natural disasters unrelated to climate change, such as earthquakes, typhoons, or tsunamis, can cause injury or loss of life, property and infrastructure damage, and disruption of socio-economic activities (UNDP 2009). Risk is determined not just by the hazard itself (i.e. its magnitude and frequency), but also by exposure and vulnerability to the hazard event (Cardona et al. 2012). Exposure to risk refers to the inventory of elements in an area in which hazard events may occur (Cardona 1990; UNISDR 2004). This can be influenced by patterns of human settlement within hazard zones. In the Netherlands, for example the location of urban settlements in the City of Rotterdam has been influenced by their accessibility to the sea and navigable inland waterways. As a result, the city's urban population, infrastructure, commercial establishments, and public institutions are exposed to flooding, since they are within the dikes which lie along the main rivers. On the other hand, vulnerability to risk refers to the propensity of people, assets, and ecosystems that are exposed to natural hazards to be susceptible to, and unable to cope with, adverse effects of natural hazards. Hence, the vulnerability of a BioCity to risk is influenced by its ability to absorb, recover, and prepare for future unexpected shocks and stresses. BioCities must uphold environmental justice by mitigating differential exposure and vulnerability to risks resulting from demographic distinctions and associated settlement patterns.

8 Links to European Policies and Actions

The enabling environment of future BioCities will be seeded by a number of global, international, national, regional, and local R&D + I policies and actions. Paramount for its massive scale and cross-cutting missions, HEU is the EU's key funding programme for 2021–2027 with a budget of €95.5 billion.

The structure of HEU has three programme pillars, of which the second is *Global Challenges and European Industrial Competitiveness.* Under this pillar, programmes are grouped into six clusters, with three directly pertaining to the development of BioCities: *Cluster 2—Culture, Creativity and Inclusive Society; Cluster 5—Climate, Energy, and Mobility;* and *Cluster 6—Food, Bioeconomy, Natural Resources, Agriculture and Environment.* These programmes will host multiple tenders and provide resources for advancing sustainable forest management, engineered timber processing, and the decentralised/distributed construction of diverse engineered timber buildings necessary for the realisation of BioCities (EC 2018).

The HEU also incorporates research and innovation missions to target programme funding. Two of the five mission areas of HEU (*Adaptation to Climate Change, Including Societal Transformation*; *Climate-Neutral and Smart Cities*) are closely aligned with the goals of BioCities, and a third (*Soil Health and Food*) is relevant as well, given the ramifications of underlying forestry practices.

Another programme within HEU expected to influence the enabling environment of BioCities is the *New European Bauhaus* (NEB). NEB specifically targets the intersection of art, culture, inclusivity, science, and technology (NEB 2021). Prior to HEU, the EU research and innovation funding scheme H2020 featured similarly intersectional priorities as reflected in projects like *Build-in-Wood*, a €10.3 million action running from 2019 to 2023, with the goal of drastically increasing the proportion of timber construction (Build-in-Wood 2020). H2020 also co-funded the €1 billion set of Green Deal tenders with the ambition of making Europe the first climate-neutral continent by 2050 (EC 2019).

The link between cities and nature has been highlighted in several EU policy documents on the environment, clarifying that urban areas and natural areas have shared rather than opposite destinies. The EU *Biodiversity Strategy for 2030* considers the greening of cities as an axis of the EU *Nature Restoration Plan* in the same way as the restoration of natural ecosystems (EC 2020a). The strategy aims to systematically integrate healthy ecosystems, green infrastructure, and NBS into urban planning. To bring nature back to cities, the EC calls on European cities to develop ambitious urban greening plans. To facilitate this work, the EC will set up an EU *Urban Greening Platform* under the *Green City Accord* (Ahlman 2020; EC 2020a). The new *Urban Greening Platform* will also facilitate urban tree planting, including under the LIFE programme. The recently launched EU *Strategy on Adaptation to Climate Change* further considers the development of urban green spaces and the installation of green roofs and walls as an NBS for adaptation (EC 2021a, b). The strategy indicates that buildings can contribute to adaptation through local water retention with green roofs and walls, which reduces the urban heat island effect. On the other hand, the future EU *Forest Strategy* will stress that the significance of the rural–urban nexus in shaping the future of forests, and that improved communication and dialogues on forests and their roles are needed between the two sides of this interface (EEA 2021). The future EU *Forest Strategy* additionally identifies urban and peri-urban areas as a potential for extending forest and tree coverage in the EU.

Jointly, these frameworks will support satisfaction of the UN SDGs, notably SDG 3—*Good Health and Well-Being*; SDG 7—*Affordable and Clean Energy*; SDG 9—*Industry, Innovation and Infrastructure*; SDG 10—*Reduced Inequalities*; SDG 11—*Sustainable Cities and Communities*; SDG 12—*Responsible Consumption and Production*; SDG 13—*Climate Action*; and SDG 15—*Life on Land* (UN 2015).

9 Gaps and Perspectives

The enabling environment for BioCities requires a thorough understanding of the challenges, barriers, and trade-offs that will be faced in cities wishing to transition to a BioCity. The key factors to be identified are the social, ecological, and economic issues that constrain or support this transition. In this context, a major consideration is the variability of challenges (e.g. demographic and cultural trends, geography, and climate) and factors (e.g. funding, resources) faced by cities in Europe and beyond, due to different geographical and social-political contexts. BioCities should be understood as models for sustainable growth, as they follow the principles of natural systems. In view of that fact, they should aspire to reach sustainable development objectives, such as the UN SDGs. Conversely, action plans for reaching sustainable development objectives will affect the transition towards BioCities. An understanding of these interlinkages is imperative. A core component for improving the knowledge of the necessary enabling environment is to ensure that the follow-up on transitional processes includes reliable monitoring systems and sets of indicators. Having a detailed understanding of the research gaps enabling the transition, as well as of priority research areas related to the concept of BioCities, is necessary. Further details on both issues can be found in the associated *BioCities Research Agenda*.

10 Take-Home Messages

- Future BioCities must adopt the philosophy of missions (i.e. systemic public policies that draw on frontier knowledge to attain specific goals), in order to foster achievement through bottom-up innovation, inclusive participatory processes, and valuable spill over.
- The enabling environment for BioCities should address five broad strategies: (1) governance; (2) policies and legal framework; (3) investment, collaboration,

and partnership; (4) social inclusion and participation; and (5) risks and their management.

- A 'mosaic government approach' is to be promoted by BioCities in order to balance trade-offs between co-benefits and co-harms of implemented NBS amongst diverse stakeholders.
- BioCities should leverage quantification of the real value and impact of natural capital (NCA) and ecosystem services (ESS), along with so-called externalities, to inform decision and policy making which follows the principles of natural systems to promote life, and implement viable strategies and metrics for measuring progress towards that vision. However, it is fundamental that NCA is not allowed to 'crowd out' alternative conservation efforts that may provide stricter protections.
- In complement to public sector initiatives, BioCities must also prioritise 'crowding in' private sector investment, whilst nurturing PPPs, interdisciplinary collaboration, and co-creation practices.
- In all planning, design, and implementation processes, BioCities are obliged to respect the mandate of 'leaving-no-one-behind', ensuring representativeness through participatory processes and environmental justice.
- Learning from the failures and successes of the past, as well as from reliable data from the present and robust predictions of future trends, BioCities should holistically integrate technical, social, cultural, behavioural, and economic tools and understandings to anticipate, mitigate, and adapt to risks posed by climate change, natural hazards, demographic changes, resource depletion, biodiversity loss, supply chain variations, and other potential disruptions.

11 Case Study

Launched in 2012, *Smart Citizen* is an exemplar of a 'citizen science' platform for the generation of social participatory processes in urban areas (IAAC 2012). By connecting data, people, and knowledge, the *Smart Citizen* platform serves as a node for building productive and open urban environmental indicators (Fig. 3). *Smart Citizen* achieves this goal with distributed tools, leading to the collective construction of the city by and for its own inhabitants. In this way, *Smart Citizen* creates 'an integrated governance ecosystem incorporating bottom-up decision making and

Fig. 3 Smart Citizen Kit and Platform. © Smart Citizen (Seeed Studio 2021)

communal rights, in which local residents and communities are actively engaged in self-determining the realities of their BioCity, coming to see its spaces as shared property, and deep local understandings lead to insightfully adapted nature-based interventions', as prompted by chapter "Towards the Development of a Conceptual Framework of BioCities". In a similar way, *Smart Citizen* also advances 'The Universal BioCity' criteria of ensuring the 'involvement of citizens is natural at all levels, from locally founded activities and management to planning and policy-making'. *Smart Citizen* is operationalised by the *Smart Citizen Kit*, an open-source hardware and software bundle allowing diverse urban dwellers to easily measure and collect data such as air and noise pollution from their environment, then visualise it whilst sharing with thousands of other users in the dedicated online space. Accordingly, *Smart Citizen* additionally supports the principle of 'The BioCity as a Forest' by seeing the BioCity as an urban system that 'does not emit carbon dioxide (CO_2) and other greenhouse gases (GHGs) which trap heat in the earth's atmosphere, but rather absorbs them, as forests do'. This and comparable tools, both technological and human-facing, will be instrumental in the transition process from the modern city to the BioCity.

At present, the *Smart Citizen Kit* measures: air temperature, relative humidity, noise levels, ambient light, barometric pressure, CO_2e, VOCs, and particulate matter (Seeed Studio 2021). To better support the transition to BioCities, the online platform could be coupled with geographic information systems (GIS) and geospatial data provided by public agencies in order to better evidence correlations between GI, NBS, ESS, urban demographics, and the metrics monitored by the *Smart Citizen Kit*.

References

Ahlman, VRJ Group, Green Academy Aarhus, OKNygaard, ITTEC Irrigation and Turf Technologies, VOŠ a SZeŠ Benešov, European Landscape Contractors Association, Wellant College (2020) European Platform for Urban Greening—Connecting Centres of Vocational Excellence (CoVEs) as part of a KA3 project co-funded by the Erasmus+ programme of the EU. https://www.platformurbangreening.eu/partners

Anguelovski I, Shi L, Chu E, Gallagher D, Goh K, Lamb Z, Reeve K, Teicher H (2016) Equity impacts of urban land use planning for climate adaptation: critical perspectives from the global north and south. J Plan Educ Res 36(3):Article 3. https://doi.org/10.1177/0739456X16645166

Artz B, Bitler Davis D (2017) Green care: a review of the benefits and potential of animal-assisted care farming globally and in rural America. Animals: Open Access J MDPI 7(4):Article 4. https://doi.org/10.3390/ani7040031

Bockarjova M, Botzen WJW, Koetse MJ (2020) Economic valuation of green and blue nature in cities: a meta-analysis. Ecol Econ 169:1–13. https://doi.org/10.1016/j.ecolecon.2019.106480

Boorsma B (2017) A new digital deal: beyond smart cities. How to Best Leverage Digitalisation for the Benefit of our Communities

Build-in-Wood (2020) Why timber is better than building law thinks it is. Build-in-Wood, December 3. https://www.build-in-wood.eu/post/building-regulations

Campbell-Lendrum D, Manga L, Bagayoko M, Sommerfeld J (2015) Climate change and vector-borne diseases: what are the implications for public health research and policy? Philos Trans R Soc B: Biol Sci 370(1665):Article 1665. https://doi.org/10.1098/rstb.2013.0552

Cardona OD (1990) Terminología de Uso Común en Manejo de Riesgos (AGID No. 13; Issue 13). Escuela de Administración, Finanzas, y Tecnología. https://www.ipcc.ch/apps/njlite/srex/njlite_download.php?id=6129

Cardona OD, van Aalst MK, Birkmann J, Fordham M, McGregor G, Perez R, Pulwarty RS, Schipper ELF, Sinh BT (2012) Determinants of risk: exposure and vulnerability (a special report of working groups I and II of the intergovernmental panel on climate change (IPCC), pp 65–108). https://www.ipcc.ch/site/assets/uploads/2018/03/SREX-Chap2_FINAL-1.pdf

David D, Anbumozhi V (2018) A Comparative Study on the Role of Public–Private Partnerships and Green Investment Banks in Boosting Low-Carbon Investments (Issue 870). Asian Development Bank. https://www.adb.org/publications/comparative-study-role-ppp-green-investment-banks-boosting-low-carbon

Díaz S, Settele J, Brondízio ES, Ngo HT, Agard J, Arneth A, Balvanera P, Brauman KA, Butchart SHM, Chan KMA, Garibaldi LA, Ichii K, Liu J, Subramanian SM, Midgley GF, Miloslavich P, Molnár Z, Obura D, Pfaff A, Zayas CN et al (2019) Pervasive human-driven decline of life on Earth points to the need for transformative change. Science 366(6471):eaax3100. https://doi.org/10.1126/science.aax3100

Dick K, Samanfar B, Barnes B, Cober ER, Mimee B, Tan LH, Molnar SJ, Biggar KK, Golshani A, Dehne F, Green JR (2020) PIPE4: fast PPI predictor for comprehensive inter- and cross-species interactomes. Sci Rep 10:1390. https://doi.org/10.1038/s41598-019-56895-w

EC (2013a) Social innovation [text]. European Commission. https://ec.europa.eu/growth/industry/policy/innovation/social_en

EC (2013b) Social investment: commission urges member states to focus on growth and social cohesion. European Commission: Employment, Social Affairs & Inclusion, February 20. https://ec.europa.eu/social/main.jsp?langId=en&catId=1044&newsId=1807&furtherNews=yes

EC (2016) EUR-Lex—52016PC0157—EN - EUR-Lex. EUR-Lex. https://eur-lex.europa.eu/legal-content/EN/TXT/?uri=CELEX%3A52016PC0157

EC (2018) Horizon Europe [Text]. European Commission, January 3. https://ec.europa.eu/info/research-and-innovation/funding/funding-opportunities/funding-programmes-and-open-calls/horizon-europe_en

EC (2019) A European Green Deal [European Commission]. European Commission - European Commission, December 10. https://ec.europa.eu/info/strategy/priorities-2019-2024/european-green-deal_en

EC (2020a) Green City Accord. European Commission https://ec.europa.eu/environment/topics/urban-environment/green-city-accord_en

EC (2020b) Biodiversity strategy for 2030. European Commission, May 20. https://ec.europa.eu/environment/strategy/biodiversity-strategy-2030_en

EC (2020c) 100 climate-neutral cities by 2030 – by and for the citizens [text]. European Commission. https://ec.europa.eu/info/publications/100-climate-neutral-cities-2030-and-citizens_en

EC (2021a) EU adaptation strategy [text]. European Commission, February 24. https://ec.europa.eu/clima/policies/adaptation/what_en

EC (2021b) Overview of natural and man-made disaster risks the European Union may face. European Commission. https://ec.europa.eu/echo/sites/default/files/overview_of_natural_and_man-made_disaster_risks_the_european_union_may_face.pdf

EEA (2021) A new EU Forest strategy—European Environment Agency [policy document]. European Environment Agency https://www.eea.europa.eu/policy-documents/the-eu-forest-strategy-com

Elmqvist T, Andersson E, Frantzeskaki N, McPhearson T, Olsson P, Gaffney O, Takeuchi K, Folke C (2019) Sustainability and resilience for transformation in the urban century. Nat Sustain 2(4): Article 4. https://doi.org/10.1038/s41893-019-0250-1

Ergas H (1987) Does technology policy matter? Technology and Global Industry: Companies and Nations in the World Economy 191–245

Figueira A, Trevisan R (2019). Baugruppen: the German model of cohousing and its constitutive variables

Fischer F (2017) Participatory environmental governance: civil society, citizen engagement, and participatory policy expertise. In: Fischer F (ed) Climate crisis and the democratic prospect: participatory governance in sustainable communities. Oxford University Press. https://doi.org/10.1093/oso/9780199594917.003.0007

Fisher D, Svendsen E, Connolly J (2015) Urban environmental stewardship and civic engagement: how planting trees strengthens the roots of democracy. Routledge https://doi.org/10.4324/9781315857589

GB (2021a) Der Stadtnatur zuliebe—Baumscheiben begrünen in Wien. Gebietsbetreuungen Stadterneuerung. https://www.gbstern.at/news/baumscheiben-begruenen/

GB (2021b) Nachbarschaftsgärten. Gebietsbetreuungen Stadterneuerung. https://www.gbstern.at/themen-projekte/urbanes-garteln/nachbarschaftsgaerten/

Gould KA, Lewis TL (2018) From green gentrification to resilience gentrification: an example from Brooklyn. City & Community 17(1):Article 1. https://doi.org/10.1111/cico.12283

Grundel I, Dahlström M (2016) A quadruple and quintuple helix approach to regional innovation systems in the transformation to a forestry-based bioeconomy. J Knowl Econ 7(4):Article 4. https://doi.org/10.1007/s13132-016-0411-7

Helbling T (2017) Externalities: prices do not capture all costs. Finance and Development. International Monetary Fund. https://www.imf.org/external/pubs/ft/fandd/basics/external.htm

IAAC (2012) Smart Citizen—Institute for Advanced Architecture of Catalonia (IAAC). https://iaac.net/project/smart-citizen/

Johnson LR, Johnson ML, Aronson MFJ, Campbell LK, Carr ME, Clarke M, D'Amico V, Darling L, Erker T, Fahey RT, King KL, Lautar K, Locke DH, Morzillo AT, Pincetl S, Rhodes L, Schmit JP, Scott L, Sonti NF (2020) Conceptualising social-ecological drivers of change in urban forest patches. Urban Ecosyst 24:633. https://doi.org/10.1007/s11252-020-00977-5

Kotchen MJ, Powers SM (2006) Explaining the appearance and success of voter referenda for open-space conservation. J Environ Econ Manag 52(1):373–390. https://doi.org/10.1016/j.jeem.2006.02.003

Krajter Ostoić S, Vuletić D, Planinšek Š, Vilhar U, Japelj A (2020) Three decades of urban forest and green space research and practice in Croatia and Slovenia. Forests 11(2):Article 2. https://doi.org/10.3390/f11020136

Krull W, Berry P, Bauduceau N, Elmqvist T, Fernandez M, Hartig T, Mayerhofer E, Naumann S, Noring L, Raskin K, Roozen E, Sutherland W, Tack J, Vandewoestijne S, Boissezon B (2015) Towards an EU research and innovation policy agenda for nature-based solutions and re-naturing cities. Final report of the Horizon 2020 expert group on 'Nature-based solutions and re-naturing cities'. https://doi.org/10.2777/765301

Kullenberg C, Kasperowski D (2016) What is citizen science? A scientometric meta-analysis. PLOS One 11(1):Article 1. https://doi.org/10.1371/journal.pone.0147152

Lachapelle P (2008) A sense of ownership in community development: understanding the potential for participation in community planning efforts. Community Dev 39(2):Article 2. https://doi.org/10.1080/15575330809489730

Lawrence A, De Vreese R, Johnston M, Konijnendijk van den Bosch CC, Sanesi G (2013) Urban forest governance: towards a framework for comparing approaches. Urban Forestry Urban Green 12(4):Article 4. https://doi.org/10.1016/j.ufug.2013.05.002

Liyanage SIH, Netswera FG (2021) Greening universities with mode 3 and quintuple helix model of innovation – production of knowledge and innovation in knowledge-based economy, Botswana. J Knowl Econ 13:1126. https://doi.org/10.1007/s13132-021-00769-y

Lorenzoni I, Nicholson-Cole S, Whitmarsh L (2007) Barriers perceived to engaging with climate change among the UK public and their policy implications. Glob Environ Chang 17(3): 445–459. https://doi.org/10.1016/j.gloenvcha.2007.01.004

Ludvig A, Weiss G, Sarkki S, Nijnik M, Živojinović I (2018) Mapping European and forest related policies supporting social innovation for rural settings. Forest Policy Econ 97(C), Article C

Ludvig A, Braun M, Hesser F, Ranacher L, Fritz D, Gschwantner T, Jandl R, Kindermann G, Ledermann T, Pölz W, Schadauer K, Schmid BF, Schmid C, Schwarzbauer P, Weiss G, Wolfslehner B, Weiss P (2021) Comparing policy options for carbon efficiency in the wood value chain: evidence from Austria. J Clean Prod 292:125985. https://doi.org/10.1016/j.jclepro.2021.125985

Mazzucato M (2018) Mission-oriented research and innovation in the European Union. European Commission 36

Mazzucato M (2019) Governing missions in the European Union. European Commission 32

NEB (2021) New European Bauhaus: beautiful, sustainable, together. New European Bauhaus, May 18. https://europa.eu/new-european-bauhaus/index_en

Nelson A, Pruetz R, Woodruff D (2011) The TDR handbook. Island Press, Washington, DC. https://islandpress.org/books/tdr-handbook

Newig J, Challies E, Jager NW, Kochskaemper E, Adzersen A (2018) The environmental performance of participatory and collaborative governance: a framework of causal mechanisms. Policy Stud J 46(2):269–297. https://doi.org/10.1111/psj.12209

Ordonez C, Threlfall CG, Livesley SJ, Kendal D, Fuller RA, Davern M, van der Ree R, Hochuli DF (2020) Decision-making of municipal urban forest managers through the lens of governance. Environ Sci Policy 104:136–147. https://doi.org/10.1016/j.envsci.2019.11.008

Rodriguez DGP, Prestvik AS (2020) The need for stakeholder engagement and participative governance to promote bioeconomy. In: Nagothu US (ed) The bioeconomy approach. Routledge, London

Salbitano F, Borelli S, Conigliaro M, Yujuan C (2016) Guidelines on urban and peri-urban forestry. FAO Forestry Paper (FAO) Eng No. 178. http://www.fao.org/3/a-i6210e.pdf

Sarabi SE, Han Q, Romme AGL, de Vries B, Wendling L (2019) Key enablers of and barriers to the uptake and implementation of nature-based solutions in urban settings: a review. Resources 8(3):3. Scopus. https://doi.org/10.3390/resources8030121

Seeed Studio (2021) Smart Citizen Starter Kit. https://www.seeedstudio.com/Smart-Citizen-Starter-Kit-p-2865.html

Shokry G, Connolly J, Anguelovski I (2019) Understanding climate gentrification and shifting landscapes of protection and vulnerability in green resilient Philadelphia. Urban Clim 31: 100539. https://doi.org/10.1016/j.uclim.2019.100539

UN (2017) Natural capital accounting and valuation of ecosystem services project | system of environmental economic accounting. System of Environmental Economic Accounting. https://seea.un.org/home/Natural-Capital-Accounting-Project

UN DESA. (2015) THE 17 GOALS | sustainable development. https://sdgs.un.org/goals

UNDP (2009) UNDP Bureau for crisis prevention and recovery 2009 annual report. UNW WRD Knowledge Hub. https://wrd.unwomen.org/explore/library/undp-bureau-crisis-prevention-and-recovery-2009-annual-report

UNEP, WEF, ELD, Vivid Economics (2021) State of Finance for Nature. UN Environment Programme, May 27. http://www.unep.org/resources/state-finance-nature

UNFCCC (2015) The Paris Agreement | UNFCCC. United Nations Framework Convention on Climate Change, December 12. https://unfccc.int/process-and-meetings/the-paris-agreement/the-paris-agreement

UNISDR (2004) Living with risk: a global review of disaster reduction initiatives. United Nations. https://www.un.org/press/en/2004/iha922.doc.htm

van der Schalk J (2018) Accounting for natural capital. FreedomLab, November 30. https://freedomlab.org/accounting-for-natural-capital/

Waag Society, IAAC, University of Dundee, Peer Educators Network, Joint Research Centre, Fab Lab Network (2017) Making Sense. Making Sense. http://making-sense.eu/

Walsh K, Wallace A, Pavis M, Olszowy N, Griffin J, Hawkins N (2021) Intellectual property rights and access in crisis. Int Rev Intell Property Competition Law 52(4):Article 4. https://doi.org/10.1007/s40319-021-01041-1

Walters K (2021) LONGREAD: Easing Australia's Housing Pain, September 13. https://bluenotes.anz.com/posts/2018/05/longread%2D%2Deasing-australias-housing-pain

Wilkes-Allemann J, Ludvig A, Hogl K (2020) Innovation development in forest ecosystem services: a comparative mountain bike trail study from Austria and Switzerland. Forest Policy Econ 115:102158. https://doi.org/10.1016/j.forpol.2020.102158

Towards BioCities: The Pathway to Transition

Clive Davies, Fabio Salbitano, Giuseppe E. Scarascia-Mugnozza, and Simone Borelli

1 Forest Ecosystems as an Analogue for the BioCity

In creating a manifesto for BioCities, this book proposes that a city acting as a nature-based socio-ecological system will also be building resilience to climate change and other stresses and risks (Biggs et al. 2015). This thinking has been adapted into four basic principles for a BioCity (Fig. 1). Nature embedded means that the BioCity should include a wide variety of species, ecosystems and habitats which are planned and sustained at all spatial and temporal levels. In a BioCity the circular bioeconomy is essential as it ensures that biological material are wholly integrated into products, development processes and that waste is regarded as a renewable resource. BioCities do not live apart from the wider region beyond the municipal boundary, which is not only local but regional and global too. The supply chain for the city is vast and closer to the BioCity as green infrastructure networks and urban forests extend into peri-urban areas and beyond, an instance among many that speak of the urban rural nexus. Finally, adaptive management ensures that policy and planning of forest based solutions are reviewed and revised based on actual observation in a socio-ecological learning paradigm.

C. Davies (✉)
School of Architecture, Planning, and Landscape (SAPL), Newcastle University, European Forest Institute, London, UK
e-mail: clive.davies@newcastle.ac.uk

F. Salbitano
University of Sassari, Sassari, Italy

G. E. Scarascia-Mugnozza
University of Tuscia, Viterbo, Italy

S. Borelli
United Nations Food and Agriculture Organisation (UN-FAO), Rome, Italy

G. E. Scarascia-Mugnozza et al. (eds.), *Transforming Biocities*, Future City 20,
https://doi.org/10.1007/978-3-031-29466-2_12

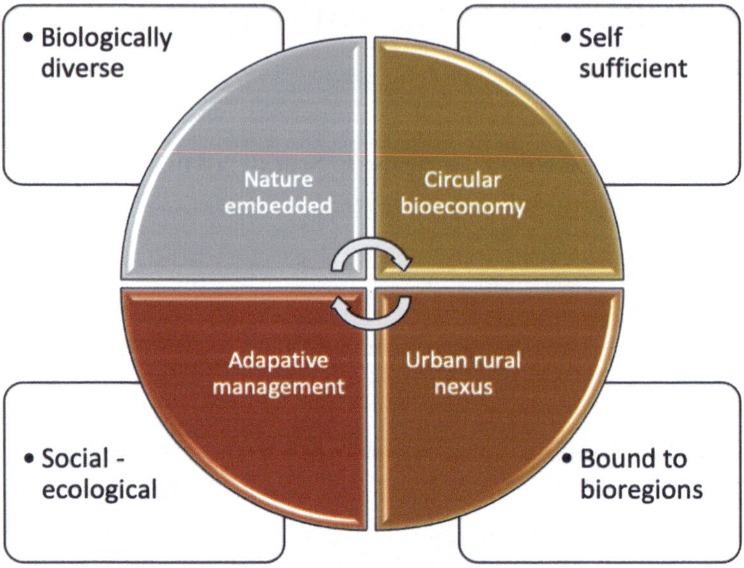

Fig. 1 The four principles of the BioCity

The key idea to emerge from these principles is to envision the planning and management of a BioCity as a 'forest analogue'. When considering the properties of this forest analogue, and hence how it might link to a city (or an urban area of any scale), it is useful to consider the forest as an ecosystem, or a community with countless interrelated pieces. In fact, the Convention on Biological Diversity (SCBD 2001) describes forest ecosystems as a '*dynamic complex of plant, animal and microorganism communities, and their abiotic environment, that interact as a functional unit that reflects the dominance of ecosystem conditions and processes by trees. Humans, with their cultural, economic and environmental needs, are an integral part of many forest ecosystems*'.

Forest ecosystems are highly dynamic and, in a mature state, represent a highly sustainable self-renewing community. They are complex biological systems that are vertically and horizontally stratified, much more so than other terrestrial ecosystems. The urban analogue to forest ecosystems is that not only should the city be home to different species, but also to a diverse range of humanity. This complex biological community applies equally at the human scale as it does at the biodiversity scale. Of notable importance is that forest ecosystems also contain substantive abiotic elements. In view of this, the interaction between the biological (living) environment and the physical (abiotic) infrastructure in the BioCity is analogous to the forest ecosystem. In many instances cities can look to nature-based approaches when they renew their physical infrastructure, hence rebalancing the biotic–abiotic nexus. Over time in the BioCity, the biotic quotient will increase and the abiotic quotient will decrease. Both forest ecosystems and BioCities, however, should not be seen solely as biological.

Forest ecosystems are a key element of global green infrastructure (GI). The GI concept has gained traction over the last few decades, including its key ideas of multifunctionality and connectivity. Returning to the analogue, forest ecosystems are highly multifunctional and well-connected at both the micro and macro scales, and directly relate to GI. Through appropriate city planning and policy, the same principles of multifunctionality and connectivity should be embedded in the creation of the BioCity.

Humans continue to play a major role in forest ecosystems. Foresters managing forests in both urban and rural areas often play the role of inter-generational conservators of forest ecosystems. For centuries, foresters have been a vital part of the cultural, economic, and environmental benefits that forests have provided to the community. Current foresters build on the knowledge and experience of their predecessors and provide a segue to the next generation. The city analogue to the forester legacy is to ensure that the same long-term principles exist in policymaking for city planning and management, which is a notable challenge given that short-termism is rife in urban politics.

Forest ecosystems are managed not just by foresters, however, but by many different types of professionals (e.g. geologists, hydrologists, wildlife biologists, and civil engineers). They are an exemplary example of a transdisciplinary approach. The professions are by no means limited to those working in either forestry or arboriculture, but also include planners, economists, sociologists, and many more. In much the same way, managing the BioCity should be transdisciplinary.

2 Framing the BIOCITY: The Role of the Sustainable Development Goals (SDGs)

The Sustainable Development Goals (SDGs) are a universal set of goals, targets, and indicators that United Nations (UN) member states have committed to use. These goals frame both domestic and international development policies in order to pursue a broad agenda that encompasses the social, environmental, and economic aspects of sustainable development. Seventeen (17) goals and 169 targets address critical issues facing the world today, including the eradication of extreme poverty, tackling global inequality, and climate change, promoting sustainable urbanisation and industrial development, protecting natural ecosystems, and fostering the growth of peaceful and inclusive communities and governing institutions (Kanuri et al. 2016). According to ICLEI, the SDGs are an unprecedented opportunity for local governments to develop practical solutions to the challenges of sustainable development, and offer cities a blueprint for action. For mayors and local leaders that are working to improve the quality of life in urban environments, the SDGs provide a roadmap for more balanced and equitable urban development. All cities must now aim to increase prosperity, promote social inclusion, and enhance resilience and environmental sustainability. In this way, the SDGs capture large parts of the existing political agenda in virtually every city. When aligned with existing planning frameworks and development priorities, they can strengthen development outcomes and provide additional resources for local governments. Along the lines of SDGs, the BioCity approach aims to address the issues of sustainable urban development by challenging all the different sectors to act in an integrated manner in order to make transformative changes in the way we work, live, and play in the cities of today and, most importantly, in the cities of tomorrow (Fig. 2).

3 Pathways to Transition

Transitioning to a BioCity is a process. One way to consider this is through a series of 'pathways' which, if followed, will enable transition. Others have already sought such pathways and an exemplar of this approach is ICLEI's 15 pathways of the Basque Declaration (ICLEI 2019). This approach is linked to the UN SDGs (Fig. 2) and makes a virtue of considering them as an integrated whole, due to their 'interrelated nature'. ICLEI considers that these 15 pathways provide local government and other local stakeholders with inspiration for their own initiatives, and that 'every pathway utilised will mean a step in the right direction'. The 15 pathways of the Basque Declaration are grouped together and offer guidance for socio-cultural (pathways 1–5), socio-economic (pathways 6–10), and technological (pathways 11–15) transformation of societies, and which contribute to the implementation of the SDGs at the local level. These pathways are particularly relevant and can, with adaptation, become key pathways for a BioCity too. They are, of course, a general framework, and each city will need to determine their own. As is often the case, the journey is at least as important as the destination. Figure 3 lists seven areas where pathways are especially important.

Determining where to start the journey commences with an audit of processes already underway, and is best described as a purposeful audit to identify the baseline. Just as it is unlikely that any city can claim to already be a 'wholly formed' BioCity, neither is it unlikely that any given city has not already got processes underway that contribute towards it. Knowing where these are and the extent of current practice, is critical to the start of the journey. After the baseline audit, a conversation is needed to establish realistic timeframes for the pathway, including milestones, and to establish a monitoring and evaluation framework. Ideally, these elements should be locally derived and 'owned' by all sectors. Opportunities for twinning between agencies and sectors can be very useful to affirm processes, steps, milestones, and the appropriateness of locally adopted pathways. There is also an opportunity for experimentation when city districts are undergoing regeneration and/or renewal. They can be considered as Bio-Neighbourhoods, on a scale below the BioCity, acting as 'living laboratories' and function as a basis for upscaling and ground truthing.

Modest | Significant | High

SDG	Target	BioCities contribution to the UN Sustainable Development Goals
1 NO POVERTY	1.5	Urban and rural forests provide essential services to those suffering from impoverishment and trees can be a source of fuel and food in a crisis. Urban Forests also create employment and provide a resource for entrepreneurs. BioCities invest in poverty alleviation and ensure that everyone has a pathway out of poverty both economically and environmentally consistent with the delivery of ecosystem services.
2 ZERO HUNGER	2.3 2.4	Urban gardens, communal growing areas, urban agriculture and allotments bring benefits in terms of local food production and contribute to a more sustainable and self-efficient BioCity if it is done at a sufficiently large scale. Urban forests are direct sources of food and support healthy eating by providing edible foods and affordable wood fuel.
3 GOOD HEALTH AND WELL-BEING	3.4 3.9	BioCities build on the integration of natural and human processes in urban design, with natural biotic and abiotic factors and processes integrated with the development of constructed features to provide for human well-being. BioCities reduce environmental contamination in air, water and soil. They also conserve and expand natural green spaces which promote mental health and well-being.
4 QUALITY EDUCATION	4.4 4.7	BioCities promote good quality education for all ages and abilities. This encompasses all socio-economic backgrounds and includes the formal system, citizen science, volunteerism, universities, and research. The key focus is to substantially increase the pathways towards green employment and provide education for sustainable development and sustainable lifestyles.
5 GENDER EQUALITY	5.5	BioCities practice co-governance crucially involving a wide range of stakeholders at all stages of decision-making because it is their "right" to participate regardless of their status, age, gender or ethnicity. BioCities encourage women to take a full and effective role in leadership of transition in public, private and civil society.
6 CLEAN WATER AND SANITATION	6.3 6.5 6.6	BioCities manage their water systems sustainably, ensuring zero pollution, ample supply of potable water, and separate their wastewater before using biological based treatments. Water leakage is managed, and water captured for reuse. BioCities recognise that the water footprint for the city extends beyond its immediate boundary to the watershed and beyond.
7 AFFORDABLE AND CLEAN ENERGY	7.2	The BioCity produces locally the derivative resources it needs for its operation. It produces energy through its own renewable systems, extracts water from its own natural basins or subsoils, and grows food and biomass (in the BioCity or the associated BioRegion) for its own population. Urban and peri urban forests within and close to the city are a source of biofuels.
8 DECENT WORK AND ECONOMIC GROWTH	8.3 8.4 8.9	Urban policy, planning and management processes in BioCities can create jobs for urban dwellers and their inclusion secures higher qualities in terms of ecosystem services, from better air quality to recreational activities & access to green spaces. A BioCity creates an attractive environment for tourism and business investment, improves home values and rental rates and provides low carbon materials for development.
9 INDUSTRY, INNOVATION AND INFRASTRUCTURE	9.2 9.3 9.4 9.5	BioCities encourage reinvention and create ample aligned job opportunities through the use and development of local bio-based materials and recycled materials to manufacture, maintain and improve the products required for city functioning. A key feature is entrepreneurship leading to the growth of small-scale enterprise and product innovation.
10 REDUCED INEQUALITIES	10.4	In BioCities co-creation between planners and diverse stakeholder groups can avoid the current exclusion of specific societal strata and clusters by benefiting from all the advantages of future urban settings.
11 SUSTAINABLE CITIES AND COMMUNITIES	11.3 11.4 11.6 11.7	By following the principles of natural systems to promote life Biocities look to nature and ecological functioning to promote their sustainability. Biomimicry promotes a new way of approaching territories, starting from the implementation of Nature-Based Solutions, the optimisation of ecosystem services and the advancement of the circular bioeconomy.
12 RESPONSIBLE CONSUMPTION AND PRODUCTION	12.2 12.5 12.6	In Bio-economies products are designed to be repurposed thereby ensuring their constituent matter remains within the manufacturing ecosystem as long as possible. Otherwise, they are designed to be compostable, ending up as necessary nutrients for building sound and productive soils.
13 CLIMATE ACTION	13.1 13.2 13.3	Urban Forests and production forests help tackle many climate issues, such as urban cooling, air pollution, energy costs, sequestering carbon and contributing wood-based alternatives. A substantial increase in urban tree cover is needed in most cities along with suitable levels of investment and management.
14 LIFE BELOW WATER	14.1	BioCities ensure healthy ecosystems exist in their rivers, reservoirs and lakes and use nature-based solutions to manage their flood waters. Water courses can be used for recreation as well as encouraging biodiversity. BioCities ensure zero waste and that biproducts are not transferred through natural systems or human systems in a way that can pollute the marine environment.
15 LIFE ON LAND	15.1 15.2 15.5 15.9	In BioCities nature provides multiple ecosystem functions, including photosynthesis, respiration, transpiration, decomposition and nutrient cycling, these support other ecosystem services useful for human being. Urban forests help create and enhance habitats and act as a pool of diversity and improve urban soil quality.
16 PEACE, JUSTICE AND STRONG INSTITUTIONS	16.7	BioCities reflect the commitment to leaving no one behind and therefore strive to achieve equal and fair distribution and access to greenspaces and green jobs through inclusive and participatory urban planning and management. BioCIties support participative and representative decision-making at all levels.
17 PARTNERSHIPS FOR THE GOALS	17.14 17.16 17.17	BioCities place multi-stakeholder partnerships and policy and institutional coherence at the centre of their planning. The sharing of sustainable development knowledge especially in terms of twinning and sharing of technologies on a north-south, south-south, regional and international basis.

Fig. 2 The SDG targets mapped against BioCity characteristics showing modest, significant, and high relationships

TRANSITION PATHWAYS TO THE BIOCITY
1. BioCity should provide universal access to ecosystem goods and services for all sections of society including underrepresented groups such as women and children, older persons and persons with disabilities.
2. That a BioCity fully involves its citizens in the different steps of the planning, design and management of it through co-production, co-design and co-creation.
3. The public sector, private companies and citizens work together to achieve common objectives in a BioCity.
4. A BioCity promotes social innovation and the engagement of civil society to achieve social inclusion of marginalised groups through access to nature and its benefits.
5. Collaborative planning, design and management processes combining the ideas of entrepreneurship, civic engagement and societal transformation in a BioCity
6. The BioCity funds its natural capital through innovative approaches such as blended finance, tax incentives, crowd-funding or micro-contributions.
7. A BioCity steers its development towards a circular economy to reduce consumption of natural resources and the production of waste.

Fig. 3 Pathways describe journeys to an end goal, in this case, transition to a BioCity. Seven pathways are listed which are critical to the BioCity journey. Based on ICLEI's 15 pathways from the Basque Declaration (ICLEI 2016)

4 Relationships with Other Networks

The concept of BioCities is not presented as an alternative to other approaches to long-term city sustainability, but as an addition to a 'family of approaches'. It is believed that by offering city policy-makers a choice, they can determine through deep and wide citizen engagement the approach that works best for them. The BioCity will be attractive to those who wish to use the analogue of the powerful forest ecosystem as the basis to move forward. The global environmental crisis coupled with a global health emergency has focused attention on the quality of life in our cities and about their functioning and planning. A wide series of networks, platforms, and projects have evolved in Europe and in all other continents, focusing on various aspects of an improved quality of life. Whilst the authors of this book are eager to advocate for the BioCity approach, it is only fair to acknowledge and demonstrate an awareness of the role of other networks and concepts which underline and attest to the effort being made to improve living conditions of urban areas. To reiterate, the idea of BioCities is not intended as a replacement for other networks, which would be unrealistic and unfair, but rather offered as a new member of the family working towards sustainable development and resilience, using the principles of forest ecosystem functioning and management as its guiding principle.

Networks and platforms planning to evolve towards becoming better places to live and work can be grouped according to three main aspects: (1) smart and sustainable planning; (2) green infrastructure and nature-based approaches; and (3) climate-resilient approaches (Table 1). The contribution and added value of the BioCities vision lies in its unifying concept of the city as a forest ecosystem; one where the functions of that ecosystem analogy are applied. Guiding principles include:

- Continuous flow of renewable energy.
- Recycling of nutrients.
- Reuse of timber.
- Minimisation of waste.
- Sustainable production and use of bio-resources, including timber, as a renewable material for construction and the long-term storage of carbon.
- Influence of forests and trees on the local and global climate, and on the quality of air, water, and soil.
- Enhance biodiversity conservation, including the regeneration of trees in the urban forest, living structures, and ecological niches.

5 Measuring Success

According to Glaeser et al. (2021), modern cities face many new challenges as a result of the 'digital age', in particular, the uncertain promise of high urban benefits from the widespread use of communication technology in a smart city context. They introduce the theme of 'Shared Spaces in Smart Places', describing it as the connection between information technology and urban community. Would this be defined differently, however, if the theme was changed to 'BioCities are Sustainable Places'? Certainly for BioCities to work, the pathways to transition need to be accompanied by new measures of progress in the same way that has occurred in smart city thinking; noting that pathways for BioCities are not likely to be linear and that reversals are not only possible, but likely. Furthermore, as Farley and Smith (2013) point out, if sustainability is everything, is it nothing? These authors offer a new approach, which they describe as neo-sustainability, to help guide policies and practices that respect the primacy of the environment, including its natural limits, and its relationship with social and economic systems. This too has implications for measuring success in BioCities, in that neo-sustainability is conceptually much closer than more traditional approaches to the sustainability discourse.

The starting point for measuring success normally commences with a baseline audit. This audit should establish the baseline, in both policy and practice, in the fields of:

- Urban planning (territorial and development)
- Governance (including co-governance)
- Landscape (macro and micro)

Table 1 Regional and global networks and platforms of sustainable and resilient cities

Group	Network	Description	Link
SMART AND SUSTAINABLE CITIES	United for Smart Sustainable Cities (U4SSC)	Joint initiative of 16 United Nations agencies and programmes to assess the contribution of information and communication technology to the creation of smarter and more sustainable cities.	https://u4ssc.itu.int
	EBRD Green Cities	The programme is based on (1) Green City Action Plans through policy interventions; (2) Sustainable infrastructure investment to stimulate public or private green investments; (3) Capacity-building to provide technical support to stakeholders.	https://www.ebrdgreencities.com
	Commonwealth Sustainable Cities	Supporting a more effective use of IC technologies for building smarter cities, strengthening urban democracy and inclusiveness, and enhancing economic development.	https://www.clgf.org.uk
GREEN INFRASTRUCTURE AND NATURE-BASED APPROACHES	European Green Capitals	Their vision embraces adaptation to climate change and 100% renewable energy, circular economies, and green spaces integrated in urban planning.	ec.europa.eu/environment/europeangreencapital
	Green Cities of Europe	Aims at increasing awareness and investments in green infrastructures in public urban areas together with emphasis on biodiversity conservation, climate, and public health.	https://thegreencities.eu
	Tree Cities of the World	Focus on adoption of common standards for urban trees and forests, their inventory, planting, and careful management.	https://treecitiesoftheworld.org
	CitiesWithNature	Shared platform for cities and their partners to enhance the value of	https://citieswithnature.org

(continued)

Table 1 (continued)

Group	Network	Description	Link
		nature in and around cities across the world.	
	Biophilic Cities	Global network of partner cities working to conserve nature in all its forms and celebrate the benefits for their inhabitants from biodiversity and wild urban spaces.	https://www.biophiliccities.org
CLIMATE RESILIENT CITIES	C40 Cool Cities	Network of cities taking ambitious, collaborative, and urgent climate action that aligns with science-backed targets.	https://www.c40.org
	Resilient Cities	Global city network focusing on mobilising resources to invest in pilot projects for the benefit of vulnerable communities, including also crucial issues as circular economy and migration.	resilientcitiesnetwork.org/network/
	One Planet City Challenge	Cities joining the WWF ongoing mission to enable people to thrive in balance with nature; OPCC supports cities in accelerating their climate transformation, assessing the achievement of the Paris Agreement goals, and showcasing participants' best practices.	https://wwf.panda.org/projects/one_planet_cities/

- Social community (insiders and outsiders)
- Public and private finances (including procurement, non-market benefits, and investment decision-making)
- Status of the circular bioeconomy (if it is recognised and active)
- Status of the urban forest (trees and woodland and other tree-based habitats)
- Status of nature-based solutions (NBS), amongst others.

Transition monitoring requires the measurement of change in terms of flows and ebbs. There should be a net positive change, but it is imperative to recognise that there will be setbacks too, which should be addressed. The use of key performance indicators (KPIs) is the likely tool for measuring this movement. Normally associated with the performance of a single organisation, in the case of the transition to BioCities, KPIs should meet multiple requirements not all of which are organisational. A multilevel approach is needed which meets not just universal requirements, but cultural ones too. As such, the adoption of KPIs must be locally appropriate. In pursuit of BioCity KPIs, a combination of quantitative and qualitative indicators are necessary. Whilst quantitative indicators are largely self-explanatory, the qualitative measures should be linked to societal goals, as measures of success that include co-design principles and co-governance.

One approach, allied to the 'milestone concept' is 'growth awards'. An example of this is the Tree City USA Growth Awards (Arbor Day Foundation 2022). These awards recognise major milestones and annual activities in five categories that combine to build a sustainable (community forestry, in this example) programme over the long term. Each activity is valued between 1 and 10 points, and communities must select and then document activities that total at least 10 points—from any of the subcategories—to receive the Growth Award for the calendar year. By extension, these categories could be replaced by those needed to transition to a BioCity over the longer term, with communities including 'all the actors', such as local authorities, civil society, and business. Furthermore, the list of growth indicators can evolve with time and change to meet new needs and circumstances. Responsibility for such a process of development could be embedded in a dedicated global BioCity facility.

There are also existing KPIs that can be adopted in the BioCity with little or no adaptation. For example, United for Smart Sustainable Cities Initiative (U4SSC) has developed KPIs for Smart Sustainable Cities (Smiciklas 2022). These are broken down based on a conventional framework model of sustainable development (SD) variously described as economy, environment, society, and culture. They are broken down into two categories referred to as 'core KPIs' and 'advanced KPIs'. Examples that are highly relevant to the BioCity concept include 'the length of bicycle paths and lanes per 100,000 population' or 'percentage of local food supplied from within 100 km of the urban area'.

There are also valuable resources to be drawn upon such as the CitiesWithNature (2022) tools and resources, which include measured variables such as 'biodiversity', 'ecosystem', and 'resilience'. These accessible tools help to monitor development with locally appropriate KPIs. Another notable monitoring tool is the City Biodiversity Index (Chan et al. 2021), which is described as a 'self-assessment tool for cities to benchmark and monitor the progress of their biodiversity conservation efforts against their own individual baselines'.

The adoption of existing KPIs to the specific needs of BioCities is resource efficient, especially if a candidate BioCity is already using them. Since the development of KPIs includes a long-drawn-out process of consultation and ground truthing, there is a danger of duplication of effort. The use of existing KPIs can shortcut a process that might otherwise absorb time and resources when the urgency for transition is great.

6 Global Inspiration: Case Studies

1. *Achieving Harmony Amongst People, Water, and the City: Wuhan Sponge City Programme, China*
   The aim of a 'Sponge City' is to allow it to solve its water management concerns, such as urban flooding, storage of water, management of discharges, and improvement to overall water quality, whilst also managing the urban heat island (UHI) effect, and to do so in a sustainable way using green and grey infrastructure. In a Sponge City, the protection and restoration of original water-based ecosystems are considered highly important and is supported and operationalised through technical measures including infiltration, retention, storage, purification, utilisation, and discharge. There is a social purpose too, which is to build better relationships between the city, people, and water. A good example of the 'Sponge City' approach is Wuhan, China, where a target of 80% of the city should meet locally derived 'sponge city' targets by 2030. The required infrastructure is supported by a national technical guide that includes an array of nature-based solutions (NBS), including rain gardens, green roofs, grass swales, and bioretention facilities, accompanied by contemporary grey infrastructures including permeable pavements and rainwater storage modules.
2. *Green City, Clean Waters—Philadelphia's (USA) 21st Century Green Stormwater Infrastructure Programme*
   Philadelphia's 25-year *Green City, Clean Waters Plan* (PWD 2022) is designed to protect and enhance city watersheds by managing stormwater with innovative green infrastructure. It was launched by the Philadelphia Water Department in 2011, and was driven by a 'triple bottom line' analysis (economic, social, and environmental) that showed how investments in green infrastructure at a watershed scale can meet state and federal regulations for reducing stormwater runoff and sewer overflows, at less cost and with greater public benefit than engineered solutions. The success of the plan is measured in several ways, the most prominent of which is the number of 'greened acres' established. A greened acre, established using green infrastructure elements such as plants, soil, stones, and water-absorbing pavements, can soak up and filter over 100,000 l of stormwater. The aim is to take pressure off of Philadelphia's sewer system and add new landscaped green spaces to neighbourhoods. As of 2021, the city has already greened over 2000 acres, with an objective to reach 10,000 by 2036.
3. *Governance and Economic Development—Singapore*
   Much of Singapore's greenness today can be attributed to its active urban greening programme. But the programme also has a special value because it was initiated over 50 years ago when Singapore was still a young developing nation, with a wide range of socio-economic challenges. The late founding Prime Minister of Singapore, Lee Kuan Yew, said in his memoirs that 'greening is the most cost-effective project I have launched'. Although Singapore is now already a green city, the budget for greening has steadily increasing. The reason is simple: the city has benefited immensely from modest expenses invested in greenery and

its management. Greening and biodiversity conservation programmes have already paid for themselves by attracting investment, and will continue to pay into the future under the Green Plan 2030 (Singapore 2022).

4. *Liveability and Well-Being—Ljubljana, Slovenia*

   In 2016, Ljubljana, Slovenia, was the recipient of the European Commission's Green Capital Award, which rewards cities that have a consistent record of achieving high environmental standards. The city continues to make the environment a cornerstone of future development plans (CoL 2022), in particular, the goal to become a sustainable city (a city living in harmony with its natural environment). Various strategies and activities reflect the determination to reach this goal. In 2010, the city declared about 1200 hectares of natural forest, mostly located in and around the city centre, as special purpose forests. Today, Ljubljana residents enjoy a high quality of life, partly because of the strong interaction between the constructed and natural environments, and the great diversity and easy accessibility of valuable natural features. Involving residents in the upkeep and improvement of the city's forests is one of the ways through which the city administration aims to ensure the continued beauty of their 'beloved' city.

5. *Land Reclamation and Cultural Regeneration—Landscape Park Duisburg Nord, Germany*

   The Ruhrgebeit is a polycentric urban region located in the state of North Rhine Westphalia, Germany. Although still a manufacturing powerhouse, much has changed since the German 'economic miracle' of the 1950s. Heavy industry has been substantially replaced by new technologies, services, and elite engineering. A creative approach has been taken to a culture-led regeneration of the Ruhrgebeit. The former Meiderich Ironworks is a key example of this and forms the core of the Landscape Park Duisburg Nord. Since 1994, nature and industrial heritage have been combined in what the British daily newspaper 'The Guardian' called one of the 10 most beautiful urban oases in the world. Industrial structures have been converted into attractions across the extensive 180-hectare site, which features gardens, meadows, water courses, and an extensive 'wild urban forest' created largely through natural regeneration.

6. *Cool Urban Spaces—Phoenix, Arizona, USA*

   Global warming and the urban heat island effect combine to make living in many cities a challenge. One of the most affected cities is Phoenix, USA, where living can be difficult in the warm summer months. Phoenix has responded by launching initiatives to tackle this problem. One of these is called Nature's Cooling System (Messerschmidt 2022), which involves redesigning badly impacted neighbourhoods. These and other initiatives involve close collaboration with communities. At the centre of this thinking is a Nature's Cooling Systems (NCS) partnership involving the Nature Conservancy, Arizona State University, and the Maricopa County Department of Public Health. The use of shade trees is one such intervention, but design is important. Drought-tolerant native trees are being interspersed with leafy thirstier trees under which people can congregate in the shade. Phoenix is ramping up efforts to meet a 20-year goal of achieving 25%

urban forest canopy coverage, which is projected to reduce temperatures by nearly 8 degrees (F), when compared with more open areas.

7. *TreeTown—Freetown, Sierra Leone*

   Freetown, Sierra Leone, has been devastated by the aftermath of massive deforestation that resulted in catastrophic flooding and mudslides. Faced with this level of destruction Freetown City Council (FCC), working along with the World Bank, The Environmental Foundation for Africa (www.EFASL.org), and Greenstand (www.greenstand.org), have launched a programme to implement and monitor a multi-year tree planting programme (Transform Freetown 2022). Sierra Leone remains amongst the world's poorest countries, ranking 180th out of 187 countries in the United Nations Human Development Index in 2011. With a focus on raising funds, especially in the United States, the aim is to undertake a wide range of improvements. In 2020/2021, the focus was on tree planting following years of catastrophic tree loss due to rapid development, overharvesting of timber, and crash-and-burn agriculture. Symbolically, FCC launched a campaign called #OneMillionTreesInFreetown, with a goal to plant one million trees by the end of 2021. A key focus has been the involvement of schools and the wider community.

7 The Role of the European Forest Institute (EFI)

The BioCity approach, with its deep commitment to forest-based solutions and to the improvement of urban sustainability, is well embedded in future European policies and strategies. The *New EU Forest Strategy for 2030* (EC 2021) fully acknowledges that sustainable afforestation and tree planting, urban parks, trees on public and private property, green buildings and infrastructure, and urban gardens, are effective in climate change and disaster risk mitigation and good for people's physical and mental health. Furthermore, the strategy mandates the EU Commission to 'develop a 2050 roadmap for reducing whole life-cycle carbon emissions in buildings and, in the context of the revision of the *Construction Products Regulation*, to develop a standard, robust and transparent methodology to quantify the climate benefits of wood construction products and other building materials'. At the same time, the EU *Biodiversity Strategy for 2030* (EC 2022) pledges 'to plant at least 3 billion additional trees by 2030 in full respect of ecological principles' and to 'raise societal awareness and commitment to biodiversity restoration and to the circular economy'.

In line with these European strategies addressing the use of nature-based thinking and of woody renewable material in urban and peri-urban areas, the European Forest Institute (EFI) has issued its own *Strategy Implementation Plan* (EFI 2022) that clearly identifies the development of circular BioCities as a new urban reality, to reshape 'cities through the circular bioeconomy lenses, based on a new and synergistic relationship between urban economy and ecology'. The EFI's aim is 'to provide new knowledge on the potential of forest-based solutions in creating sustainable and resilient cities'. In fact, it provides that a holistic approach, 'from adapted trees and urban forests to green design and wood construction, as emphasised also in the New European Bauhaus initiative (EU 2022), is urgently needed'. The role of EFI in the evolution of BioCities, therefore, is to develop and advance scientific knowledge in forest-based solutions, as well as to provide scientifically based evidence and information for the benefit of political discussions and decision-making at the European and national policy levels.

Key issues for EFI research and policy support in the field of BioCities are as follows:

- The potential benefits of urban forests for human health and well-being, climate change mitigation, and biodiversity, as well as how these forests should be governed and managed.
- The role that wood material could play in reimagining cities locally and globally, in view of their decarbonisation capabilities and the need for circular urban systems.
- Developing the planning strategies, tools (including indicators), and policy instruments necessary to facilitate the transition to circular BioCities, whilst addressing the interrelationship between cities and (rural) hinterlands.

References

Biggs R, Schluter M, Schoon ML (eds) (2015) Principles for building resilience: sustaining ecosystem services in social-ecological systems. Cambridge University Press

Chan L, Hillel O, Werner P, Holman N, Coetzee I, Galt R, Elmqvist T (2021) Handbook on the Singapore Index on Cities' Biodiversity (also known as the City Biodiversity Index). Secretariat of the Convention on Biological Diversity and Singapore: National Parks Board, Montreal, 70 p

CitiesWithNature Tools and Resources. Accessed February 21, 2022., from https://citieswithnature.org/tools-and-resources/

Farley HM, Smith ZA (2013) Sustainability: if it's everything, is it nothing? 1–177. https://doi.org/10.4324/9780203799062

Glaeser E, Kourtit K, Nijkamp P (2021) New urban challenges: Shared spaces in smart places – Overview and positioning. Land Use Policy 111:art. no. 105672

ICLEI BRIEFING SHEET - Cities and the Sustainable Development Goals - Urban Issues, No. 02, 2019

Kanuri C, Revi A, Espey J, Kuhle H (2016) Getting started with the SDGs in cities. Sustainable Development Solutions Network, New York

Secretariat of the Convention on Biological Diversity (2001) The Value of Forest Ecosystems. Montreal, SCBD, 67 p. (CBD Technical Series no. 4)

Smiciklas J for U4SSC. Accessed February 21, 2022., from https://www.itu.int/en/ITU-D/Regional-Presence/CIS/Documents/Events/2019/02_Minsk/Presentations/Training-S1-and-S2-Pres2-SmiciklasJohn-U4SSC_KPIs-John-Smiciklas.pdf

Glossary

Adaptive capacity/adaptive management Adaptive capacity—Urban systems with the capacity to be resilient whilst responding and adapting to a variety of chronic stresses and acute shocks, reducing vulnerability to climate change and extreme events (EEA 2020).Adaptive management—A systematic and cyclic process for continuously improving management policies and practices by learning from the outcomes of previously employed policies and practices.

Agro-Forestry Land use management system where trees and shrubs are integrated with crops and animal farming to create environmental, economic, and social benefits.

BioCity An urban settlement that follows the principles of natural ecosystems to promote life, particularly by developing their network interactions such as the harnessing and flow of renewable energy, the storage of carbon, the cycling of bio-materials or other matter, and the conservation of evolutionary information as biodiversity, a fundamental feature of ecological as well as sociological systems.

Biodiversity The variety and variability of life on Earth, at the genetic, species, and ecosystem levels.

Bioregion A complex social-ecological, low-entropy release system, at regional scale, that includes the BioCity and its surroundings and is formulated and defined by natural and social interconnections (e.g. watersheds, foodsheds, wood and biomass supply systems) rather than administrative and economic boundaries.

Brownfield Any derelict, abandoned harbour, industrial or commercial areas, typically located in urban settlements; brownfields are not necessarily part of a formal planning, designing, and management regime, but still may be flexibly managed for long-term natural development.

Carbon footprint/carbon sequestration/carbon sink Carbon footprint—The total amount of greenhouse gases (expressed as carbon dioxide equivalent) that are generated by our actions or to produce a given commodity or service.Carbon sequestration—A natural or artificial process by which carbon dioxide is removed

G. E. Scarascia-Mugnozza et al. (eds.), *Transforming Biocities*, Future City 20,
https://doi.org/10.1007/978-3-031-29466-2

from the atmosphere and held in solid or liquid form. Carbon dioxide is naturally captured from the atmosphere through biological, chemical, and physical processes.Carbon sink—Any component of the Earth system, natural or otherwise, that accumulates and stores some carbon-containing chemical compound for an indefinite period and thereby removes carbon dioxide from the atmosphere. Globally, the two most important carbon sinks are vegetation and the ocean.

Circular bioeconomy A model of production and consumption, based on renewable natural resources and bio-based materials to produce food, energy, products, processes, and services, including related knowledge, science, technology, and innovation, that aims to maintain the value of products and resources for as long as possible by returning them to the product cycle at the end of their use whilst minimising the generation of waste (European Parliament 2018).

City A large and densely populated urban settlement. A city is defined in relation to a political level (administrative boundary) and a densely populated 'urban centre' (population > 50,000) (EC 2012 and EC 2016).

Climate change/climate change mitigation/climate change adaptation Climate change—The man-made (anthropogenic) forcing of climate that is considered to be causing an increase in global temperatures, driven by emissions of gases, known as greenhouse gases (National Academies 2023).Climate change mitigation—The reduction of the flow of heat-trapping greenhouse gases into the atmosphere, either by reducing sources of these gases (for example, the burning of fossil fuels for electricity, heat, or transport) or enhancing the "sinks" that accumulate and store these gases (such as the oceans, forests, and soil).Climate change adaptation—Any action that reduces the negative impact of climate change, whilst taking advantage of potential new opportunities.

Ecology/Urban ecology Ecology—The science that studies the relationships between living organisms, including humans, and their physical environment. Urban ecology—The study of ecosystems that include humans living in cities and urbanising landscapes (Indiana University 2023).

Ecosystem service Any environmental, economic, social, and cultural benefits (as outputs, conditions, or processes) provided by nature and its ecosystems to human society.

Ecological network Interconnected system of habitats whose biodiversity needs to be safeguarded. The geometry of the network has a structure based on the recognition of core areas, buffer zones, and ecological corridors that allow the exchange of individuals to reduce the extinction risk of local populations (ISPRA 2023).

Enabling environment *Ensemble* of relatively high-level factors, policies, or technologies that, based on their level of availability, facilitate (drivers) or hinder (barriers) the transition towards urban environmental sustainability (EEA 2023).

Entropy The measure of a system's thermal energy per unit temperature that is unavailable for doing useful work. The amount of entropy is also a measure of the molecular disorder, or randomness, of a system.

Green (and blue) infrastructure A strategically planned network of natural and semi-natural areas with other environmental features designed and managed to deliver a wide range of ecosystem services. Green infrastructure is present in both rural and urban settings. In urban areas, many different features may be part of green infrastructure (e.g. park, garden, grassy verge, bioswale, green wall, or green roof) in so far as they are part of an interconnected network and are delivering multiple ecosystem services. These green urban elements (or blue if aquatic ecosystems are concerned) may be found within the city and in its peri-urban area (EC 2013; EEA 2017).

Greenhouse gases A group of gases contributing to global warming and climate change, such as carbon dioxide, methane, and nitrous oxide. Converting them to carbon dioxide (or CO_2) equivalents makes it possible to compare them and to determine their individual and total contributions to global warming.

Green urban mobility Solutions for promoting green, smart, and inclusive movement of people and goods in cities. Allowing people and goods to move freely and safely in towns and cities whilst respecting the environment, which is crucial both for quality of life and for the health of the economy.

Governance The interaction between the formal institutions and those in civil society whereby those actors wield power, authority, and influence and enact policies and decisions concerning public life and social upliftment. Public engagement, soft governance, transparency, accountability, and integrated decision-making processes involving all relevant sectors, stakeholders (e.g. civil society platforms), and levels of government are crucial to support urban sustainability transitions (Global Development Research Centre 2021).

Landscape A section or expanse of land scenery, usually extensive, that can be seen from a single viewpoint, which usually integrates natural and man-made features.

Nature-based solutions Actions to protect, sustainably manage, and restore natural and modified ecosystems that address societal challenges effectively and adaptively, simultaneously benefiting people and nature (IUCN 2023).

Nexus The interlinkages and interrelationships between two or more systems (e.g. food and energy, BioCity, and BioRegion) or policy areas relevant to urban environmental sustainability (FAO 2014).

One health approach A collaborative, multisectoral, and transdisciplinary approach—working at the local, regional, national, and global levels—with the goal of achieving optimal health outcomes recognising the interconnection between people, animals, plants, and their shared environment (CDC 2023).

Park/urban park Park—An area of natural, semi-natural, or planted space reserved set aside for the protection of 'wild nature' or for human enjoyment and recreation.Urban park—A green space set aside for recreation inside towns and cities.

Planetary boundary Environmental threshold or limit within which humanity can survive, develop, and thrive for generations to come.

Resilience The capacity of individuals, communities, institutions, businesses, and systems to reduce their exposure to, prepare for, cope with, recover better from, adapt to and transform, as necessary, in response to the impacts of climate change (Resilient Cities Network 2021).

Social-ecological systems Complex adaptive systems, in which human societies are embedded in nature. The social component refers to all human activities that include economy, technology, politics, and culture. On the other hand, the ecological component refers to the biosphere, that is, the part of the planet on which life develops (SARAS 2023).

Sustainable buildings Buildings that have high levels of energy and resource efficiency and reduce environmental impacts across their life cycle. Their users enjoy better health and well-being and productivity gains. In turn this translates into cost savings (EC 2016).

Sustainable urban growth The way in which cities and national governments can foster more growth that protects environmental quality and creates thriving, low-carbon, and climate-resilient communities that promote economic vitality, health, well-being, and social inclusion (EEA 2020).

Transitions (towards BioCities) The fundamental and structural changes in urban systems through which persistent environmental and societal challenges are addressed (EEA 2019).

Urban areas Areas, including cities but also smaller urban settlements and suburban areas, developed for residential, industrial, or recreational purposes.

Urban agriculture The cultivation, processing, and distribution of agricultural products in urban and suburban areas. Community gardens, rooftop farms, hydroponic, aeroponic, and aquaponic facilities, and vertical production are all examples of urban agriculture (USDA 2022).

Urban environmental sustainability Sustainable perspective is achieved by focusing on environmental issues in urban areas, such as air and water pollution, green spaces providing space for people and nature, biodiversity loss, resource efficiency, and mitigation measures to reduce greenhouse gas emissions and manage the impacts of climate change (World Bank 2018).

Urban forest Networks or systems comprising all woodlands, groups of trees, and individual trees located in urban and peri-urban areas; they include, therefore, forests, street trees, trees in parks and gardens, and trees in derelict corners. Urban forests are the backbone of the green infrastructure, bridging rural and urban areas and ameliorating a city's environmental footprint (FAO 2022).

Urban planning The planning discipline dealing with the physical, social, economic, and environmental development of metropolitan regions, municipalities, and neighbourhoods. The expression 'urban planning' covers developing land use and building plans as well as local building and environmental regulations (Council of Europe 2007).

Urban sprawl The physical pattern of the low-density expansion of large urban areas into the surrounding agricultural areas under certain market conditions. Sprawl lies in advance of the principal lines of urban growth and implies little planning control of land subdivision. Development is patchy, scattered, and strung out, with a tendency to discontinuity because it leap-frogs over some areas, leaving agricultural enclaves.

Urban sustainability An adaptive process of addressing economic (e.g. economic equity), social (e.g. resilience to climate change impacts), environmental (e.g. reduced air pollution), and governance (e.g. ensuring citizens' active participation in carrying out urban functions) issues in an integrated way within and beyond urban areas (World Bank 2018).

Urbanisation Urbanisation is a long-term process characterised by both an increasing share of the population living in towns and cities and the growth of urban areas (Council of Europe 2007).

Water cycle/water basin Water cycle—The continuous movement of water within the Earth and atmosphere as a complex system that includes many different processes. Liquid water evaporates into water vapour, condenses to form clouds, and precipitates back to earth in the form of rain and snow. Water in different phases moves through the atmosphere (transportation). Liquid water flows across land (runoff), into the ground (infiltration and percolation), and through the ground (groundwater). Groundwater moves into plants (plant uptake) and evaporates from plants into the atmosphere (transpiration) (NOAA 2019).Water basin (or watershed)—The area of land that catches rain and snow and drains or seeps into a marsh, stream, river, lake, or groundwater.

Wood A structural tissue found in the stems and roots of trees and other woody plants. It is an organic material—a natural composite of cellulose fibres that are strong in tension and embedded in a matrix of lignin that resists compression. It can be used for fuel, as energy source, or timber for construction and other wood products.

Index

G. E. Scarascia-Mugnozza et al. (eds.), *Transforming Biocities*, Future City 20,
https://doi.org/10.1007/978-3-031-29466-2

The manufacturer's authorised representative in the EU is Springer Nature Customer Service Centre GmbH, Europaplatz 3, 69115 Heidelberg, Germany. If you have any concerns regarding our products, please contact ProductSafety@springernature.com

Printed and bound by CPI Group (UK) Ltd, Croydon, CR0 4YY
05/10/2026
02228122-0010